Springer Theses

Recognizing Outstanding Ph.D. Research

For further volumes:
http://www.springer.com/series/8790

Aims and Scope

The series "Springer Theses" brings together a selection of the very best Ph.D. theses from around the world and across the physical sciences. Nominated and endorsed by two recognized specialists, each published volume has been selected for its scientific excellence and the high impact of its contents for the pertinent field of research. For greater accessibility to non-specialists, the published versions include an extended introduction, as well as a foreword by the student's supervisor explaining the special relevance of the work for the field. As a whole, the series will provide a valuable resource both for newcomers to the research fields described, and for other scientists seeking detailed background information on special questions. Finally, it provides an accredited documentation of the valuable contributions made by today's younger generation of scientists.

Theses are accepted into the series by invited nomination only and must fulfill all of the following criteria

- They must be written in good English.
- The topic should fall within the confines of Chemistry, Physics, Earth Sciences, Engineering and related interdisciplinary fields such as Materials, Nanoscience, Chemical Engineering, Complex Systems and Biophysics.
- The work reported in the thesis must represent a significant scientific advance.
- If the thesis includes previously published material, permission to reproduce this must be gained from the respective copyright holder.
- They must have been examined and passed during the 12 months prior to nomination.
- Each thesis should include a foreword by the supervisor outlining the significance of its content.
- The theses should have a clearly defined structure including an introduction accessible to scientists not expert in that particular field.

Sebastian Will

From Atom Optics to Quantum Simulation

Interacting Bosons and Fermions in Three-Dimensional Optical Lattice Potentials

Doctoral Thesis accepted by
the Johannes Gutenberg University of Mainz, Germany

Springer

Author
Dr. Sebastian Will
Massachusetts Institute of Technology
Cambridge, MA
USA

Supervisor
Prof. Dr. Immanuel Bloch
Quantum Many-Body Systems Division
MPI of Quantum Optics
Garching b. München
Germany

ISSN 2190-5053 ISSN 2190-5061 (electronic)
ISBN 978-3-642-33632-4 ISBN 978-3-642-33633-1 (eBook)
DOI 10.1007/978-3-642-33633-1
Springer Heidelberg New York Dordrecht London

Library of Congress Control Number: 2012948588

Printed on acid-free paper

Springer is part of Springer Science+Business Media (www.springer.com)

More is different.

P. W. Anderson[1]

[1] Quote from: P. W. Anderson. *More Is Different*. Science, **177**, 393 (1972). Abbreviations: **SF** = superfluid, **MI** = Mott insulator, **AF** = antiferromagnet.

meiner Familie,
Roswitha, Josef, Elisa,
Maximilian und Isabell

Supervisor's Foreword

Over the past years, ultracold quantum gases have proven to be versatile model systems, in which complex many-body effects can be studied in a very controlled environment. One of the main research fields in this direction has been the realization of artificial lattice structures made out of light, into which degenerate bosonic or fermionic quantum gases are loaded. Such optical lattice structures can be controlled in geometry, dimensionality, and strength by simply interfering several laser beams of selected intensity. By loading a degenerate Bose or Fermi gas into such optical lattices, one has been able to e.g., realize strongly correlated Mott insulating phases of matter, which form the basis of many intriguing and challenging condensed matter phenomena. Ultracold atoms in optical lattices therefore hold the potential to shed new light on longstanding open questions in the field of strongly correlated electronic materials, where interpretation of experimental results is often complicated by the complex interplay of many effects and undesired impurities in the system. The origin of high-T_c superconductivity, spin orbit coupled many-body systems, non-equilibrium dynamics, and the realization of strong effective magnetic fields are only a few of numerous topics that have come into the focus of this novel interdisciplinary research field at the interface of quantum optics, atomic, and condensed matter physics.

The thesis of Sebastian Will explores ultracold atomic quantum gases of bosons and fermions in optical lattices. The highly controllable experimental settings discussed in this work have enabled new insight into static and dynamical properties of interacting quantum matter. A crucial technical innovation demonstrated in this thesis, is the implementation of an optical lattice that allows for independent control of tunneling, interactions, and overall system size. This has paved the way for new classes of experiments with interacting bosonic and fermionic atoms.

As one of the highlights of the thesis, the reader is introduced to quantum revival spectroscopy. This novel time-resolved spectroscopic technique allows for precision measurements of interactions and correlations in atomic quantum many-body systems. By following the intriguing collapse and revival dynamics of a bosonic many-body system after a quantum quench, Sebastian Will has shown how the characteristic frequencies of the dynamics precisely reveal the energies of atom number states that play a crucial role in the underlying Hamiltonian. For example, quantum revival spectroscopy has revealed the presence of multi-body interactions in optical lattice systems and has been shown to yield information on the bosonic number statistics on individual lattice sites. Extending the technique to mixtures of bosonic and fermionic atoms, Sebastian has studied the interaction of a single fermionic impurity with a coherent bosonic field. He observed how the interactions among bosons can be modified by the presence of a single interacting impurity.

The observation of emerging multi-body interactions in lattice structures has triggered both experimental and theoretical investigations. In recent experiments, Sebastian's findings have been confirmed and multi-body interactions have been used to manipulate individual lattice sites depending on the number of atoms occupying them. Theoretical studies have explored the effects of multi-body interactions on the phases of many-body quantum systems and led to predictions of novel strongly correlated quantum phases, e.g., with topological order and exotic ground-state properties.

Sebastian Will's thesis not only contains exciting and topical results that are of interest to a broad audience ranging from atomic physics over quantum optics to condensed matter physics. It also concisely covers the fundamentals of ultracold bosons and fermions in optical lattices, it is written in a very clear style, and is accompanied by a series of excellent figures. This will allow even a reader from outside the research field to rapidly learn about the exciting topics investigated in this thesis.

Garching b. München, Germany, August 2012 Immanuel Bloch

Acknowledgments

My time as a Ph.D. student at the Johannes Gutenberg-Universität Mainz and the Ludwig-Maximilians-Universität München was truly amazing. Not only on the scientific, but also on the personal level I have learned for life. I am indebted to many, many people whom I met, whom I discussed with, and whom I learned from.

First of all, I want to thank my supervisor Immanuel Bloch. A number of years back, when I was an undergraduate in the fifth semester, he caught my attention for ultracold atoms with his fantastic atomic physics class. After a few months as an intern in his newly started labs in Mainz, he generously supported my plans for a research stay with Wolfgang Ketterle at MIT, where I carried out my diploma thesis. Among Immanuel's most amazing gifts are his ability to produce streams of creative ideas that never run dry, and to explain physics with very intuitive pictures. His strive for scientific accuracy and completeness is legendary. This trait is challenging for a young student, but at the same time constitutes a role model for a good scientist. I am very grateful for his support and his trust into my work right from the start. It has been a fantastic collaboration.

Experiments as those reported in this thesis are obviously not the work of a single person. They require the contributions of many people. I am thankful to all the members of our Fermion team: Thorsten Best, Lucia Hackermüller, Dries van Oosten, Tim Rom, Ulrich Schneider, and Martin Zwierlein. In particular, during my first year of graduate school we had to move mountains. The challenge to make the new apparatus work so well, that it allowed us to set out to unknown scientific shores, often seemed daunting. However, each of us contributed a great deal to the final success. Within less than two years before the move to Munich we took the data for an amazing number of six papers that made it into Physical Review Letters, Science and Nature! An enormous thank you goes to Thorsten and Tim for laying the solid foundations of this wonderful apparatus. And again thank you, Thorsten, for teaching me electronics, caringly introducing me to the nuts and bolts of the experimental setup and for being a great office mate. A new generation of talented students has taken over the Fermion apparatus, Simon Braun, Jens Phillip Ronzheimer, and Michael Schreiber. I wish you the best of luck and wise decisions. Build wings to the machine and let it fly again. Thank you, Simon, for enjoyable measurement nights, in which "Lafer! Lichter! Lecker!" kept us hungry and awake. And thank you, Michael, for great outdoor adventures, in particular our fantastic tour de force to the top of Zugspitze.

After the move to Munich, I shared the office with Stefan Trotzky. We had a fabulous time, enjoying good discussions about physics, good food, good music and, last but not least, sharing a severe coffee addiction. Thank you for having that thing about cleanliness in our office, for introducing me to Viking chess, and for countless seemingly foolish ideas.

I thank all the members of the rapidly growing Bloch group and the neighboring research groups in Mainz and Munich. In particular, I want to mention Martin Boll, Patrick Cheinet, Yuao Chen, Christoph Clausen, Manuel Endres, Simon Fölling, Fabrice Gerbier, Tatjana Gericke, Carolin Hahn, Fabian Heidrich-Meisner, Stefan Kuhr, Herwig Ott, Arno Rauschenbeutel, Ute Schnorrberger, Jacob Sherson, Ariane Stiebeiner, Jeff Thompson, Christof Weitenberg, and Artur Widera.

Thank you to all colleagues and friends in theoretical physics, who helped us to understand what we observed in the experiments. Especially, I want to thank Dirk-Sören Lühmann from the University of Hamburg. Through our discussions about multi-orbital physics in optical lattices the idea of quantum revival spectroscopy took shape. Furthermore, I am grateful to Ulf Bissbort, Michael Buchhold, and Walter Hofstetter from the University of Frankfurt for a great and fruitful collaboration.

I would also like to thank Christine Best and Elvira Stuck-Kerth as well as Zohra Hauck and Marianne Kargl, our secretaries in Mainz and Munich. Your assistance in all organizational matters made the lives of all team members so much easier. Thank you for making the research group feel like a big family!

I am very grateful for the generous support by the Johannes Gutenberg-Universität Mainz. Being a member of the materials science graduate school (MAINZ) was an important experience. The school offered a wonderful platform for interdisciplinary exchange of ideas. At the student meetings, seminars, and workshops I have learned a great deal about how chemists and physicists think about and solve scientific problems. And, aside from all scientific benefit, the graduate school was a great place to socialize.

It has been a great honor to be a fellow of the Gutenberg Academy at the University of Mainz. This wonderful group of people, students and professors, half and half across all departments of the university, really allowed me to peek beyond the rim of my teacup. The monthly round tables were very inspiring, broadening everybody's view and allowing for intriguingly deep discussions. The annual excursions of the academy's junior fellows, which led me to Edinburgh, Bologna, Budapest, and Istanbul, were no less than legendary. The mentoring program of the Gutenberg Academy was definitely a stroke of luck for me: I am indebted to Jürgen Gauss, who has been and still is a fantastic academic mentor, giving me precious advice. His deliberate and honest way of caring about science and personal matters is absolutely exceptional. Also, I am very grateful to my external mentor Jürgen Kluge. Being a physicist by training and a very successful economic leader at the same time, he has opened up my view to the worlds of business and politics. His broad knowledge, his good judgement, his grounded and cheerful temper have left a deep impression on me.

I thank my many friends, who have supported and encouraged me a lot during graduate school. You have suffered my repetitive words that this thesis is going to be finished soon… and I am more than happy that you did not lose patience! My special thank you goes to Elisabeth Balss, Andreas Dechange, Dominik Höhler, Alexander Klaus, Carmen Mergen, Dominik Schultes, and Gerd Schupp for your unconditional friendship, which I hope will last for a lifetime.

Mein größter Dank gilt meiner Familie, der ich diese Arbeit widme. Ohne Euren Glauben an mich, ohne Eure Unterstützung wäre all das nicht möglich…

Contents

Chapter 1
Introduction

The first creation of atomic Bose-Einstein condensates in 1995 [1–5] was the starting point for the research field of ultracold atoms, which since then has made advances at a breathtaking pace. In the beginning, the experiments focussed on weakly interacting Bose-Einstein condensates (BECs), in which up to several million atoms undergo a quantum phase transition. They occupy a single macroscopic wavefunction and form a quantum degenerate Bose gas. It was possible to confirm the predictions of Bogoliubov's mean-field theory [6, 7], to demonstrate the coherence properties via matter wave interference [8] and to prove superfluidity via the creation of vortices [9–11]. The first quantum degenerate gas of fermionic atoms was created in 1999 [12–14]. Spin-polarized fermions do not undergo a quantum phase transition on the way to ultra low temperatures, but fermionic spin mixtures of two atomic hyperfine states soon turned out to be a meaningful analog to spin up and spin down electrons in condensed matter.

Reaching beyond mean-field physics, fast progress was made to enter the regime of strong correlations with ultracold atoms. In this regime, many-particle systems are no longer described by a single macroscopic wavefunction. Instead, strong interactions between the particles lead to quantum mechanical correlations in the many-body wavefunctions, which give rise to much richer physics and complex quantum phases.

The advent of Feshbach resonances [15, 16] made it possible to tune and enhance the interactions between ultracold atoms over enormous ranges. With this novel tool, in particular, degenerate Fermi gases started to unfold their full experimental potential. In a Feshbach resonance an external magnetic field is used to tune a molecular level into resonance with the energy of the colliding atom pair [16]. This effectively controls the interatomic scattering length and thereby the interaction between atoms. However, it bears the risk of enhanced three-body losses in the vicinity of the resonance. Being a fortunate coincidence of nature, such losses proved to be strongly suppressed for fermionic spin mixtures giving access to an extremely fruitful branch of research: The study of the crossover between molecular Bose-Einstein condensation and Bardeen-Cooper-Shrieffer (BCS) superfluidity. The breakthroughs include the creation of BECs of molecules on the repulsive side of interactions [17–20],

S. Will, *From Atom Optics to Quantum Simulation*, Springer Theses,
DOI: 10.1007/978-3-642-33633-1_1,

the observation of pairing on the BCS side [21, 22], the proof of BCS-superfluidity via the creation of vortices [23] as well as studies of imbalanced Fermi mixtures [24–27].

For bosonic atoms the approach of strong correlations via Feshbach resonances turned out to be hamperd by inelastic three-body losses. However, a different, but no less successful route towards strong interactions and strongly correlated phases was found: In 1998, Dieter Jaksch and collaborators [28] proposed to trap bosons in a so-called optical lattice to experimentally realize the Bose-Hubbard Hamiltonian, which was so far only known as a model system in condensed matter theory [29]. An optical lattice is usually formed by the interference of counterpropagating laser beams, creating a regular array of intensity minima and maxima with simple cubic symmetry. Based on the optical dipole force, atoms are trapped in this array that resembles the structure of ions in a solid crystal [30]. When the intensity of the lasers is increased, the tunneling of atoms through the artificial crystal is more and more suppressed. At the point where the interaction energy dominates over the kinetic energy, the atoms get localized at individual lattice sites and the many-body state corresponds to a Mott insulating phase. Soon after the observation of number-squeezed states in an one-dimensional optical lattice [31], the field had a seminal breakthrough in 2002. Markus Greiner and collaborators [32] observed the quantum phase transition from a bosonic superfluid to a Mott insulator in a three-dimensional (3D) optical lattice. In the following years optical lattices have been used to create Mott insulators in 1D and 2D [33, 34], a Tonks-Girardeau gas [35, 36] as well as a band insulator of spin-polarized fermions [37]. Recently, also more complex lattice geometries such as superlattices [38–40] and hexagonal lattices [41] have been realized.

The enormous experimental progress that has been made with fermions at Feshbach resonances and bosons in optical lattices has fueled the prospect that ultracold atoms can be used to study problems of quantum many-body physics with high precision. This idea follows Richard P. Feynman's vision of a quantum simulator, which he presented in 1981 [42]. Feynman argued on general grounds that, on a classical computer, the accuracy of simulations of quantum mechanical systems must have an intrinsic limit. Therefore he proposed to use a well-controlled quantum system for the simulation of quantum mechanical problems.

In this thesis I report on the realization, characterization and analysis of many-body quantum states of ultracold atoms in an optical lattice. The studies with ensembles of interacting bosonic ^{87}Rb and fermionic ^{40}K atoms are enabled by a versatile setup, that integrates many of the aforementioned experimental concepts. A key innovation is the first implementation of a so-called blue-detuned optical lattice. In combination with a red-detuned dipole trap, this provides independent control of the lattice depth and the underlying trapping potential. Thereby, it was possible to realize the first homogeneous optical lattice. The combination of these unprecedented capabilities with the interaction control offered by Feshbach resonances has opened the door to new classes of equilibrium and nonequilibrium experiments.

The research of this thesis has lead to several important achievements: First, it has been possible to implement the Fermi-Hubbard model in an optical lattice with an ultracold spin mixture of fermionic atoms. Metallic and insulating phases have been identified, including the fermionic Mott insulator. Second, a new atom optical technique has been developed to measure the on-site interactions of bosonic atoms with high precision. This has allowed to demonstrate the presence of effective multi-body interactions pointing towards important multi-band effects in optical lattices. Third, the first lattice Bose-Fermi mixtures with tunable interspecies interactions have been created. The influence of fermionic “impurities” on the bosonic superfluid to Mott insulator transition has been studied and several nontrivial interaction effects have been revealed. In the following, a brief motivation and outline of the specific experiments is given:

1.1 Metallic and Insulating Phases of Fermionic Spin Mixtures

In condensed matter physics the Mott insulator phase is a paradigmatic manifestation of strong correlations between electrons in solids [43, 44]. When interactions between electrons are exceedingly strong, a Mott insulator can emerge in materials, in which the unit cell of the crystal is on average occupied by one electron. Under these circumstances individual electrons (either with spin up or spin down) get localized at the atomic sites and the material is not able to conduct electrical current, although the conduction band is only half-filled. For crystals with simple cubic symmetry the electron spins in the Mott insulator are antiferromagnetically ordered, if the temperature of the material is below the Néel-temperature T_{N}. This temperature is determined by the energy scale of the Heisenberg exchange interaction J_{ex}, which describes the energy cost for neighboring spins to exchange places. Up to this point, the physics of Mott insulators is described by the famous Fermi-Hubbard model. It is the simplest model system that captures the competition between interactions and kinetic energy in electronic materials [45]. However, the known territory is soon departed when Mott insulating materials are doped: Some of them quickly turn into superconductors with surprisingly high transition temperatures. These are the famous high-T_c superconductors [46], which are typically found in cuprates, chemical compounds consisting of two-dimensional copper-oxide layers. It is a long-standing open question, reaching back to the discovery of high-T_c materials in 1986 [47], whether the corresponding superconducting many-body state also emerges from within the Fermi-Hubbard model. So far, conclusive explanations for the mechanism of high-temperature superconductivity could not be given. This is — cast in simple words — the reason, why quantum simulation of the Fermi-Hubbard Hamiltonian in a clean and defect free model system is a highly relevant research topic.

In the experiment, we implement the Fermi-Hubbard model by loading an equal mixture of fermionic spin up and spin down ^{40}K atoms into a three-dimensional optical lattice. The quantum phases of the system are explored by changing the strength of the external confining potential. Leaving the overall atom number fixed, this effectively changes the system size and the local filling at the lattice sites. Such a procedure is virtually impossible in real materials. The response of our model quantum system is probed by monitoring the in-situ cloud size via phase-contrast imaging, which allows to extract the compressibility of the quantum many-body state. We first show

how a noninteracting spin mixture evolves from a dilute compressible metal into an incompressible band insulating state with doubly occupied sites (one spin up and one spin down atom) when the confinement is increased. For strong interactions and intermediate confinement, we observe the emergence of an incompressible Mott insulator state in the center of the trap. As an additional confirmation, we also observe a strong suppression of doubly occupied lattice sites as a result of the localization of single atoms.

Given the above outline of the status in condensed matter physics, it is the natural next step to direct experimental efforts to the realization and detection of an antiferromagnetically ordered Mott insulator. To reach this aim, the entropy per particle should lie below $k_B \ln(2)$, which corresponds to staying below the Néel-temperature. Current experiments, including ours, lie a factor of two above this value. Schemes for the reduction of entropy in optical lattices are intensely pursued with the prospect of studying spin order and quantum magnetism with ultracold atoms.

1.2 Quantum Revival Spectroscopy and Multi-Body Interactions

When a Bose-Einstein condensate is loaded into a shallow optical lattice, the atoms macroscopically occupy a quantum state that is delocalized across the whole system. What happens when in this situation the lattice depth is abruptly increased? The atoms will remain in the delocalized state, although it is not the ground state of the system. If the lattice depth is chosen sufficiently deep, the tunnel coupling between lattice sites is very small and the many-body quantum state does not equilibrate. Taking now a glance at an individual lattice site, the on-site quantum state can be well described as a coherent matter wave field that is formed by a Poissonian superposition of atom number states (Fock states). Such coherent states have originally been introduced by Glauber [48] to describe coherent fields of light formed by a Poissonian superposition of photon number states. Coherent states can show intriguing quantum dynamics [49, 50]: When the phase evolution of the individual Fock states is nonlinear in the particle number, the coherent state undergoes a series of collapses and revivals! For atomic coherent states, this quantum evolution is induced by interatomic collisions and the periodicity of the revivals is inversely proportional to the interaction energy of a colliding atom pair. This direct conversion of the interaction energy into an observable frequency makes the observation of quantum revivals a precise measurement tool for atomic interactions.

For the experimental realization of this intriguing atom optical phenomenon, we load a Bose-Einstein condensate of ^{87}Rb atoms into a shallow optical lattice, giving rise to an enormously large array of atomic coherent states. An abrupt, nonadiabatic increase of the lattice depth initiates the quantum dynamics of all the coherent states in parallel. Simultaneously with the increase of the lattice depth we cancel the underlying confinement of the lattice. This minimizes mutual energy offsets between lattice sites and boosts the coherence time of the quantum evolution by more than a factor of ten compared to previous experiments [51, 52]. The atomic interference patterns after time-of-flight expansion show more than forty revival cycles in the temporal evolution of the contrast of time-of-flight interference patterns.

The resulting high spectral resolution allows us to perform the most precise measurements of atomic Fock state energies to date. The spectral analysis of the quantum evolution reveals that the collisions of atoms in an optical lattice cannot be exclusively described by two-body interactions. In fact, virtual transitions of atoms to excited on-site orbitals generate effective multi-body interactions as higher-order corrections to the single orbital two-body interaction. Our precision measurement of multi-body interaction energies signifies the impact of multi-orbital effects in optical lattices and provides crucial input for the comparison of optical-lattice quantum simulators with many-body quantum theory.

The observation of quantum revivals additionally reveals information on the atom number statistics at individual lattice sites. Due to the effect of multi-body interactions, each Fock state appears with a characteristic frequency in the experimental spectra. The corresponding spectral weight encodes the occupation of individual Fock states, which allows to monitor how the on-site quantum state evolves from coherent, for shallow lattice depths, to highly number-squeezed, for deeper lattices. Underlining the intriguing analogies between quantum and atom optics, our technique is similar to foundational experiments in cavity quantum electrodynamics that yield the statistics of the cavity photon field [53].

1.3 Equilibrium Phases of Bose-Fermi Mixtures in Optical Lattices

Multi-component systems play an important role in quantum many-body physics. From interacting atoms and photons to electrons and phonons the interplay of interactions in binary mixtures gives rise to intriguing many-body phenomena such as superradiance, BCS-superfluidity or polaronic effects. However, a scenario, in which particles of different quantum statistics occupy the same lattice structure and interact with each other, is somewhat unusual. A novel quantum system of this kind becomes available when loading ultracold clouds of bosonic and fermionic atoms into an optical lattice. The complex interplay of Bose-Bose and Bose-Fermi interactions gives rise to a physical richness that has inspired numerous theoretical investigations. They led to the prediction of a plethora of intriguing quantum phases, including exotic superfluids [54], charge-density waves [55] and polaron-like quasiparticles [56]. Even a supersolid phase of the bosonic component has been conjectured [57].

Being an initial step on the way towards the creation of complex quantum phases, we have realized the first lattice Bose-Fermi mixture with tunable interspecies interactions using ^{87}Rb and ^{40}K. The investigation of the coherence of the bosonic species shows that the superfluid to Mott insulator transition is shifted towards lower lattice depths for negative interspecies scattering lengths. This shift can be attributed to an effective deepening of the optical lattice owing to the presence of attractively interacting fermions. The resulting renormalization of bosonic interactions and tunneling reaches beyond the single-band Hubbard model and requires the inclusion of higher lattice bands. On the side of repulsive interspecies interactions we do not see a notable shift of the Mott insulator transition suggesting that bosons and fermions do not occupy the same lattice sites. This behavior is consistent both with a trivial global phase separation and a local anticorrelation of the species, as expected for a charge-density wave or a supersolid. However, with the present techniques we cannot distinguish between these scenarios, yet.

1.4 Few-Body Bose-Fermi Systems on Individual Lattice Sites

In order to gain further insight into the interaction effects of Bose-Fermi mixtures in optical lattices, we carried out the first application of quantum revival spectroscopy. Specifically, we create an array of few-body systems, which consist of a single fermion and a bosonic coherent state. In this ensemble of "microlabs" we induce collapse and revival dynamics and deduce the interaction effects between the fermion and the individual bosonic Fock states. For the first time, the absolute strength of the Bose-Fermi interactions is directly measured as a function of the interspecies scattering length. Furthermore, it is demonstrated that the presence of an interacting fermion mediates a modification of the Bose-Bose interactions. Indeed, the data shows an increase of the repulsion among the bosons when an attractively interacting fermion is present, which is consistent with the observed shift of the superfluid to Mott insulator transition.

Additionally, we have identified an intriguing interference effect in quantum revival spectroscopy that allows us to selectively infer the mean fermionic filling on those sites of the lattice array, in which bosons and fermions overlap. With this method it is possible to show that the local fermionic filling indeed varies as a function of the interspecies interactions. We observe a marked increase of the on-site fermion density for interspecies attraction, while for repulsive interactions again a separation between bosons and fermions is detected.

Our studies on tunable Bose-Fermi mixtures in optical lattices constitute early steps in the investigation of this extremely rich quantum system. In fact, we are presented with a promising starting point to investigate tunable disorder [58], polaron physics [59] and complex quantum phases [57]. On the few-body level, Bose-Fermi systems on individual lattice sites may be a useful platform to study impurity physics or effective field theories that are relevant in the description of atomic nuclei [60].

1.5 Outline

In Chap. 2 we give a brief overview of the physics of weakly interacting quantum gases, which is followed by the description of the two main routes to enter the strongly correlated regime: Optical lattice potentials and Feshbach resonances.

Chapter 3 is concerned with the description of Hubbard models for ultracold atoms in optical lattices. We derive the Hubbard Hamiltonian and indicate the approximations that are made with respect to the full many-body Hamiltonian. This illustrates that also physics beyond the usual single-band Hubbard model can be expected in realistic lattice systems. The Bose- and Fermi-Hubbard models are discussed as well as the expected phases in implementations of the models with ultracold atoms in optical lattices.

Chapter 4 provides theoretical background on the methods for the detection of ultracold atoms that are used in this thesis: Absorption imaging and phase-contrast imaging. The latter is of crucial importance for the reliable detection of fermionic in-situ density distributions in Chap. 6. Furthermore, we discuss relevant physical observables that can be extracted from in-situ and time-of-flight images.

Chapter 5 gives a brief overview of the experimental apparatus focussing on the main innovations that are crucial for the presented experiments. In particular, we describe the phase-contrast imaging system as well as alignment and characterization methods of the combined setup of blue-detuned optical lattice and red-detuned dipole trap.

In Chap. 6 the experimental realization of the Fermi-Hubbard model with fermionic spin mixtures is presented. We describe the experimental route towards the observation of metallic and insulating phases in repulsively interacting mixtures, including the demonstration of a fermionic Mott insulator. Furthermore, the observation of a counterintuitive effect, the anomalous expansion of attractively interacting mixtures at finite temperature is briefly outlined. The outlook discusses strategies to cool fermionic spin mixtures below the Néel-temperature as well as detection methods for antiferromagnetic order.

Chapter 7 reports on the observation of coherent multi-body interactions in a bosonic lattice quantum gas by means of quantum revival spectroscopy. We introduce the theory of collapse and revival dynamics of a matter wave field and show that it can be used to measure the interaction energy of atomic Fock states. The experimental part reports on long-lived quantum phase revival dynamics in a homogeneous lattice and the resulting precision measurement of Fock state energies and effective multi-body interactions. We further work out a proposal for the experimental detection of a Schrödinger cat state that arises during the collapse of a matter wave field.

Chapter 8 outlines the realization of a Bose-Fermi mixture with tunable interspecies interactions in a three-dimensional optical lattice. It is demonstrated that the superfluid to Mott insulator transition shifts towards lower lattice depths for attractive interspecies interactions, while evidence for phase separation is found for interspecies repulsion.

Chapter 9 describes the application of quantum revival spectroscopy to few-body systems consisting of a single fermion and a small coherent bosonic field. We report on the direct measurement of absolute Bose-Fermi interaction energies and demonstrate that the Bose-Bose interaction energy is modified by the presence of an interacting fermion. Additional unpublished data shows the observation of collapse and revival dynamics in the fermionic component.

1.6 Publications

Several papers have been published during my PhD work. The most relevant ones for the content of this thesis are marked:

- • M. Buchhold, U. Bissbort, S. Will, and W. Hofstetter, *Creating Exotic Condensates via Quantum-Phase-Revival Dynamics in Engineered Lattice Potentials*, Physical Review A, **84**, 023631, (2011)
- ► S. Will, Th. Best, S. Braun, U. Schneider, and I. Bloch, *Coherent Interaction of a Single Fermion with a Small Bosonic Field*, Physical Review Letters, **106**, 115305, (2011)
- • U. Schneider, L. Hackermüller, J. P. Ronzheimer, S. Will, S. Braun, Th. Best, I. Bloch, E. Demler, S. Mandt, D. Rasch, and A. Rosch, *Fermionic Transport and Out-of-Equilibrium Dynamics in a Homogeneous Hubbard Model with Ultracold Atoms*, Nature Physics, **8**, 213-218, (2012)
- ► S. Will, Th. Best, U. Schneider, L. Hackermüller, D.-S. Lühmann, and I. Bloch, *Time-Resolved Observation of Coherent Multi-Body Interactions in Quantum Phase Revivals*, Nature, **465**,197-201, (2010)
- • L. Hackermüller, U. Schneider, M. Moreno-Cardoner, T. Kitagawa, Th. Best, S. Will, E. Demler, E. Altman, I. Bloch, and B. Paredes, *Anomalous Expansion of Attractively Interacting Fermionic Atoms in an Optical Lattice*, Science, **327**,1621-1624, (2010)
- • S. Will, B. Paredes, L. Hackermüller, U. Schneider, Th. Best, M. Moreno-Cardoner, and I. Bloch, *Strongly Correlated Bosons and Fermions in Optical Lattices*, Proceedings of the XIX International Conference on Laser Spectroscopy (ICOLS), 191, World Scientific, (2009)
- ► Th. Best, S. Will, U. Schneider, L. Hackermüller, D.-S. Lühmann, D. van Oosten, and I. Bloch, *Role of Interactions in* ^{87}*Rb*-^{40}*K Bose-Fermi Mixtures in a 3D Optical Lattice*, Physical Review Letters, **102**, 030408, (2009)
- ► U. Schneider, L. Hackermüller, S. Will, Th. Best, I. Bloch, T. Costi, R. Helmes, D. Rasch, and A. Rosch, *Metallic and Insulating Phases of Repulsively Interacting Fermions in a 3D Optical Lattice*, Science, **322**, 1520-1525, (2009)

Earlier work:

- • C. Christensen, S. Will, M. Saba, G.-B. Jo, Y.-I. Shin, W. Ketterle, and D. Pritchard, *Trapping of Ultracold Atoms in a Hollow-Core Photonic Crystal Fiber*, Physical Review A, **78**, 033429, (2008)
- • G.-B. Jo, Y.-I. Shin, S. Will, T. Pasquini, M. Saba, W. Ketterle, D. Pritchard, M. Vengalattore, and M. Prentiss, *Long Phase Coherence Time and Number Squeezing of Two Bose-Einstein Condensates on an Atom Chip*, Physical Review Letters, **98**, 030407, (2007)

References

1. M.H. Anderson, J.R. Ensher, M.R. Matthews, C.E. Wieman, E.A. Cornell, Observation of Bose-Einstein condensation in a dilute atomic vapor. Science **269**, 198 (1995)
2. K.B. Davis, M.O. Mewes, M.R. Andrews, N.J. van Druten, D.S. Durfee, D.M. Kurn, W. Ketterle, Bose-Einstein Condensation in a gas of sodium atoms. Phys. Rev. Lett. **75**, 3969 (1995)
3. F. Dalfovo, S. Giorgini, L.P. Pitaevskii, S. Stringari, Theory of Bose-Einstein condensation in trapped gases. Rev. Mod. Phys. **71**, 463 (1999)
4. W. Ketterle, D. Durfee, D. M. Stamper-Kurn. Making, probing and understanding Bose-Einstein condensates, in *Bose-Einstein Condensation in Atomic Gases*, ed. by S. Stringari, M. Inguscio, C. E. Wieman (IOS Press, Amsterdam, 1999), pp. 67–176
5. C. Pethick , H. Smith, *Bose-Einstein Condensation in Dilute Gases* (Cambridge University Press, Cambridge, 2008)
6. M.R. Andrews, D.M. Kurn, H.-J. Miesner, D.S. Durfee, C.G. Townsend, S. Inouye, W. Ketterle, Propagation of sound in a Bose-Einstein condensate. Phys. Rev. Lett. **79**, 553 (1997)
7. D.M. Stamper-Kurn, A.P. Chikkatur, A. Görlitz, S. Inouye, S. Gupta, D.E. Pritchard, W. Ketterle, Excitation of phonons in a Bose-Einstein condensate by light scattering. Phys. Rev. Lett. **83**, 2876 (1999)
8. M.R. Andrews, C.G. Townsend, H.-J. Miesner, D.S. Durfee, D.M. Kurn, W. Ketterle, Observation of interference between two Bose condensates. Science **275**, 637 (1997)
9. M.R. Matthews, B.P. Anderson, P.C. Haljan, D.S. Hall, C.E. Wieman, E.A. Cornell, Vortices in a Bose-Einstein condensate. Phys. Rev. Lett. **83**, 2498 (1999)
10. K.W. Madison, F. Chevy, W. Wohlleben, J. Dalibard, Vortex Formation in a Stirred Bose-Einstein Condensate. Phys. Rev. Lett. **84**, 806 (2000)
11. J.R. Abo-Shaeer, C. Raman, J.M. Vogels, W. Ketterle, Observation of vortex lattices in Bose-Einstein condensates. Science **292**, 476 (2001)
12. B. DeMarco, D.S. Jin, Onset of Fermi degeneracy in a trapped atomic gas. Science **285**, 1703 (1999)
13. S. Giorgini, L.P. Pitaevskii, S. Stringari, Theory of ultracold atomic Fermi gases. Rev. Mod. Phys. **80**, 1215 (2008)
14. W. Ketterle, M.W. Zwierlein, Making, probing and understanding ultracold Fermi gases, in *Ultracold Fermi Gases, Proceedings of the International School of Physics "Enrico Fermi", Course CLXIV, Varenna, 20–30 June 2006*, ed. by M. Inguscio, W. Ketterle, C. Salomon (IOS Press, Amsterdam, 2008)
15. S. Inouye, M.R. Andrews, J. Stenger, H.-J. Miesner, D.M. Stamper-Kurn, W. Ketterle, Observation of Feshbach resonances in a Bose-Einstein condensate. Nature **392**, 151 (1998)
16. C. Chin, R. Grimm, P. Julienne, E. Tiesinga, Feshbach resonances in ultracold gases. Rev. Mod. Phys. **82**, 1225 (2010)
17. C.A. Regal, C. Ticknor, J.L. Bohn, D.S. Jin, Creation of ultracold molecules from a Fermi gas of atoms. Nature **424**, 47 (2003)
18. M. Greiner, C. Regal, D. Jin, Emergence of a molecular Bose-Einstein condensate from a Fermi gas. Nature **426**, 537 (2003)
19. S. Jochim, M. Bartenstein, A. Altmeyer, G. Hendl, S. Riedl, C. Chin, J. Hecker Denschlag, R. Grimm, Bose-Einstein condensation of molecules. Science **302**, 2101 (2003)
20. M.W. Zwierlein, C.A. Stan, C.H. Schunck, S.M.F. Raupach, S. Gupta, Z. Hadzibabic, W. Ketterle, Observation of Bose-Einstein condensation of molecules. Phys. Rev. Lett. **91**, 250401 (2003)
21. C.A. Regal, M. Greiner, D.S. Jin, Observation of resonance condensation of fermionic atom pairs. Phys. Rev. Lett. **92**, 040403 (2004)
22. M.W. Zwierlein, C.A. Stan, C.H. Schunck, S.M.F. Raupach, A.J. Kerman, W. Ketterle, Condensation of Pairs of Fermionic Atoms near a Feshbach Resonance. Phys. Rev. Lett. **92**, 120403 (2004)

23. M. Zwierlein, J. Abo-Shaeer, A. Schirotzek, C. Schunck, W. Ketterle, Vortices and superfluidity in a strongly interacting Fermi gas. Nature **435**, 1047 (2005)
24. M.W. Zwierlein, A. Schirotzek, C.H. Schunck, W. Ketterle, Fermionic superfluidity with imbalanced spin populations. Science **311**, 492 (2006)
25. Y. Shin, M.W. Zwierlein, C.H. Schunck, A. Schirotzek, W. Ketterle, Observation of phase separation in a strongly interacting imbalanced Fermi gas. Phys. Rev. Lett. **97**, 030401 (2006)
26. G. B. Partridge, W. Li, R. I. Kamar, Y.-a. Liao, R. G. Hulet. Pairing and phase separation in a polarized fermi gas. Science **311**, 503 (2006)
27. Y. Liao, A. Rittner, T. Paprotta, W. Li, G. Partridge, R. Hulet, S. Baur, E. Mueller, Spin-imbalance in a one-dimensional Fermi gas. Nature **467**, 567 (2010)
28. D. Jaksch, C. Bruder, J.I. Cirac, C.W. Gardiner, P. Zoller, Cold Bosonic atoms in optical lattices. Phys. Rev. Lett. **81**, 3108 (1998)
29. M.P.A. Fisher, P.B. Weichman, G. Grinstein, D.S. Fisher, Boson localization and the superfluid-insulator transition. Phys. Rev. B **40**, 546 (1989)
30. I. Bloch, J. Dalibard, W. Zwerger, Many-body physics with ultracold gases. Rev. Mod. Phys. **80**, 885 (2008)
31. C. Orzel, A.K. Tuchman, M.L. Fenselau, M. Yasuda, M.A. Kasevich, Squeezed states in a Bose–Einstein condensate. Science **291**, 2386 (2001)
32. M. Greiner, O. Mandel, T. Esslinger, T. Hänsch, I. Bloch, Quantum phase transition from a superfluid to a Mott insulator in a gas of ultracold atoms. Nature **415**, 39 (2002)
33. T. Stöferle, H. Moritz, C. Schori, M. Köhl, T. Esslinger, Transition from a strongly interacting 1D superfluid to a mott insulator. Phys. Rev. Lett. **92**, 130403 (2004)
34. I.B. Spielman, W.D. Phillips, J.V. Porto, Mott-insulator transition in a two-dimensional atomic Bose gas. Phys. Rev. Lett. **98**, 080404 (2007)
35. B. Paredes, A. Widera, V. Murg, O. Mandel, S. Fölling, I. Cirac, G. Shlyapnikov, T. Hänsch, I. Bloch, Tonks-Girardeau gas of ultracold atoms in an optical lattice. Nature **429**, 277 (2004)
36. T. Kinoshita, T. Wenger, D.S. Weiss, observation of a one-dimensional Tonks-Girardeau gas. Science **305**, 1125 (2004)
37. M. Köhl, H. Moritz, T. Stöferle, K. Günter, T. Esslinger, Fermionic atoms in a three dimensional optical lattice: Observing fermi surfaces, dynamics, and interactions. Phys. Rev. Lett. **94**, 080403 (2005)
38. S. Fölling, S. Trotzky, P. Cheinet, M. Feld, R. Saers, A. Widera, T. Müller, I. Bloch, Direct observation of second-order atom tunnelling. Nature **448**, 1029 (2007)
39. M. Anderlini, P. Lee, B. Brown, J. Sebby-Strabley, W. Phillips, J. Porto, Controlled exchange interaction between pairs of neutral atoms in an optical lattice. Nature **448**, 452 (2007)
40. S. Trotzky, P. Cheinet, S. Fölling, M. Feld, U. Schnorrberger, A.M. Rey, A. Polkovnikov, E.A. Demler, M.D. Lukin, I. Bloch, Time-resolved observation and control of superexchange interactions with ultracold atoms in optical lattices. Science **319**, 295 (2008)
41. P. Soltan-Panahi, J. Struck, P. Hauke, A. Bick, W. Plenkers, G. Meineke, C. Becker, P. Windpassinger, M. Lewenstein, K. Sengstock, Multi-component quantum gases in spin-dependent hexagonal lattices. Nat. Phys. **7**, 434 (2011)
42. R. Feynman, Simulating physics with computers. Int. J.Theor. Phys. **21**, 467 (1982)
43. N.W. Ashcroft, N.D. Mermin, *Solid state physics* (Saunders College Publishing, Fort Worth, 1976)
44. A. Altland, B. Simons, *Condensed Matter Field Theory* (Cambridge University Press, Cambridge, 2006)
45. G. Kotliar, D. Vollhardt, Strongly correlated materials: insights from dynamical mean-field theory. Phys. Today **57**, 53 (2004)
46. P.A. Lee, N. Nagaosa, X.-G. Wen, Doping a Mott insulator: physics of high-temperature superconductivity. Rev. Mod. Phys. **78**, 17 (2006)
47. J.G. Bednorz, K.A. Müller, Possible High T_c Superconductivity in the Ba-La-Cu-O System. Zeitschrift für Physik B - Condens. Matter **64**, 189 (1986)
48. R.J. Glauber, coherent and incoherent states of the radiation field. Phys. Rev. **131**, 2766 (1963)
49. M. Scully and S. Zubairy, *Quantum Optics* (Cambridge University Press, Cambridge, 1997)

50. S. Haroche, J.-M. Raimond, *Exploring the Quantum: Atoms, Cavities, and Photons (Oxford Graduate Texts)*, 1st edn. (Oxford University Press, USA, 2006)
51. M. Greiner, O. Mandel, T. Hänsch, I. Bloch, Collapse and revival of the matter wave field of a Bose-Einstein condensate. Nature **419**, 51 (2002)
52. J. Sebby-Strabley, B.L. Brown, M. Anderlini, P.J. Lee, W.D. Phillips, J.V. Porto, P.R. Johnson, Preparing and probing atomic number states with an atom interferometer. Phys. Rev. Lett. **98**, 200405 (2007)
53. M. Brune, F. Schmidt-Kaler, A. Maali, J. Dreyer, E. Hagley, J.M. Raimond, S. Haroche, Quantum Rabi oscillation: a direct test of field quantization in a cavity. Phys. Rev. Lett. **76**, 1800 (1996)
54. M. Lewenstein, L. Santos, M.A. Baranov, H. Fehrmann, Atomic Bose-Fermi mixtures in an optical lattice. Phys. Rev. Lett. **92**, 050401 (2004)
55. D.-W. Wang, M.D. Lukin, E. Demler, Engineering superfluidity in Bose-Fermi mixtures of ultracold atoms. Phys. Rev. A **72**, 051604 (2005)
56. L. Mathey, D.-W. Wang, W. Hofstetter, M.D. Lukin, E. Demler, Luttinger liquid of polarons in one-dimensional Boson-Fermion mixtures. Phys. Rev. Lett. **93**, 120404 (2004)
57. H.P. Büchler, G. Blatter, Supersolid versus phase separation in atomic Bose-Fermi mixtures. Phys. Rev. Lett. **91**, 130404 (2003)
58. U. Gavish, Y. Castin, Matter-wave localization in disordered cold atom lattices. Phys. Rev. Lett. **95**, 020401 (2005)
59. A. Privitera, W. Hofstetter, Polaronic slowing of fermionic impurities in lattice Bose-Fermi mixtures. Phys. Rev. A **82**, 063614 (2010)
60. L. Platter, Low-energy universality in atomic and nuclear physics. Few-Body Syst. **46**, 139 (2009)

Chapter 2
Towards Strongly Interacting Bosons and Fermions

This chapter sets the stage from which this thesis departs into the realm of strongly interacting bosons and fermions. Several theoretical concepts that are needed for a basic understanding of the experiments in later chapters are briefly summarized.

The first section introduces theoretical background on ultracold quantum gases. After a fast journey through second quantization, quantum statistics and basics of statistical mechanics, ideal fermions and weakly interacting bosons are discussed in the Thomas-Fermi limit. The consequences of interactions in ultracold quantum gases are exemplarily illustrated in a mean-field analysis of an interacting Bose-Fermi mixture in a harmonic trap. The section concludes with a discussion of possible routes towards strongly interacting quantum systems. The second section is dedicated to the theory of optical lattices. After discussing the landscape of experimental lattice potentials, we turn to the band structure of a simple cubic lattice and introduce the Wannier basis. In the third section, the concept of Feshbach resonances is presented. After a brief summary of the basics of quantum mechanical scattering theory, resonance scattering, the emergence of bound molecular states and the problem of two interacting atoms in a tight harmonic potential are discussed. Finally, the specific Feshbach resonances that are used in the experiments of this thesis are introduced.

2.1 Ultracold Quantum Gases

We start with a reminder of quantum statistics [1] and many-body quantum theory in second quantization [2]. Second quantization is the language of choice to formulate the many-body Hamiltonians for bosons and fermions in optical lattice potentials. It is heavily used throughout the thesis. Then, basic formalisms for bosonic and fermionic quantum gases in harmonic trapping potentials are introduced, including the important Thomas-Fermi approximation [3–5]. The bosonic and fermionic formalisms are both applied in a self-consistent mean-field calculation for an interacting Bose-Fermi mixture in a harmonic trap, revealing a marked influence of interspecies interactions on the phases of the mixture. Finally, we discuss under which

S. Will, *From Atom Optics to Quantum Simulation*, Springer Theses,
DOI: 10.1007/978-3-642-33633-1_2,

conditions a quantum system can be regarded as strongly interacting [6, 7] and identify fundamental routes to reach this regime with ultracold atoms.

2.1.1 Bosons and Fermions

Indistinguishability of particles is a fundamental concept of quantum theory. We consider a system of N indistinguishable particles. The single-particle states in the system are described by a basis of single-particle wavefunctions $\{\psi_E(x)\}$, where x denotes the collection of spatial and spin coordinates and E uniquely labels the state by representing a complete set of single-particle quantum numbers. For example, for free fermions in a box E denotes the momentum $\mathbf{k}$ and the z-component of the spin, s_z. Using the single-particle wavefunctions, the most general many-particle wavefunction $\Psi(x_1, \ldots x_N)$ for the whole N-particle system can be constructed by

$$\Psi(x_1, \ldots, x_N) = \sum_{E_1,\ldots,E_N} C(E_1, \ldots, E_N)\, \psi_{E_1}(x_1) \cdots \psi_{E_N}(x_N), \tag{2.1}$$

where each E_k in the sum runs over the complete set of quantum numbers. Because quantum particles are fundamentally indistinguishable, a single-particle wavefunction cannot be strictly assigned to a certain particle. This implies the invariance of observables under the exchange of particles. Particularly, the probability density $|\Psi|^2$ must be unchanged:

$$|\Psi(\ldots, x_k, \ldots, x_l, \ldots)|^2 = |\Psi(\ldots, x_l, \ldots, x_k, \ldots)|^2. \tag{2.2}$$

From this we conclude that there are two possibilities for the sign of the many-particle wavefunction when particle coordinates are exchanged:

$$\Psi(\ldots, x_k, \ldots, x_l, \ldots) = \pm\Psi(\ldots, x_l, \ldots, x_k, \ldots). \tag{2.3}$$

Therefore, the principle of indistinguishability suggests that quantum particles are grouped into two fundamental classes: First, the bosons, for which the wavefunction transforms symmetrically (upper sign). Second, the fermions, for which the wavefunction transforms antisymmetrically (lower sign). Indeed, the famous spin-statistics theorem in quantum field theory states, that particles with integer spin are bosons and particles with half-integer spin are fermions [8]. According to Eq. 2.3 two fermions cannot occupy the same quantum state, because then Ψ would be equal to $-\Psi$ implying, that the wavefunction must vanish

$$\Psi(\ldots, x_k, \ldots, x_k, \ldots) = 0. \tag{2.4}$$

This is Pauli's exclusion principle for fermions.

Quantum statistics has a profound impact on the construction of many-particle wavefunctions. For the case of bosons, a single-particle quantum state can be occupied by many particles. Accordingly, the general many-particle wavefunction for N bosons is expanded in terms of a basis of completely symmetrized wavefunctions

$$\Phi^{\mathrm{B}}_{n_1,\ldots,n_\infty}(x_1,\ldots,x_N) = \left(\frac{n_1!\cdots n_\infty!}{N!}\right)^{1/2} \sum_{\substack{E_1,\ldots,E_N \\ (n_1,\ldots,n_\infty)}} \psi_{E_1}(x_1)\cdots\psi_{E_N}(x_N). \quad (2.5)$$

Here, the sum runs over all possibilities, in which N particles are distributed over the single-particle states, such that n_1 particles are in state ψ_{E_1}, n_2 particles are in state ψ_{E_2} and so on. In total, there are $N!/(n_1!n_2!\cdots n_\infty!)$ possibilities, which explains the normalization.

In the case of fermions, Pauli's principle restricts the occupation of single-particle states to either $n_k = 0$ or 1. The basis states of the general many-particle wavefunction for N fermions are given by the normalized Slater determinants

$$\Phi^{\mathrm{F}}_{n_1,\ldots,n_\infty}(x_1,\ldots,x_N) = \frac{1}{\sqrt{N!}} \begin{vmatrix} \psi_{E_1}(x_1) & \psi_{E_1}(x_2) & \cdots & \psi_{E_1}(x_N) \\ \psi_{E_2}(x_1) & \psi_{E_2}(x_2) & \cdots & \psi_{E_2}(x_N) \\ \vdots & \vdots & \ddots & \vdots \\ \psi_{E_N}(x_1) & \psi_{E_N}(x_2) & \cdots & \psi_{E_N}(x_N) \end{vmatrix}. \quad (2.6)$$

The mathematical properties of the determinant ensure, that the basis wavefunctions transform antisymmetrically under exchange of any two particles.

Second Quantization

Second quantization offers a convenient way to capture the symmetry properties of bosons and fermions without explicitly writing down the above many-particle basis states during calculations. The underlying formalism implicitly takes care of maintaining the appropriate symmetry of the many-particle wavefunctions. For example the completely symmetrized bosonic N-particle state 2.5 is represented by an occupation number state $|\Phi^{\mathrm{B}}_{n_1,\ldots,n_\infty}\rangle = |n_1, n_2, \ldots\rangle \equiv |n_1\rangle|n_2\rangle\cdots|n_\infty\rangle$ with the occupation numbers n_i as defined above.

For bosons, the whole range of integer occupation numbers is allowed, including zero. In order to construct and manipulate the occupation number states, the creation and annihilation operators $\hat{a}_k^\dagger$ and $\hat{a}_k$ are introduced for each single-particle state k. The operators obey the bosonic commutation relations

$$[\hat{a}_k, \hat{a}_l]_- = 0, \quad [\hat{a}_k^\dagger, \hat{a}_l^\dagger]_- = 0 \quad \text{and} \quad [\hat{a}_k, \hat{a}_l^\dagger]_- = \delta_{kl}, \quad (2.7)$$

where $[A, B]_- = AB - BA$. These relations determine all properties of the operators. They imply, that the creation operator $\hat{a}_k^\dagger$ raises and the annihilation operator $\hat{a}_k$

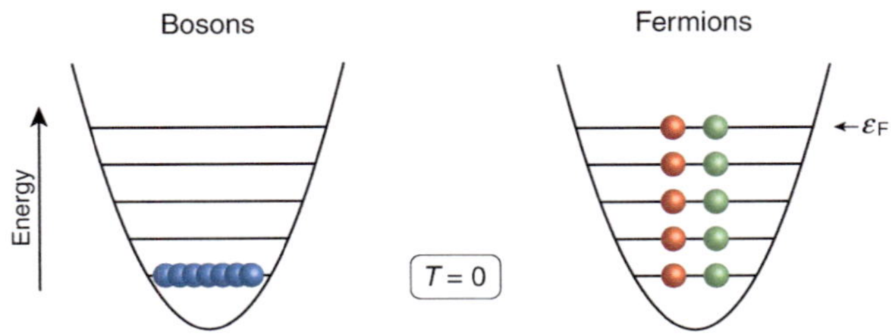

Fig. 2.1 Bosonic and fermionic quantum statistics. At zero temperature bosons form a Bose-Einstein condensate, while fermions arrange in a Fermi sea. The energy of the highest occupied state in the Fermi sea is the Fermi energy ϵ_F. *Green* and *red balls* correspond to two spin states (spin up and spin down). The *color* coding introduced in this figure is kept throughout the thesis: *blue balls* indicate bosons, *red* and *green balls* indicate fermions

lowers the number of particles in the single-particle state ψ_{E_k} by one according to $\hat{a}_k^\dagger|n_k\rangle = \sqrt{n_k+1}\,|n_k+1\rangle$, $\hat{a}_k|n_k\rangle = \sqrt{n_k}|n_k-1\rangle$, and especially $\hat{a}_k|0\rangle = 0$, where $|0\rangle$ is the vacuum state. Furthermore, it is easy to show that the eigenvalues of the operator $\hat{n}_k \equiv \hat{a}_k^\dagger\hat{a}_k$ correspond to the number of particles occupying the single-particle state ψ_{E_k}, which suggests the name number operator for $\hat{n}_k$.

For fermions, the occupation numbers are restricted to $n_k = 0$ or 1 and the many-particle wavefunction must be antisymmetric. Those requirements are automatically taken care of by defining the anticommutation relations for the fermionic creation and annihilation operators $c_k^\dagger$ and c_k according to

$$[\hat{c}_k, \hat{c}_l]_+ = 0, \quad [\hat{c}_k^\dagger, \hat{c}_l^\dagger]_+ = 0 \quad \text{and} \quad [\hat{c}_k, \hat{c}_l^\dagger]_+ = \delta_{kl}, \tag{2.8}$$

where $[A, B]_+ = AB + BA$. With these relations one can show, that $\hat{c}_k^\dagger\hat{c}_k^\dagger|0\rangle = 0$, which prevents double occupation of a single quantum state (Pauli's exclusion principle), as well as $c_k^\dagger|0\rangle = |1\rangle$, $c_k^\dagger|1\rangle = 0$, $c_k|1\rangle = |0\rangle$ and $c_k|0\rangle = 0$.

Within the formalism of second quantization it is simple to write down many-particle ground states. For N noninteracting, spinless bosons with the single-particle ground state ψ_{E_0}, the many-particle ground state can be directly constructed via

$$|\Phi_B\rangle = |N, 0, 0, \ldots\rangle \equiv \frac{1}{\sqrt{N!}}(\hat{a}_0^\dagger)^N|0, 0, 0, \ldots\rangle. \tag{2.9}$$

In this state, all particles occupy the same single-particle state, which is the defining property of a Bose-Einstein condensate (see Fig. 2.1).

In a fermionic N-particle system each single-particle state can only be occupied by a single fermion at most. Therefore, the zero-temperature many-particle ground state is realized, when the N single-particle states with lowest energy are filled from bottom up as expressed by

$$|\Phi_{\rm F}\rangle = |\underbrace{1,1,\ldots,1}_{N \text{ times}},0,0,\ldots,0\rangle = \prod_{k\le k_{\rm F}} \hat{c}_k^\dagger |0,0,0,\ldots\rangle. \tag{2.10}$$

The indices are ordered in such a way, that a lower index corresponds to a lower energy. The index $k_{\rm F}$ is defined such, that $\sum_{k\le k_{\rm F}} = N$. The energy $E_{k_{\rm F}} \equiv \epsilon_{\rm F}$ of the highest occupied single-particle state is called the Fermi energy (see Fig. 2.1).

Operators in Second Quantization

As shown above, second quantization offers a concise way to express many-particle states. However, the formalism unfolds its full potential, when it is used to rewrite N-particle Hamiltonians in a way that allows for elegant diagonalization formalisms. In real space, a N-particle Hamiltonian generally has the form

$$\begin{aligned} H = \sum_{k=1}^{N} T(x_k) + \sum_{k\ne l=1}^{N} V_2(x_k, x_l) \\ + \sum_{k\ne l\ne m=1}^{N} V_3(x_k, x_l, x_m) + \ldots \end{aligned} \tag{2.11}$$

Here $T(x_k)$ denotes an operator acting on a single particle (e.g. kinetic or potential energy), $V_2(x_k, x_l)$ an operator acting on two particles (e.g. two-body interaction between particles) and $V_3(x_k, x_l, x_m)$ an operator acting on three particles (e.g. three-body interaction between particles). Generally, also operators involving a larger number of particles can play a role as indicated by the dots. We will see in Chap. 7 that higher particle terms can happen to be relevant not only theoretically, but also experimentally. In second quantization the Hamiltonian takes the form [2]

$$\begin{aligned} \hat{H} = \sum_{ij} \hat{a}_i^\dagger \langle i|T|j\rangle \hat{a}_j + \frac{1}{2}\sum_{ijkl} \hat{a}_i^\dagger \hat{a}_j^\dagger \langle ij|V_2|kl\rangle \hat{a}_l \hat{a}_k \\ + \frac{1}{6}\sum_{\substack{ijk\\lmn}} \hat{a}_i^\dagger \hat{a}_j^\dagger a_k^\dagger \langle ijk|V_3|lmn\rangle \hat{a}_n \hat{a}_m \hat{a}_l + \ldots \end{aligned} \tag{2.12}$$

Here, only the case of bosonic operators is shown for brevity. The identical expression holds for fermions, when the operators $\hat{a}_k$ are replaced by $\hat{c}_k$. However, it is very important to keep the ordering of the indices as changes affect the overall sign. The matrix elements are complex numbers that are calculated by integration over the generalized coordinate x, for example $\langle i|T|j\rangle = \int dx\, \psi_{E_i}^*(x)\, T(x)\, \psi_{E_j}(x)$. We come across Hamiltonians of this kind several times in this thesis. Nevertheless, it is often convenient to use Hamiltonian 2.12 in a slightly different form that is obtained by introducing the field operators $\hat{\psi}(x) = \sum_k \psi_k(x)\hat{a}_k$ and $\hat{\psi}^\dagger(x) = \sum_k \psi_k^*(x)\hat{a}_k^\dagger$.

A simple calculation yields

$$\hat{H} = \int dx\, \hat{\psi}^\dagger(x)\, T(x)\, \hat{\psi}(x) + \frac{1}{2}\int dx\, dx'\, \hat{\psi}^\dagger(x)\hat{\psi}^\dagger(x')\, V_2(x, x')\, \hat{\psi}(x')\hat{\psi}(x)$$
$$+ \frac{1}{6}\int dx\, dx'\, dx''\, \hat{\psi}^\dagger(x)\hat{\psi}^\dagger(x')\hat{\psi}^\dagger(x'')\, V_3(x, x', x'')\, \hat{\psi}(x'')\hat{\psi}(x')\hat{\psi}(x) + \ldots \tag{2.13}$$

2.1.2 Bose-Einstein and Fermi-Dirac Distribution

In the preceding section, we have derived the effects of quantum statistics on an elementary level. However, in systems of practical importance the particle number is often very large and, even more importantly, such systems generally have a finite temperature. Therefore it is crucial to consider the impact of quantum statistics on statistical mechanics and thermodynamics.

According to the fundamental postulate of statistical mechanics, a macroscopic system in thermodynamic equilibrium is equally likely to be in any of the states that satisfy the macroscopic conditions [1]. This means, that the system is a member of an ensemble, in which the total energy E, the particle number N and the volume V are fixed. It is called the *microcanonical* ensemble. Remarkably, those conditions are quite closely met in experiments with ultracold atoms.[1] However, for calculations it is often more practical to consider the system being in contact with a large reservoir, which allows for the exchange of particles and energy. The corresponding ensemble is called the *grand canonical ensemble*, in which the temperature $T = 1/(k_B\beta)$ and the chemical potential μ, which is the energy cost to add a particle, are fixed. The probability of the system to be in any state with particle number N and total energy E is determined by the Boltzmann factor $e^{-\beta(E-\mu N)}/Z$, where Z is the partition function, which we are going to derive for ideal gases in the following. The partition function has crucial importance for the calculation of ensemble averages of physical observables (Fig. 2.2).

We assume an ideal gas consisting of noninteracting particles, where the eigenenergies of the single-particle states are denoted by ϵ_i and the many-particle states are given by $|n_1, \ldots, n_\infty\rangle$ with the total energy $\sum_i \epsilon_i n_i$. The grand canonical partition function is then given by [2]

$$Z = \prod_{i=1}^{\infty} \mathrm{Tr}_i\, e^{-\beta(\epsilon_i-\mu)\hat{n}_i} = \prod_{i=1}^{\infty}\sum_n \left(e^{-\beta(\epsilon_i-\mu)}\right)^n. \tag{2.14}$$

[1] This is true under the assumption that ultracold atom systems are truly in thermodynamic equilibrium. In practice, an ultracold sample under investigation has typically undergone a sequence of parameter changes, which can only be adiabatic to a certain degree. For example, changes of E and V are induced by variation of the interparticle interactions or the trapping potential.

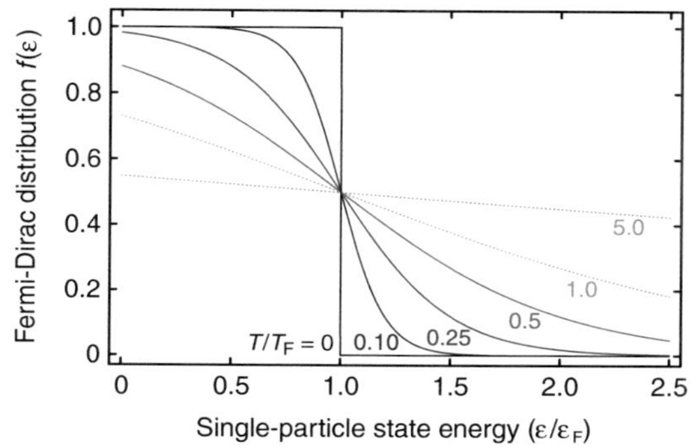

Fig. 2.2 Fermi-Dirac distribution for several dimensionless temperatures T/T_{F} as a function of the dimensionless energy of single-particle quantum states $\epsilon/\epsilon_{\mathrm{F}}$. *Solid curves* indicate the quantum degenerate regime

We can further evaluate this expression by taking into account quantum statistics. For bosons, the occupation numbers n are unrestricted and cover all integer numbers, which yields

$$Z_{\mathrm{B}} = \prod_i \frac{1}{1 - e^{-\beta(\epsilon_i - \mu)}} \tag{2.15}$$

and for fermions, n can either take the value 0 or 1, such that

$$Z_{\mathrm{F}} = \prod_i (1 + e^{-\beta(\epsilon_i - \mu)}). \tag{2.16}$$

Now, statistics can be connected to thermodynamics via the fundamental relation [2]

$$\Omega(T, V, \mu) = -\frac{1}{\beta} \ln Z = \pm\frac{1}{\beta} \sum_i \ln(1 \mp e^{\beta(\mu - \epsilon_i)}), \tag{2.17}$$

which is the grand canonical potential that allows to calculate all macroscopic thermodynamic properties in equilibrium. The upper (lower) sign refers to bosons (fermions) in this and the following equations. The mean total atom number of the ideal gas is given by

$$\langle N \rangle = -\frac{\partial \Omega}{\partial \mu} = \sum_i \langle n_i \rangle \quad \text{with} \quad \langle n_i \rangle = \frac{1}{e^{\beta(\epsilon_i - \mu)} \mp 1} = f_{\mathrm{B/F}}(\epsilon_i), \tag{2.18}$$

which are the famous Bose-Einstein and the Fermi-Dirac distributions for the mean occupation of individual quantum states in noninteracting systems. Additionally the mean entropy is given by [1]

$$\langle S \rangle = -\frac{\partial \Omega}{\partial T} = k_B \sum_i \frac{\beta(\epsilon_i - \mu)}{e^{\beta(\epsilon_i - \mu)} \mp 1} \mp k_B \ln\left(1 \mp e^{-\beta(\epsilon_i - \mu)}\right) \tag{2.19}$$

It is important to note that the results derived here are generally valid for all systems of noninteracting bosonic and fermionic particles. However, for practical calculations it often proves useful to replace the discrete summations over the state index i, which appears in Eqs. 2.17–2.19, by a continuous integral. This leads to the concept of the density-of-states capturing the effects of the trapping potential on the energy levels.

2.1.3 Quantum Degenerate Fermionic Gases

A gas of fermionic atoms in a single spin state at low temperature is the "most ideal" gas one can think of. As we will discuss later in this chapter, low energy collisions between fermionic atoms are largely suppressed due to Pauli's principle (see Sect. 2.3). We consider the gas to be trapped in a three-dimensional harmonic potential

$$V(\mathbf{r}) = \frac{1}{2}m\left(\omega_x^2 x^2 + \omega_y^2 y^2 + \omega_z^2 z^2\right), \tag{2.20}$$

where m is the mass of the particles and the ω_α denote the trapping frequencies in each direction $\alpha = x, y, z$. The eigenenergies of the single-particle states are given by [9]

$$\epsilon_{n_x,n_y,n_z} = \hbar\sum_\alpha \omega_\alpha\left(n_\alpha + \frac{1}{2}\right). \tag{2.21}$$

This single-particle spectrum of the harmonic oscillator gives rise to the density-of-states[2]

$$g(\epsilon) = \frac{\epsilon^2}{2(\hbar\bar{\omega})^3}, \tag{2.22}$$

where $\bar{\omega} = (\omega_x\omega_y\omega_z)^{1/3}$ is the geometric mean of the trapping frequencies. With Eq. 2.22 and the Fermi-Dirac distribution the total particle number N in the system can be expressed by

$$N = \int_0^\infty d\epsilon\,\frac{g(\epsilon)}{e^{\beta(\epsilon-\mu)}+1} = -\left(\frac{k_BT}{\hbar\bar{\omega}}\right)^3 \mathrm{Li}_3(-e^{\beta\mu}) \xrightarrow{\beta\to\infty} \int_0^\infty d\epsilon\, g(\epsilon)\Theta(\mu-\epsilon). \tag{2.23}$$

At a certain temperature T and total particle number N, this relation implicitly fixes the chemical potential μ. When the temperature approaches zero, the Fermi-Dirac distribution becomes a step function, indicating that the single-particle states are filled from bottom up.[3] Under these conditions the chemical potential is called the

---

[2] The continuum approximation of the quantum mechanically discrete eigenenergies is strictly speaking only valid, when the discreteness is not resolved, for example, due to finite temperature. For the harmonic oscillator, the condition justifying the use of a density-of-states requires that $k_BT \gg \hbar\omega_\alpha$ ($\alpha = x, y, z$).

[3] $\Theta(x)$ is defined as 0 for $x < 0$ and 1 for $x \geq 0$.

Fermi energy ϵ_{F}, denoting the energy of the highest occupied single-particle state. Using Eq. 2.23 we obtain

$$\epsilon_{\mathrm{F}} = \hbar\bar{\omega}(6N)^{1/3}, \tag{2.24}$$

which allows for the definition of the Fermi temperature $T_{\mathrm{F}} = \epsilon_{\mathrm{F}}/k_B$ and the Fermi momentum $k_{\mathrm{F}} = \sqrt{2m\epsilon_{\mathrm{F}}/\hbar^2}$. Furthermore, using the Eqs. 2.19 and 2.22 the total entropy of the system can be calculated [10, 11], which to the lowest order in the temperature reads

$$\frac{S}{k_B} = \pi^2 N \frac{T}{T_{\mathrm{F}}} + \mathcal{O}\left[\left(\frac{T}{T_{\mathrm{F}}}\right)^2\right]. \tag{2.25}$$

Thomas-Fermi Approximation

The calculation of finite temperature properties of an ideal Fermi gas, such as the real space density distribution, is considerably simplified by taking a semi-classical approach that is called the Thomas-Fermi approximation. This approach is quantum in so far, that the Fermi-Dirac distribution is used, but classical in so far, that the energies of the single-particle states ϵ are approximated by the classical Hamiltonian $H(\mathbf{r}, \mathbf{p})$, according to

$$f_{\mathrm{F}}(\mathbf{r}, \mathbf{p}) = \frac{1}{e^{\beta\left(\frac{\mathbf{p}^2}{2m} + V(\mathbf{r}) - \mu\right)} + 1}. \tag{2.26}$$

The semi-classical approach is valid in the limit of large particle numbers and corresponds to a local-density approximation [5]. The volume of a single quantum state viewed in the classical phase space $(\mathbf{r}, \mathbf{p})$ is $(2\pi\hbar)^3$. Accordingly, the real-space density distribution in an arbitrary potential $V(\mathbf{r})$ is obtained by integration over all momenta [4, 5]

$$n_{\mathrm{F}}(\mathbf{r}) = \frac{1}{(2\pi\hbar)^3} \int d\mathbf{p}\, f_{\mathrm{F}}(\mathbf{r}, \mathbf{p}) = -\frac{1}{\lambda_{\mathrm{dB}}^3} \mathrm{Li}_{3/2}\left(-e^{\beta(\mu - V(\mathbf{r}))}\right), \tag{2.27}$$

where $\lambda_{\mathrm{dB}} = \sqrt{2\pi\hbar^2/mk_B T}$ is the de Broglie wavelength. $\mathrm{Li}_n(z)$ denotes the polylogarithm of nth order.[4]

[4] The polylogarithm is defined by a series expansion $\mathrm{Li}_n(z) = \sum_{k=1}^{\infty} x^k/k^n$ that can also be written as [4]

$$\mathrm{Li}_n(z) = \frac{1}{\pi^n} \int d^{2n}r \frac{1}{e^{\mathbf{r}^2}/z - 1},$$

where $\mathbf{r}$ denotes a vector in $2n$ dimensions. Note the limiting values $\mathrm{Li}_n(z) \xrightarrow{z \ll 1} z$ and $-\mathrm{Li}_n(-z) \xrightarrow{z \to \infty} \ln^n(z)/\Gamma(n+1)$, where $\Gamma(n)$ is the Gamma function.

Assuming a three-dimensional harmonic potential $V(\mathbf{r})$ (see Eq. 2.20), it is instructive to consider the Thomas-Fermi approximation in the limits of high and zero temperature. In the first case, $T \rightarrow \infty$, the classical Boltzmann distribution is recovered

$$n_{\mathrm{cl}}(\mathbf{r}) = \frac{N}{\pi^{3/2}\sigma_x\sigma_y\sigma_z} e^{-\sum_\alpha \alpha^2/\sigma_\alpha^2} \quad \text{with} \quad \sigma_\alpha^2 = \frac{k_B T}{m\omega_\alpha^2}, \tag{2.28}$$

corresponding to a Gaussian distribution as expected for a harmonic potential ($\alpha = x, y, z$). In the second case, $T \rightarrow 0$, we obtain the profile

$$\begin{aligned} n_{\mathrm{F}}(\mathbf{r}) &= \frac{(2m)^{\frac{3}{2}}}{6\pi^2\hbar^3} \mathfrak{Re}\left[(\epsilon_{\mathrm{F}} - V(\mathbf{r}))^{\frac{3}{2}}\right] \\ &= \frac{8}{\pi^2} \frac{N}{R_{\mathrm{F}x} R_{\mathrm{F}y} R_{\mathrm{F}z}} \mathfrak{Re}\left[\left(1 - \sum_\alpha \frac{\alpha^2}{R_{\mathrm{F}\alpha}^2}\right)^{\frac{3}{2}}\right], \end{aligned} \tag{2.29}$$

keeping in mind the definition of the Fermi energy ϵ_{F} as the zero temperature chemical potential 2.24. The extension of the cloud in the directions of the harmonic potential that is possible at the energy ϵ_{F} is called the Fermi radius

$$R_{\mathrm{F}\alpha} = \sqrt{\frac{2\epsilon_{\mathrm{F}}}{m\omega_\alpha^2}} = \sqrt{\frac{\hbar}{m\omega_\alpha}} (48N)^{1/6}. \tag{2.30}$$

The density profiles $n_{\mathrm{cl}}(\mathbf{r})$ and $n_{\mathrm{F}}(\mathbf{r})$ play an important role in the thermometry of ultracold fermion clouds, which is discussed in Chap. 6 and Appendix C.

2.1.4 Quantum Degenerate Bosonic Gases

The behavior of an ideal gas of bosonic atoms differs fundamentally from the fermionic case: When the temperature is lowered, an ideal Bose gas undergoes a phase transition and the single-particle ground state of the system becomes macroscopically occupied. This phenomenon of Bose-Einstein condensation is fundamentally rooted in quantum statistics and can be identified on general grounds by examining the Bose-Einstein distribution (see Eq. 2.18)

$$f_{\mathrm{B}}(\epsilon_i) = \frac{1}{e^{\beta(\epsilon_i - \mu)} - 1}. \tag{2.31}$$

Assuming without loss of generality, that the energy of the single-particle ground state ϵ_0 vanishes, $f_{\mathrm{B}}(\epsilon_0) \xrightarrow{\beta\rightarrow\infty} \infty$ diverges, when the temperature approaches zero. Note that $\mu \leq 0$, because $f_{\mathrm{B}}(\epsilon_i)$ must assume nonnegative values. The divergence of the Bose-Einstein distribution entails a macroscopic occupation of the single-particle ground state.

Assuming a three-dimensional harmonic oscillator with the density-of-states 2.22, we obtain for the total particle number

$$N - N_0 = \int_0^\infty d\epsilon \frac{g(\epsilon)}{e^{\beta(\epsilon-\mu)} - 1} = \left(\frac{k_B T}{\hbar\bar{\omega}}\right)^3 \mathrm{Li}_3(e^{\beta\mu}) \tag{2.32}$$

where it is crucial to separate out the ground state occupation N_0 that is otherwise not properly accounted for by the integral. The maximal particle number that could be accommodated in the system at a fixed temperature without condensation, $N_0 = 0$, is reached for $e^{\beta\mu} \to 1$ since $\mathrm{Li}_3(z)$ is monotonically increasing and $0 \leq e^{\beta\mu} \leq 1$. Therefore, the chemical potential must be $\mu = 0$ at this point. In turn, if a fixed number of particles is to be accommodated in the system, there is a critical temperature T_c, below which a fraction of the atoms must occupy the ground state. The critical temperature follows from Eq. 2.32 by setting $N_0 = 0$ and $\mu = 0$

$$k_B T_c = \hbar\bar{\omega} \left(\frac{N}{\mathrm{Li}_3(1)}\right)^{1/3} \approx 0.94\hbar\bar{\omega}N^{1/3}. \tag{2.33}$$

Insertion of this result into Eq. 2.32 yields the fraction of condensed atoms as a function of temperature

$$\frac{N_0}{N} = 1 - \left(\frac{T}{T_c}\right)^3. \tag{2.34}$$

Using a semi-classical approach analogous to Eq. 2.27, it turns out that condensation sets in, when the density reaches $n_{\mathrm{B}}^{\max} = \mathrm{Li}_{3/2}(1)/\lambda_{\mathrm{dB}}^3 = 2.612/\lambda_{\mathrm{dB}}^3$. This corresponds to the intuitive argument that Bose-Einstein condensation happens, when the de Broglie wavelength reaches the same order of magnitude as the interparticle spacing.

Weakly Interacting Bose Gas

Atomic Bose gases are not as ideal as spin-polarized Fermi gases, because interactions are not suppressed at low temperatures. A realistic description needs to include interparticle interactions and this has been successfully done using the Gross-Pitaevskii Eq. (2.3) for the ground state of the many-particle system $\Phi(\mathbf{r}, t)$. This is a nonlinear Schrödinger equation

$$i\hbar\frac{\partial}{\partial t}\Phi(\mathbf{r}, t) = \left(-\frac{\hbar^2}{2m}\Delta + V(\mathbf{r}) + g|\Phi(\mathbf{r}, t)|^2\right)\Phi(\mathbf{r}, t), \tag{2.35}$$

where the interactions are included by the parameter $g = 4\pi\hbar^2 a_s/m$ with the scattering length a_s (see Sect. 2.3 for further details). With the knowledge, that in a pure Bose-Einstein condensate all atoms occupy an unique single-particle state, it is plausible to make the ansatz $\Phi(\mathbf{r}, t) = \phi(\mathbf{r})e^{-i\mu t/\hbar}$. Here, $\phi(\mathbf{r})$ is understood to be normalized to the total particle number $\int d\mathbf{r}\, |\phi(\mathbf{r})|^2 = N$ and μ is the chemical potential. Accordingly, the time independent Gross-Pitaevskii equation reads

$$\left(-\frac{\hbar^2}{2m}\triangle + V(\mathbf{r}) + g|\phi(\mathbf{r})|^2\right)\phi(\mathbf{r}) = \mu\phi(\mathbf{r}), \tag{2.36}$$

allowing to calculate the real space wavefunction $\phi(\mathbf{r})$ that is connected to the density distribution via $n_{\mathrm{B}}(\mathbf{r}) = |\phi(\mathbf{r})|^2$. In general, this equation must be solved numerically. However, it turns out that under most experimental conditions the contribution of the kinetic energy term, proportional to $\triangle$, is negligible compared to the potential and interaction energy [3]. Omitting the kinetic energy, the real space density of the weakly interacting Bose gas is simply given by

$$n_{\mathrm{B}}(\mathbf{r}) = |\psi(\mathbf{r})|^2 = \max\left[\frac{\mu - V(\mathbf{r})}{g}, 0\right]. \tag{2.37}$$

This approximation is also called Thomas-Fermi approximation due to the close analogy to Eq. 2.29.

For the case of harmonic confinement (Eq. 2.20), it is straightforward to derive for the chemical potential

$$\mu = \frac{\hbar\bar{\omega}}{2}\left(\frac{15Na_s}{\ell}\right)^{2/5}, \tag{2.38}$$

where $\ell = \sqrt{\hbar/(m\bar{\omega})}$ is the harmonic oscillator length corresponding to $\bar{\omega}$. The Thomas-Fermi radius for weakly interacting bosons in a spherically symmetric trap is given by

$$R_{\mathrm{B}} = \ell\left(\frac{15Na_s}{\ell}\right)^{1/5}. \tag{2.39}$$

2.1.5 *Interacting Bose-Fermi Mixtures in a Harmonic Trap*

In the experiment, we realize interacting mixtures of quantum degenerate bosons and fermions by sympathetic cooling in a harmonic trap. This technique requires both interactions and good overlap of the two species to ensure proper thermalization. However, when approaching the quantum degenerate regime, the densities can become high enough that interactions considerably influence the density distribution of both species inside the trap. In the case of repulsive interactions they can even phase separate.

In the following section, we will derive phases of a quantum degenerate interacting mixture of bosons and fermions in a harmonic trap. We will see that in—what is commonly called—the weakly interacting regime, interactions can still have a dramatic influence on the quantum phases. We use a mean-field approach to identify phase separation ($a_{\mathrm{BF}} \gg 0$), density enhancement ($a_{\mathrm{BF}} < 0$) and eventually the collapse of the mixture ($a_{\mathrm{BF}} \ll 0$), as a function of the Bose-Bose and Bose-Fermi interaction strength. We discuss this simple model here, first of all, because it is experimentally important to understand the density distribution of harmonically trapped Bose-Fermi mixtures prior to the lattice ramp-up. Additionally, it provides intuition to effects that are also relevant for a Bose-Fermi mixture loaded into an optical lattice. In the presence of an optical lattice the expected phases are even richer and, consequently, much less accessible by simple theoretical means.

Self-Consistent Mean-Field Calculation

We calculate the density profiles of harmonically trapped Bose-Fermi mixtures at $T = 0$ using a self-consistent mean-field theory [12–14]. Our model is based on the Thomas-Fermi approximation both for the bosonic and the fermionic component given in Eqs. 2.37 and 2.29, respectively. The interactions between bosons and fermions are accounted for by adding to the external trapping potential $V_{\mathrm{B/F}}(\mathbf{r})$, a mean-field potential $g_{\mathrm{BF}} n_{\mathrm{F}}(\mathbf{r})$ felt by the bosons and, analogously, $g_{\mathrm{BF}} n_{\mathrm{B}}(\mathbf{r})$ felt by the fermions. This results in a coupled pair of equations for the bosonic and the fermionic real-space densities

$$n_{\mathrm{B}}(\mathbf{r}) = \max\left[\frac{\mu_{\mathrm{B}} - V_{\mathrm{B}}(\mathbf{r}) - g_{\mathrm{BF}} n_{\mathrm{F}}(\mathbf{r})}{g_{\mathrm{BB}}}, 0\right], \tag{2.40}$$

$$n_{\mathrm{F}}(\mathbf{r}) = \frac{(2m_{\mathrm{F}})^{\frac{3}{2}}}{6\pi^2\hbar^3} \mathfrak{Re}\left[(\mu_{\mathrm{F}} - V_{\mathrm{F}}(\mathbf{r}) - g_{\mathrm{BF}} n_{\mathrm{B}}(\mathbf{r}))^{\frac{3}{2}}\right]. \tag{2.41}$$

The Bose-Bose and the Bose-Fermi interactions in the system are parametrized by the respective scattering lengths:

$$g_{\mathrm{BB}} = 2\pi\hbar^2 a_{\mathrm{BB}}/\mu_{\mathrm{BB}},$$
$$g_{\mathrm{BF}} = 2\pi\hbar^2 a_{\mathrm{BF}}/\mu_{\mathrm{BF}},$$

where $\mu_{\mathrm{BB}} = m_{\mathrm{B}}/2$ and $\mu_{\mathrm{BF}} = (m_{\mathrm{B}} m_{\mathrm{F}})/(m_{\mathrm{B}} + m_{\mathrm{F}})$ denote the reduced masses of a colliding atom pair. The chemical potentials are implicitly determined by the bosonic and fermionic atom numbers N_{B} and N_{F} by

$$N_{\mathrm{B/F}} = \int d\mathbf{r}\, n_{\mathrm{B/F}}(\mathbf{r}, \mu_{\mathrm{B/F}}). \tag{2.42}$$

The coupled equations are solved numerically by iteratively inserting the density distributions into the respective other equation. Conveniently, the three-dimensional problem can be reduced to one dimension in the case of a harmonic trapping potential that is equally deep (in absolute units) for the bosons and the fermions. Indeed, $V_{\mathrm{B}}(\mathbf{r}) = V_{\mathrm{F}}(\mathbf{r})$ is typically a good approximation for magnetically trapped alkali atoms in the stretched hyperfine state or atoms in a far-detuned optical dipole trap (see Sect. 2.2). Introducing rescaled units

$$\tilde{x}_\alpha = \sqrt{\frac{m_{\mathrm{B}}\omega_{\mathrm{B}\alpha}^2}{2}}\, x_\alpha = \sqrt{\frac{m_{\mathrm{F}}\omega_{\mathrm{F}\alpha}^2}{2}}\, x_\alpha, \tag{2.43}$$

where $\alpha = x, y, z$, the external trapping potential takes the convenient form $V_{\mathrm{B}}(\tilde{\mathbf{r}}) = V_{\mathrm{F}}(\tilde{\mathbf{r}}) = \tilde{\mathbf{r}}^2 = \tilde{r}^2$. In the new coordinates the problem is spherically symmetric and therefore the density distributions solely depend on $\tilde{r}$ instead of a three-dimensional vector $\mathbf{r}$. The simplified coupled equations read

$$n_{\mathrm{B}}(\tilde{r}) = \max\left[\frac{\mu_{\mathrm{B}} - \tilde{r}^2 - g_{\mathrm{BF}} n_{\mathrm{F}}(\tilde{r})}{g_{\mathrm{BB}}}, 0\right], \tag{2.44}$$

$$n_{\mathrm{F}}(\tilde{r}) = \frac{(2m_{\mathrm{F}})^{\frac{3}{2}}}{6\pi^2\hbar^3}\mathfrak{Re}\left[(\mu_{\mathrm{F}} - \tilde{r}^2 - g_{\mathrm{BF}} n_{\mathrm{B}}(\tilde{r}))^{\frac{3}{2}}\right] \tag{2.45}$$

and Eq. 2.42 takes the form

$$N_{\mathrm{B/F}} = 4\pi\left(\frac{2}{m_{\mathrm{F}}\bar{\omega}_{\mathrm{F}}^2}\right)^{\frac{3}{2}}\int_0^\infty d\tilde{r}\,\tilde{r}^2\, n_{\mathrm{B/F}}(\tilde{r}, \mu_{\mathrm{B/F}}). \tag{2.46}$$

After rescaling, the density distributions merely depend on the geometrical mean of the trapping frequencies $\bar{\omega}_{\mathrm{F}} = (\omega_{\mathrm{F}x}\omega_{\mathrm{F}y}\omega_{\mathrm{F}z})^{1/3}$ and the aspect ratio of the trap does not enter. Note that the results that are obtained from this set of equations are only valid within the requirements of the Thomas-Fermi approximations (see the preceding sections).

We solve the coupled equations by means of numerical iteration to obtain the density distributions $n_{\mathrm{B}}(\tilde{r})$ and $n_{\mathrm{F}}(\tilde{r})$. The following steps are performed in this procedure:

- **Initialization**: The starting point ($i = 0$) is the Thomas-Fermi profile of the pure Bose-Einstein condensate $n_{\mathrm{B},0}(\tilde{r})$. It is calculated using Eq. 2.44 with vanishing interspecies interactions ($g_{\mathrm{BF}} = 0$) and the chemical potential $\mu_{\mathrm{B},0}$ being implicitly given by Eq. 2.46.
- **Start of the iteration loop**: $n_{\mathrm{B},i}(\tilde{r})$ is inserted into Eq. 2.45. The resulting fermionic density profile is used to calculate the corresponding chemical potential $\mu_{\mathrm{F},i+1}$ for N_{F} fermions using Eq. 2.46. This yields the correctly normalized fermionic density $n_{\mathrm{F},i+1}(\tilde{r})$.

- **End of iteration loop**: This new fermionic density $n_{\mathrm{F},i+1}(\tilde{r})$ leads to a new bosonic density distribution $n_{\mathrm{B},i+1}(\tilde{r})$ as a result of the interspecies interactions. It is obtained by inserting $n_{\mathrm{F},i+1}(\tilde{r})$ into Eq. 2.44 and matching of the bosonic chemical potential $\mu_{\mathrm{B},i+1}$ to the number of bosons N_{B} using Eq. 2.46.

The last two steps are iteratively repeated several times (typically $i = 1$ up to 4). The iteration can yield different outcomes: Either the procedure converges and the resulting density distributions constitute the self-consistent solutions of the problem, or the procedure diverges corresponding to an increasing central density for each iteration. The latter case can happen for attractive interspecies interactions and corresponds to a collapse of the mixture [15]. Figure 2.3 shows a summary of results obtained both for attractively and repulsively interacting Bose-Fermi mixtures at typical experimental parameters.

2.1.6 Routes Towards the Strongly Interacting Regime

The ratio between the interaction energy and the kinetic energy per particle determines, whether a degenerate quantum system is in the weakly or strongly interacting regime. At quantum degeneracy the de Broglie wavelength approximately corresponds to the interparticle spacing. Therefore, the de Broglie wavelength relates to the particle density via $\lambda_{\mathrm{dB}} \sim n^{-1/3}$ and the density dependent kinetic energy can be approximated by $\epsilon_{\mathrm{kin}} = h^2 n^{2/3}/(2m)$. The density dependent interaction $\epsilon_{\mathrm{int}} = |g|n$ has already been introduced in the preceding sections, where the interaction strength reads $g = 4\pi\hbar^2 a_s/m$ and a_s is the scattering length (see Sect. 2.3). Accordingly, the ratio between interaction and kinetic energy per particle is given by the parameter [6]

$$\gamma = \frac{\epsilon_{\mathrm{int}}}{\epsilon_{\mathrm{kin}}} = \frac{|g|n}{h^2 n^{2/3}/(2m)} \approx n^{1/3}|a_s|. \tag{2.47}$$

We note, that γ corresponds to the parameter $k_{\mathrm{F}}|a_s|$ used to characterize fermionic quantum systems, because the Fermi momentum k_{F} coincides with $n^{1/3}$ in the quantum degenerate regime [4, 5].

When $\gamma < 1$, the quantum system is regarded as weakly interacting. In this regime interparticle correlations need not be taken into account. For example, weakly interacting bosonic quantum gases are remarkably well described within the effective single-particle theory of the Gross-Pitaevskii equation 2.35. This relatively simple framework has been successfully employed to describe exciting phenomena, such as interfering condensates or vortices [3, 6].

However, strongly correlated many-body quantum phases only arise, when γ exceeds unity. Equation 2.47 shows the possibilities to reach this regime: Either one increases the mass m or the interaction strength g. While these options may rather be called "impossibilities" in solid state physics, the remarkable techniques and control of atomic physics renders them possible for ultracold quantum gases.

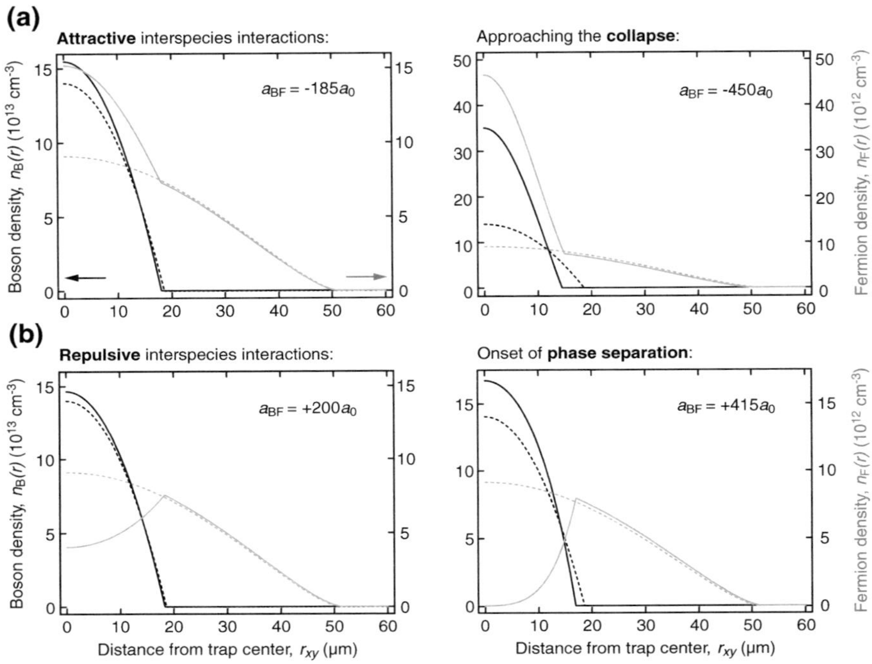

Fig. 2.3 Real space density profiles of an interacting mixture of ^{87}Rb and ^{40}K in a three-dimensional harmonic trap at four interactions. The mean-field calculation has been performed assuming $N_B = 3 \times 10^5$ bosons and $N_F = 3 \times 10^5$ fermions and trap frequencies of $\omega_{Fx} = \omega_{Fy} = 2\pi \times 40\,$Hz in the horizontal and $\omega_{Fz} = 2\pi \times 200\,$Hz in the vertical direction, similar to the experimental situation of Chap. 8. *Black* (*gray*) *lines* indicate bosonic (fermionic) profiles and $r_{xy} = \sqrt{x^2 + y^2}$. Note the factor of ten between the units of n_B and n_F. *Dashed lines* show the profiles at vanishing interactions. Strikingly, the fermionic cloud is much larger than the bosonic cloud with the same atom number; the volume of the clouds differs by a factor of about twenty! This is a remarkable manifestation of the different quantum statistics. **a** At attractive interspecies interactions the atomic density accumulates in the center until the system undergoes a mean-field collapse, when the interspecies attraction is increased beyond $a_{BF} \approx -450\ a_0$. Note that $a_{BF} = -185\ a_0$ approximately corresponds to the background scattering length between ^{87}Rb and ^{40}K (see Sect. 2.3.5). **b** In the case of repulsive interactions the fermions are pushed out from the trap center. The bosonic cloud is compressed by the surrounding fermions and shows an increased density in the trap center. When the interspecies repulsion is raised beyond $a_{BF} = +415\ a_0$ the central fermionic density vanishes. At this point the central bosonic density is enhanced by about 20 %

We discuss in the next two sections, how, on the one hand, an optical lattice potential can be employed to change the effective mass of atoms, and how, on the other hand, Feshbach resonances can serve as a direct control knob of the interaction strength g via the scattering length a_s.

Those two approaches have led to wonderful studies in ultracold atom physics within the last ten years. Seminal results have been the realization of the superfluid to Mott insulator transition in an optical lattice [16, 17] or the investigation of the BEC-BCS crossover in quantum degenerate fermionic spin mixtures [18–21]. The experiments in this thesis utilize both techniques simultaneously.

2.2 Optical Lattice Potentials

This section briefly reviews the theory of optical lattice potentials [7]. After an introduction to optical dipole forces [22], we explain how optical standing waves can be employed to create simple cubic lattice potentials for ultracold atoms. We analyze, compare and model the landscape of realistic lattice potentials, in particular focussing on the differences of using red- or blue-detuned laser light to create the underlying optical standing waves. The section concludes with a discussion of the band structure in simple cubic lattices [23].

2.2.1 Optical Dipole Potentials

A light field can act on neutral atoms both in a dissipative and a conservative way. A dissipative force arises from the absorption and subsequent reemission of photons. This process can transfer net momentum on the atoms and creates a force that is often called radiation pressure. It is used for laser cooling and magneto-optical traps, where temperatures down to 100 μK can be achieved with Doppler cooling techniques. Sub-Doppler techniques even allow to reach the microkelvin regime, at least theoretically [22, 24].[5] A conservative force, the so-called optical dipole force, originates from the interaction of a light field with the electric dipole moment that is induced in the atom by the very same light field. This interaction results in a shift of the atomic energy levels, the AC-Stark shift, that is proportional to the intensity of the field.

When neutral atoms are exposed to light generally both of the aforementioned forces are present. However, we will show in this section that the conservative part can play the dominating role, when the light field is far detuned from all atomic resonances. The large detuning suppresses scattering of photons more strongly than the interaction with the self-induced dipole. As the optical dipole potential is proportional to the intensity of the light field, an appropriately engineered intensity distribution allows for the creation of optical dipole traps [25] or optical lattice potentials [7, 23]. Typical depths of optical dipole potentials reach several microkelvin. Therefore, they are well suited to capture atoms that have been cooled by radiation pressure in a first step.

---

[5] The Doppler temperature is given by the energy scale defined by the natural linewidth Γ of the transition that is used for cooling, $T_{\mathrm{D}} = \hbar\Gamma/(2k_{\mathrm{B}})$. The minimal temperature that can principally be reached with sub-Doppler techniques is set by the recoil energy, corresponding to the recoil temperature $T_{\mathrm{SD}} = (\hbar k)^2/(2mk_B)$.

Oscillator Model

We assume an atom to be exposed to a monochromatic light field $\mathbf{E}$ that oscillates at a frequency ω. This induces an electric dipole moment proportional to the electric field $\mathbf{d}(\omega) = \alpha(\omega)\mathbf{E}$. The proportionality constant $\alpha(\omega)$ is called the polarizability. It is a function of the frequency ω and generally takes a complex value, i.e. it also contains information on a phase shift between the electric field and the induced dipole. The potential energy of the self-induced dipole is given by

$$V_{\text{dip}} \propto -\langle \mathbf{d} \cdot \mathbf{E} \rangle \propto -\Re\mathfrak{e}[\alpha(\omega)] \cdot I, \tag{2.48}$$

where $\langle \ldots \rangle$ denotes the temporal average over the fast oscillation of the light field. This is the conservative dipole potential that is proportional to the intensity $I = \epsilon_0 c E^2/2$, where the amplitude of the electric field is given by $E = |\mathbf{E}|$ and c is the speed of light. The real part of the polarizability indicates that the in-phase component of the oscillating dipole moment is responsible for the dipole force. Conversely, the imaginary part of the polarizability denotes the out-of-phase component that gives rise to the spontaneous scattering rate

$$\Gamma_{\text{sc}} \propto \Im\mathfrak{m}[\alpha(\omega)] \cdot I. \tag{2.49}$$

An expression for the polarizability can be derived using classical [26], semi-classical or fully quantized theories [27–29]. However, it turns out that for a two-level system in the limit of low saturation, also the quantum mechanical approaches yield the polarizability

$$\alpha(\omega) = 6\pi\epsilon_o c^3 \frac{\Gamma/\omega_0^2}{\omega_0^2 - \omega^2 - i(\omega^3/\omega_0^2)\Gamma}, \tag{2.50}$$

which is conveniently derived for a classical damped oscillator [25]. Here, ω_0 denotes the optical transition frequency of the atom and Γ the damping rate associated with the spontaneous decay rate of the excited level (corresponding to the line width of the transition). The limit of low saturation, i.e. negligible population in the excited level, is reached at far detuning[6] $\Delta = \omega - \omega_0$, which is typically the case for dipole traps, and also implies that $\Gamma_{\text{sc}} \ll \Gamma$ as we will see below. While the classical derivation yields an accurate description for $\alpha(\omega)$, this is not the case for the damping rate Γ. In a semi-classical derivation it turns out that the damping rate is determined by the dipole matrix element between the ground and excited state $\Gamma \propto |\langle e|\mathbf{d}|g\rangle|^2$.

Based on the above expressions, it is possible to derive the dipole potential and the scattering rate in the limit of large detuning and negligible saturation:

$$V_{\text{dip}}(\mathbf{r}) = -\frac{3\pi c^2}{2\omega^3}\left(\frac{\Gamma}{\omega_0 - \omega} + \frac{\Gamma}{\omega_0 + \omega}\right) \cdot I(\mathbf{r}), \tag{2.51}$$

[6] In the picture of the Bloch sphere this corresponds to a Bloch vector that only oscillates close to the ground state.

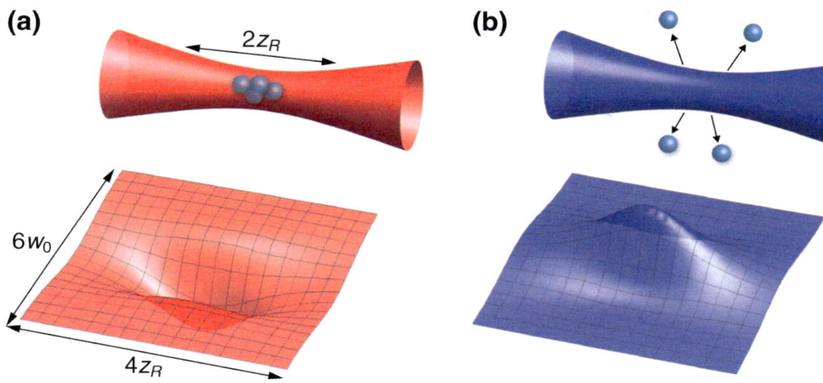

Fig. 2.4 The *lower* pictures schematically show the dipole potential in the focus of a round *red*-(**a**) and *blue*-detuned (**b**) Gaussian laser beam. The envelope area of the *upper* images is defined by the beam radius $w(z)$. A *red*-detuned beam can act as a trap for ultracold atoms, while the *blue* beam creates a repulsive potential

$$\Gamma_{\text{sc}}(\mathbf{r}) = \frac{3\pi c^2}{2\hbar\omega_0^3}\left(\frac{\omega}{\omega_0}\right)^3\left(\frac{\Gamma}{\omega_0-\omega}+\frac{\Gamma}{\omega_0+\omega}\right)^2\cdot I(\mathbf{r}). \tag{2.52}$$

For the case of large, but not too large detuning $|\Delta| \ll \omega_0$, one may neglect the terms proportional to $1/(\omega_0+\omega)$ corresponding to the often employed rotating wave approximation [26, 28]. This yields the simplified formulas

$$V_{\text{dip}}(\mathbf{r}) = \frac{3\pi c^2}{2\omega_0^3}\frac{\Gamma}{\Delta}\cdot I(\mathbf{r}), \tag{2.53}$$

$$\Gamma_{\text{sc}}(\mathbf{r}) = \frac{3\pi c^2}{2\hbar\omega_0^3}\left(\frac{\Gamma}{\Delta}\right)^2\cdot I(\mathbf{r}). \tag{2.54}$$

Those two expressions contain the physics of optical dipole potentials in a concise form. We see that V_{dip} is proportional to I/Δ, while the scattering rate Γ_{sc} scales as I/Δ^2. The sign of the detuning determines, whether the dipole potential is repulsive ($\Delta > 0$, blue detuning) or attractive ($\Delta < 0$, red detuning). Furthermore, we note that Γ_{sc} is proportional to V_{dip}/Δ. This shows that inelastic scattering can be efficiently suppressed by choosing a large detuning Δ. When optical dipole forces are employed for ultracold atoms, it is crucial to minimize inelastic scattering; the recoil energy of a single scattering process corresponds to a temperature of several 100 nK and creates strong heating compared to the motional ground state energy of the atoms. Therefore, the detuning should generally be chosen as large as possible within the limits of available laser power to ensure a conservative potential.

Although being conceptually appealing, the approximate Eqs. 2.53 and 2.54 must be taken with a grain of salt, when it comes to actual calculations. Considering the case of rubidium, the error originating from the rotating wave approximation for the dipole potential depth amounts to 3 % for $\lambda = 738$ nm (used for the optical lattice in this thesis) and even 12 % for $\lambda = 1030$ nm (used for the dipole trap in this thesis). Therefore, in most cases the rotating wave approximation must not be applied. Furthermore, for ^{40}K and ^{87}Rb, which are used in the experiment, the fine structure splitting due to the spin-orbit coupling must be taken into account [25] leading to the D_1 and D_2 line doublet, which is present in all alkali atoms. For linearly polarized light, the resulting formula for the dipole potential reads

$$V_{\text{dip}}(\mathbf{r}) = \frac{\pi c^2}{2} \left[\frac{2\Gamma_{D_2}}{\omega_{D_2}^3} \left(\frac{1}{\Delta_{D_2}} - \frac{1}{\Delta_{D_2} + 2\omega_{D_2}} \right) + \frac{\Gamma_{D_1}}{\omega_{D_1}^3} \left(\frac{1}{\Delta_{D_1}} - \frac{1}{\Delta_{D_1} + 2\omega_{D_1}} \right) \right] \cdot I(\mathbf{r}), \tag{2.55}$$

where Γ_x denotes the line width and $\Delta_x = \omega - \omega_x$ the detuning of the laser frequency ω from the respective resonance frequency ω_x of the $x = D_1, D_2$ line. The corresponding data for ^{40}K and ^{87}Rb are provided in the Appendices A.1 and A.2. For circularly polarized light, the dipole potential is also sensitive to the hyperfine splitting and depends on the quantum numbers F and m_F of Zeeman sublevels in the ground state [25]. However, using linear polarization the optical dipole force offers an elegant way to create identical potentials for different Zeeman sublevels without differential shifts. This is particularly relevant for the experiments using spin mixtures of ^{40}K reported in Chap. 6.

Red-Detuned Dipole Trap

It is conceptually simple to create a trap for ultracold atoms based on the optical dipole force discussed above. For red detuning the force is attractive and atoms are drawn towards the intensity maximum. Therefore, the focus of a single red-detuned Gaussian laser beam can be used to create a three-dimensional trapping potential for atoms (see Fig. 2.4). The intensity distribution of an elliptical Gaussian beam propagating along the z-axis, can be written as [30]

$$I(\mathbf{r}) = \frac{2P}{\pi w_x(z) w_y(z)} e^{-\frac{2x^2}{w_x^2(z)} - \frac{2y^2}{w_y^2(z)}}, \tag{2.56}$$

where P is the total power of the beam, giving rise to a peak intensity $I_0 = 2P/(\pi w_{0x} w_{0y})$. The beam radius $w_\alpha(z)$ ($\alpha = x, y$) denotes the distance from the beam center at which the intensity has dropped by a factor $1/e^2$. It is given by

$$w_\alpha(z) = w_{0\alpha} \sqrt{1 + \left(\frac{z}{z_{R\alpha}} \right)^2}, \tag{2.57}$$

where $w_{0\alpha}$ is the beam waist along the directions $\alpha = x, y$ and $z_{R\alpha} = \pi w_{0\alpha}^2/\lambda$ defines the Rayleigh length; λ is the wavelength of the laser light. The trapping potential that is created by a Gaussian beam profile can be approximated in the vicinity of the focus by

$$V_{\text{dip}}(\mathbf{r}) \approx -V_0 \left[1 - 2\left(\frac{x}{w_{0x}}\right)^2 - 2\left(\frac{y}{w_{0y}}\right)^2 - \frac{1}{2}\left(\frac{z}{z_{Rx}}\right)^2 - \frac{1}{2}\left(\frac{z}{z_{Ry}}\right)^2 \right]. \tag{2.58}$$

This corresponds to a harmonic trap with radial and axial trap frequencies

$$\omega_x = \sqrt{\frac{4V_0}{m w_{0x}^2}}, \quad \omega_y = \sqrt{\frac{4V_0}{m w_{0y}^2}} \quad \text{and} \quad \omega_z = \sqrt{\frac{V_0}{m}\left(\frac{1}{z_{Rx}^2} + \frac{1}{z_{Ry}^2}\right)}, \tag{2.59}$$

where m denotes the atomic mass. For typical parameters the radial trap frequencies ω_x and ω_y are about two orders of magnitude larger than the axial trap frequency ω_z. Therefore a Gaussian single beam trap is highly anisotropic and hard to handle experimentally without additional axial confinement.

2.2.2 *Optical Lattice Potentials*

Optical lattice potentials for ultracold atoms are created by interfering counterpropagating Gaussian laser beams. Using one, two or three of such standing waves 1D, 2D and 3D optical lattices can be formed. For blue or red detuning the atoms are either trapped in the intensity minima or maxima, respectively, which has a subtle influence on the details of the global potential landscape.

1D Optical Lattice Potential

When a Gaussian beam with a wavelength λ is retroreflected into itself, an optical standing wave with a periodicity of $\lambda/2$ forms (see Fig. 2.5). Such a standing wave can be used as a one-dimensional (1D) optical lattice potential for atoms. Typically, the axial extend of the atom cloud is much smaller than the Rayleigh length z_R, such that the axially dependent beam radius can be approximated by the constant beam waist w_z.[7] The resulting periodic potential is given by

$$V_{\text{1D}}(\mathbf{r}) = V_z\, e^{-2\frac{r_{xy}^2}{w_z^2}} \cos^2(kz) \simeq V_z \left(1 - 2\frac{r_{xy}^2}{w_z^2}\right) \cos^2(kz), \tag{2.60}$$

[7] From now on we use the convention that the index of the beam waist denotes the propagation direction of the rotationally symmetric lattice beam, i.e. w_z is the waist of a round beam propagating in z-direction.

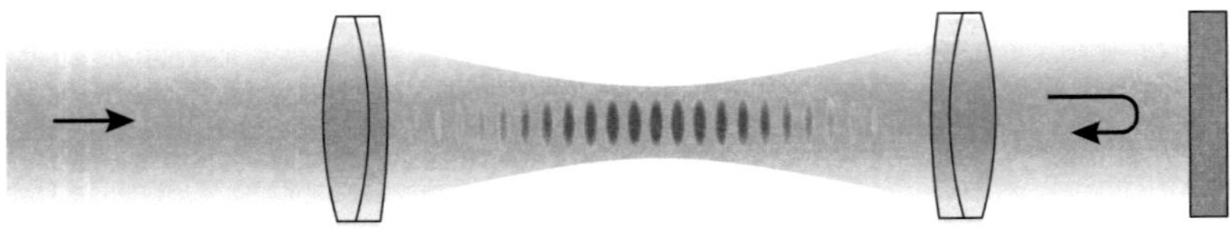

Fig. 2.5 An optical standing wave formed by retroreflection of a Gaussian laser beam creates a one-dimensional optical lattice potential. Generally, the electric field amplitude of the returning beam is reduced by a factor, which we call the effective reflection coefficient ρ

where $k = 2\pi/\lambda$ is the wavevector of the laser light, $r_{xy} = \sqrt{x^2 + y^2}$ is the radial coordinate and V_z is the depth of the optical lattice potential. We note that the depth of the lattice V_z is four times larger compared to the potential depth of the bare Gaussian beam without retroreflection. This enhancement is caused by constructive interference, which becomes obvious, when the counterpropagating beams are viewed on the level of electric fields. Usually, the depth of an optical lattice is given in units of the recoil energy $E_{\text{rec}} = \hbar^2 k^2/(2m)$. In the following, we also use the dimensionless lattice depth $s_z = V_z/E_{\text{rec}}$.

In the above description of a 1D optical lattice, we have made the implicit assumption that the retroreflected beam has the same electric field amplitude as the incoming part. In experimental realizations, however, the returning part has passed through several additional optical elements reducing the field amplitude. We account for this by introducing an effective reflection coefficient ρ_z that quantifies the ratio of the returning and the incoming electric field amplitudes at the position of the atoms. For $0 \leqslant \rho_z < 1$ the standing wave is not fully modulated anymore and reads

$$V_{\text{1D}}(\mathbf{r}) = \frac{V_z}{4} e^{-2\frac{r_{xy}^2}{w_z^2}} \left(1 + \rho_z^2 + 2\rho_z \cos\left(2kz\right)\right) \tag{2.61}$$

$$\simeq \frac{V_z}{4}\left(1 - 2\frac{r_{xy}^2}{w_z^2}\right)\left(1 + \rho_z^2 + 2\rho_z \cos\left(2kz\right)\right). \tag{2.62}$$

For perfect reflectivity ($\rho_z = 1$) Eq. 2.60 is recovered using $\cos(2kz) = 2\cos^2(kz) - 1$.

External Potential of a 1D Optical Lattice

The Gaussian beam shape and finite modulation of the standing wave give rise to a transverse underlying potential in addition to the axial modulation of the optical lattice. We aim at an effective description of the lattice potential according to

$$V_{\text{1D}}(\mathbf{r}) = V_z' \cos^2(kz) + \frac{1}{2} m\omega_{xy}^2 r_{xy}^2. \tag{2.63}$$

Here, $V_z' = \rho_z V_z$ is the effective lattice depth and $\omega_{xy} = \omega_x = \omega_y$ the transverse trap frequency of a spherically symmetric beam. We distinguish two contributions to the frequency ω_{xy}:

First, we use Eq. 2.62 to derive the transverse curvature of the lattice potential as it is felt by the atoms. It is crucial to distinguish red ($V_z < 0$) and blue detuning ($V_z > 0$) of the laser light, because in the first case the atoms are trapped in the intensity maxima for which $\cos(2kz) = 1$, while in the second case the atoms sit in the intensity minima described by $\cos(2kz) = -1$. From this, we obtain the first contribution to the transverse potential

$$\omega_{\text{pot},xy}^2 = -\frac{V_z}{m}\frac{(1+\rho_z^2 \pm 2\rho_z)}{w_z^2}, \tag{2.64}$$

where the upper (lower) sign denotes red (blue) detuning. For a positive (negative) overall sign the potential is (anti)confining (see Fig. 2.6).

The second contribution is more subtle. Viewing a single lattice well as an harmonic oscillator potential in the axial direction, that is approximating $\cos(2kz)$ harmonically, we obtain for the trap frequency in a lattice well at the center of the beam ($r_{xy} = 0$)

$$\omega_{\text{lat},z}^2 = \frac{2k^2}{m}|V_z'|. \tag{2.65}$$

Due to the Gaussian beam shape the absolute modulation depth of the standing wave radially goes down according to the factor $\exp(-2r_{xy}^2/w_z^2)$. Consequently, the trap frequency on a lattice well decreases like

$$\omega_{\text{lat},z}(r_{xy}) = \omega_{\text{lat},z}\, e^{-\frac{r_{xy}^2}{w_z^2}} \approx \omega_{\text{lat},z}\left(1 - \frac{r_{xy}^2}{w_z^2}\right) \tag{2.66}$$

in the transverse direction and so does the ground state energy $E_0(r_{xy}) = \hbar\omega_{\text{lat},z}(r_{xy})/2$ of the local harmonic oscillator. This gives rise to an additional radial anticonfinement that is independent of the sign of the laser detuning. Hence, the frequency originating from the shift in the harmonic ground state energy reads

$$\omega_{\text{ho},xy}^2 = -\frac{2}{mw_z^2}\sqrt{|V_z'|E_{\text{rec}}}. \tag{2.67}$$

Combining the two contributions and introducing the dimensionless effective lattice depth $s_z' = V_z'/E_{\text{rec}}$, the transverse trap frequency of the external potential reads

$$\omega_{xy}^2 = -\frac{E_{\text{rec}}}{mw_z^2}\left[\left(\frac{1}{\rho_z} + \rho_z \pm 2\right)s_z' - 2\sqrt{s_z'}\right], \tag{2.68}$$

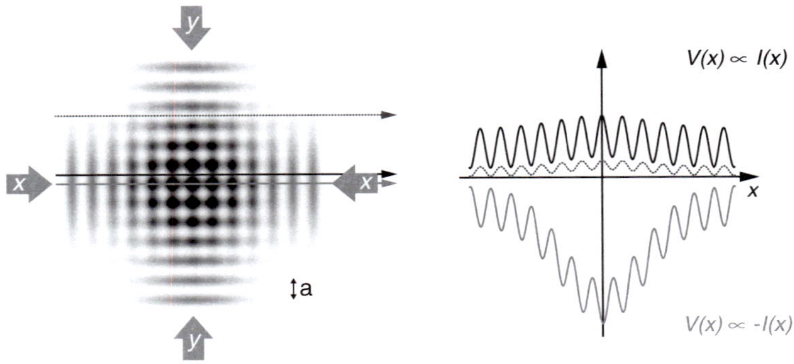

Fig. 2.6 External trapping potential for *red*- and *blue*-detuned lattices. The intensity distribution of a 3D lattice with reduced reflectivity ($\rho = 0.25$) of the retro beam is shown as a density plot. Cuts show the influence of the perpendicular beams on the lattice potential along the x axis. In the case of *blue* detuning the cut is taken for $I(x, a/2, a/2)$ (*black*) and $I(x, 7a/2, a/2)$ (*dotted*), while for *red* detuning the cut corresponds to $I(x, 0, 0)$ (*gray*). The underlying confinement of the *red* lattice is much stronger than the corresponding anticonfinement of the *blue* lattice

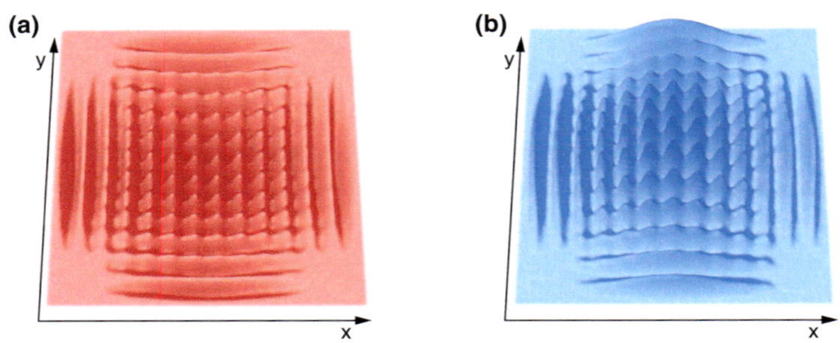

Fig. 2.7 Comparison of the lattice potentials created by a *red*- (**a**) and *blue*-detuned (**b**) laser. In a *red* lattice the potential minima are located at the intensity maxima as $V(\mathbf{r}) \propto -I(\mathbf{r})$, while in a *blue* lattice the potential minima are located at the intensity minima as $V(\mathbf{r}) \propto I(\mathbf{r})$

where the upper (lower) sign holds for red (blue) detuning. In typical experimental setups the effective reflection coefficient does not exceed the range $0.8 < \rho_z < 1$. For those values the prefactor $1/\rho_z + \rho_z \pm 2$ is very close to 4 (zero) for the case of red (blue) detuning. Therefore, a red-detuned lattice has an underlying external potential that consists of a dominant confining term (scaling as s_z') and an anticonfining term (scaling as $\sqrt{s_z'}$). In a blue-detuned lattice, however, both terms are anticonfining and the $\sqrt{s_z'}$ scaling dominates. This important result is confirmed by direct measurements of the transverse external potential in a blue-detuned optical lattice presented in Sect. 5.4 (Fig. 2.7).

3D Optical Lattice Potential

A simple cubic optical lattice potential can be created by crossing three optical standing waves orthogonally to each other. In order to suppress interference between the three standing waves it is crucial to choose mutually orthogonal linear polarizations. However, in the experiment slight deviations of the polarizations and resulting small interferences are hard to avoid. Such interferences can be rendered harmless by choosing laser frequencies of the beams that differ by some tens of MHz. This leads to rapid oscillations of spurious potential corrugations and the atoms effectively feel a smooth lattice potential due to time averaging. Therefore, we can assume independent 1D optical lattices in x, y and z direction, which create a 3D optical lattice potential of simple cubic type with a lattice constant of $\lambda/2$ and a one atom basis. Using Eq. 2.61 the 3D optical lattice potential can be written as

$$\begin{aligned} V_{3\mathrm{D}}(\mathbf{r}) = & \frac{V_x}{4} e^{-2\frac{r_{yz}^2}{w_x^2}} \Big(1+\rho_x^2+2\rho_x \cos\big(2kx\big)\Big) \\ & + \frac{V_y}{4} e^{-2\frac{r_{xz}^2}{w_y^2}} \Big(1+\rho_y^2+2\rho_y \cos\big(2ky\big)\Big) \\ & + \frac{V_z}{4} e^{-2\frac{r_{xy}^2}{w_z^2}} \Big(1+\rho_z^2+2\rho_z \cos\big(2kz\big)\Big). \end{aligned} \tag{2.69}$$

Here, V_α denote the potential depth, w_α the waist and ρ_α the effective reflectivity of the superimposed 1D standing waves ($\alpha = x, y, z$). As above, red (blue) detuning corresponds to $V_\alpha < 0$ ($V_\alpha > 0$).

External Potential of 3D Optical Lattices

It is convenient to approximate the 3D optical lattice potential in the center, for distances much smaller than the beam waists, as a sum of a homogeneous periodic modulation and an external harmonic potential according to

$$V_{3\mathrm{D}}(\mathbf{r}) \simeq V'_x \cos^2(kx) + V'_y \cos^2(ky) + V'_z \cos^2(kz) + \frac{1}{2} m \Big(\omega_x^2 x^2 + \omega_y^2 y^2 + \omega_z^2 z^2\Big). \tag{2.70}$$

Here, the effective lattice depths are denoted by $V'_\alpha = \rho_\alpha V_\alpha$ and ω_α are the effective trap frequencies of the external harmonic potential in the directions $\alpha = x, y, z$. Using the above results for the 1D case, the squared trap frequencies are readily derived and read

$$\omega_x^2 = -\frac{E_{\mathrm{rec}}}{m} \left[\frac{(1/\rho_y + \rho_y \pm 2)\, s'_y - 2\sqrt{s'_y}}{w_y^2} + \frac{(1/\rho_z + \rho_z \pm 2)\, s'_z - 2\sqrt{s'_z}}{w_z^2} \right]. \tag{2.71}$$

The corresponding expressions for ω_y^2 and ω_z^2 are obtained by cyclic permutation of the indices. Assuming an isotropic potential, for which the dimensionless effective lattice depths $s'_\alpha = s'$, the effective reflectivities $\rho_\alpha = \rho$, the waists $w_\alpha = w$ are equal in all directions $\alpha = x, y, z$ and $r^2 = x^2 + y^2 + z^2$, the entire external potential can be expressed by

$$V_{3D}^{ext}(\mathbf{r}) = -E_{rec}\left[\left(\frac{1}{\rho} + \rho \pm 2\right)s' - 2\sqrt{s'}\right]\frac{r^2}{w^2} \tag{2.72}$$

This expression contains both the effects of finite effective reflectivity as well as the anticonfinement scaling as $\sqrt{s'}$ that originates from the transverse change of the ground state energy for each lattice beam. In the case of a blue-detuned 3D optical lattice (lower sign), which is used in the experiments of this thesis, the later term is dominating the external potential of the lattice.

2.2.3 Band Structure and Bloch States

In this section we derive the eigenstates of a single particle that moves in a periodic potential. The corresponding eigenenergies form energy bands with a characteristic structure depending on the symmetries and the depth of the lattice. Irrespective whether an electron in an ionic crystal or an atom in an optical lattice is the matter of interest, the resulting physics is identical. However, we want to emphasize two caveats on the theory derived in this section:

- First, it refers to a homogeneous lattice system that extends to infinity. This is a reasonable assumption for large solid state crystals, but in the case of finite-sized optical lattices it must be handled with care. Deviations that arise in an optical lattice with an underlying confining potential are numerically addressed in Sect. 3.3.3.
- Second, what is derived in this section is a single-particle theory. Therefore, the theory itself and all implications that are drawn from it are strictly speaking only valid for single particles or noninteracting many-particle systems. Noninteracting systems can be realized with ultracold atoms using spin polarized fermions, which do not collide in the low energy limit, or by tuning the scattering length to zero using a Feshbach resonance (see Sect. 2.3). However, essentially all experiments of this thesis feature interacting particles.

Nevertheless, the single-particle band structure of homogeneous lattices is the basis to understanding the physics of more involved interacting systems. Band structure in conjunction with quantum statistics often allows to understand basic physical effects in lattices, such as the formation of metallic or band insulator states for fermions [31] (see Chap. 6).

The optical lattice used in the experiments has a three-dimensional simple cubic structure. Therefore, the movement of the atoms can be considered independently for the three coordinate axes x, y and z. It is sufficient to solve the Schrödinger equation of the one-dimensional problem:

$$\hat{H}\phi_q^{(n)}(x) = E_q^{(n)}\phi_q^{(n)}(x) \quad \text{with} \quad \hat{H} = \frac{\hat{p}^2}{2m} + V(x), \tag{2.73}$$

where $\hat{p} = -i\hbar\partial/\partial x$ is the momentum operator and $V(x)$ is assumed to be a homogeneous lattice potential with periodicity $a = \lambda/2 = \pi/k$, where λ is the wavelength of the laser creating the lattice. According to Bloch's theorem [31] each eigenstate $\phi_q^{(n)}(x)$ of this Hamiltonian can be written as a product of a plane wave with wavevector q and a function $u_q^{(n)}(x) = u_q^{(n)}(x+a)$ with the same periodicity as the lattice potential,

$$\phi_q^{(n)}(x) = e^{iqx/\hbar}\, u_q^{(n)}(x). \tag{2.74}$$

We use this wavefunction as an ansatz and insert it into Eq. 2.73 to obtain an eigenvalue problem for $u_q^{(n)}(x)$:

$$\hat{H}_q\, u_q^{(n)}(x) = \left(\frac{1}{2m}\left(\hat{p}+q\right)^2 + V(x)\right) u_q^{(n)}(x) = E_q^{(n)}\, u_q^{(n)}(x). \tag{2.75}$$

At this point we can use the fact that both the potential $V(x)$ and the functions $u_q^{(n)}(x)$ have the same periodicity. They can be expanded in discrete Fourier sums consisting of plane waves with wavevector $2k$ and the corresponding higher harmonics:

$$V(x) = \sum_r V_r e^{i2krx} \quad \text{and} \quad u_q^{(n)}(x) = \sum_s c_s^{(n,q)} e^{i2ksx}, \tag{2.76}$$

where the indices r, s run over all integer numbers. Based on the expansion of $u_q^{(n)}(x)$ it becomes apparent, that the Bloch function $\phi_q^{(n)}(x)$ can be constructed by a superposition of plane waves with wavevectors $q/\hbar + 2ks$, where $2ks$ are the reciprocal lattice vectors [31]. This implies, that the quasi-momenta q that give rise to distinct Bloch functions are restricted to the interval $[-\hbar k, \hbar k]$. This interval is called the first Brillouin zone. For each value of q we can expect an infinite number of discrete energy levels $E_q^{(n)}$ that are conveniently labeled by the positive integer index n. It will become clear below, why n is called the band index.

Inserting the Fourier sums into Eq. 2.75, we obtain for the kinetic energy

$$\frac{1}{2m}\left(\hat{p}+q\right)^2 u_q^{(n)}(x) = \sum_s \frac{(2\hbar ks + q)^2}{2m} e^{i2ksx} c_s^{(n,q)} \tag{2.77}$$

and the potential energy

$$V(x)\,u_q^{(n)}(x) = \sum_r \sum_{s'} V_r\, e^{i2k(r+s')kx}\, c_{s'}^{n,q} = \sum_r \sum_s V_r e^{iksx}\, c_{s-r}^{(n,q)}. \tag{2.78}$$

The functional form of an optical lattice potential is known to be sinusoidal, such that the corresponding Fourier expansion is simply given by

$$V(x) = V_x \cos^2(kx) = \frac{V_x}{4}\left(e^{i2kx} + e^{-i2kx} + 2\right) \tag{2.79}$$

with the only non-vanishing Fourier coefficients being $V_0 = V_x/2$ and $V_{\pm 1} = V_x/4$. This allows to write the eigenvalue problem of Eq. 2.75 in matrix form

$$\sum_{s'} H_{ss'} c_{s'}^{(n,q)} = E_q^{(n)} c_s^{(n,q)}, \tag{2.80}$$

where $H_{ss'}$ is the matrix of the Hamiltonian $\hat{H}_q$ calculated with respect to the basis of plane waves e^{i2ksx}. The entries of the matrix are given by

$$H_{ss'} = \begin{cases} \left(2s + \frac{q}{\hbar k}\right)^2 E_{\text{rec}} + V_x/2 & \text{if } |s - s'| = 0, \\ V_x/4 & \text{if } |s - s'| = 1, \\ 0 & \text{else.} \end{cases} \tag{2.81}$$

Here it becomes apparent, that the recoil energy $E_{\text{rec}} = \hbar^2 k^2/(2m)$ is the natural unit for the lattice depth V_x.

The Hamiltonian can be numerically diagonalized for a given quasi-momentum q yielding the eigenenergies $E_q^{(n)}$ and the eigenvectors $\mathbf{c}^{(n,q)} = (c_s^{(n,q)})$ that define the Bloch functions via Eqs. 2.76 and 2.74. The matrix entries for large indices $|s|$ correspond to high-energy contributions (see Eq. 2.78) and the coefficients $c_s^{(n,q)}$ become very small in the lowest bands. Therefore, it is sufficient to include the matrix entries up to a cut-off index $|s| \leq s_{\text{max}}$. For typical lattice depths up to 50 E_{rec} it is sufficient to keep the entries with $|s| \leq 7$. In principle, the numerical diagonalization within the limited Hilbert space yields results for the first $2s_{\text{max}} + 1$ bands, but the outcome is most accurate for the lowest energy bands.

The results of band structure calculations for a one-dimensional sinusoidal lattice at several depths are displayed in Fig. 2.8. For very low lattice depth the band structure does not show band gaps and corresponds to the kinetic energy of a free particle, where the dispersion parabola is reduced to the first Brillouin zone. For increasing lattice depths band gaps open up and the band width decreases exponentially in particular for the lowest lying bands. We observe that the energies $E_q^{(n)}$ for a fixed index n are a continuous function of q being bounded from below and above, which motivates the name band index for n. In very deep lattices, the individual lattice wells become more and more independent and the low energy physics can be well described in the harmonic approximation. In this case, the gap between the lowest bands is given by the energy spacing of the on-site harmonic oscillator $\hbar\omega_{\text{lat},x}$.

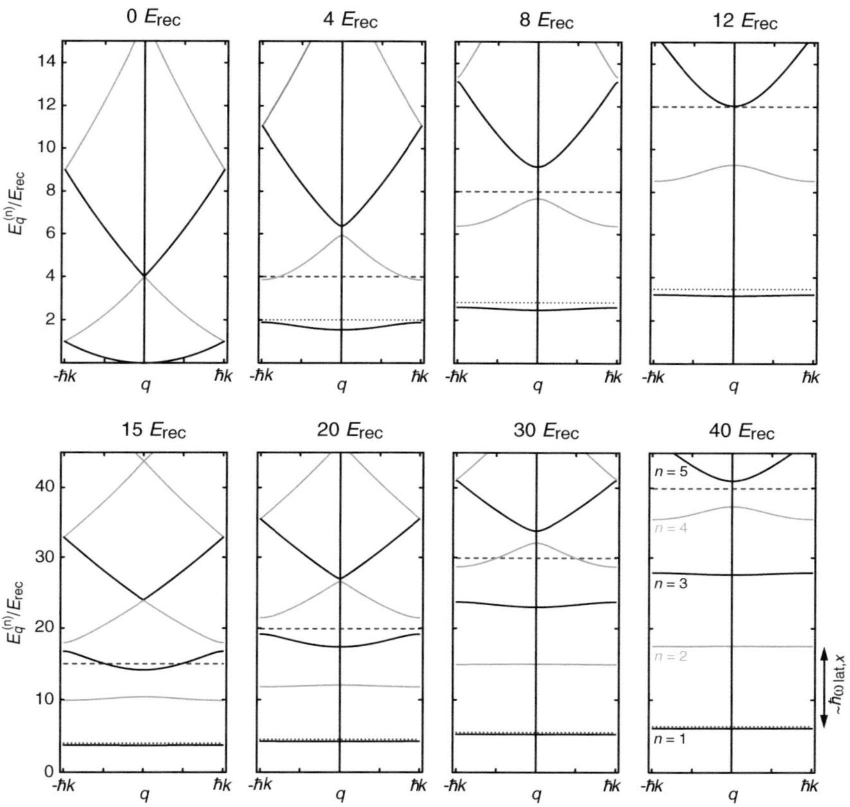

Fig. 2.8 Band structure of a one-dimensional optical lattice. The eigenenergies $E_q^{(n)}$ of Bloch states $\phi_q^{(n)}$ at quasi-momenta q are displayed for the lowest bands n at several lattice depths ranging from 0 to 40 E_{rec}. The lattice depths V_x are shown as *dashed lines*. *Dotted lines* indicate the zero point energy $\hbar\omega_{\text{lat},x}/2$, when single lattice wells are treated in harmonic approximation. For deep lattices the lowest band becomes flat and the band gap is well approximated by the level spacing of the harmonic oscillator $\hbar\omega_{\text{lat},x}$

However, we note that even in deep lattices the higher bands with energies larger than the depth of the lattice (dashed lines in Fig. 2.8) remain wide and still closely resemble the dispersion relation of a free particle. Here, the main effect of the lattice can be captured by assigning a larger effective mass $m_{\text{eff}} > m$ to the particle, which leads to a slower movement through the lattice potential. Generally, the effective mass at quasi-momentum q_0 and band index n is given by

$$m_{\text{eff}}^{(n)}(q_0) = \hbar^2 \left(\left. \frac{\partial^2 E_q^{(n)}}{\partial q^2} \right|_{q_0} \right)^{-1} . \tag{2.82}$$

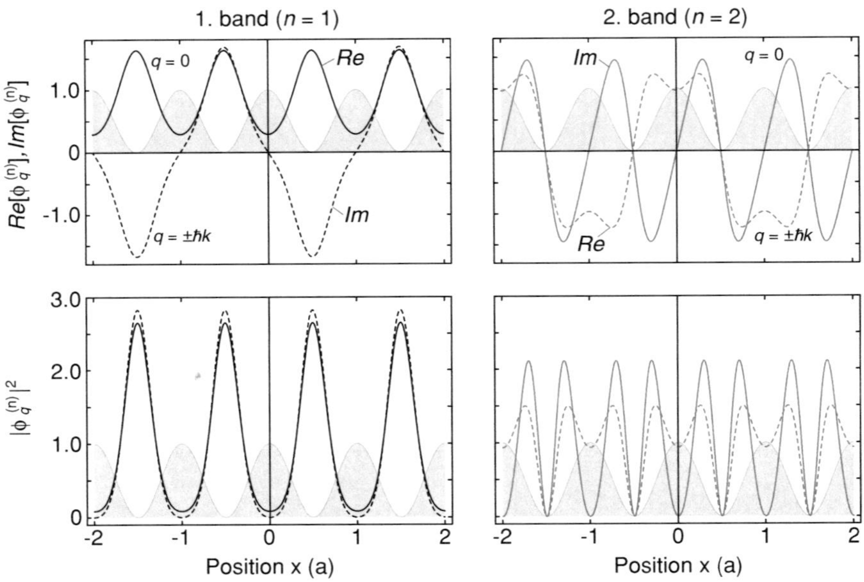

Fig. 2.9 Exemplary Bloch wavefunctions $\phi_q^{(n)}$ of a one-dimensional 8 $E_{\rm rec}$ lattice. The *upper* (*lower*) panels show the probability amplitudes (probability densities) in the first and second band for the quasi-momenta $q = 0$ (*solid lines*) and $q = \pm\hbar k$ (*dashed lines*). The periodicity of the lattice potential is indicated by the *gray* shading

Similarly, the group velocity of a wavepacket in the Bloch state $\phi_{q_0}^{(n)}$ is given by the first derivative of the dispersion relation [23]

$$v_{\rm gr}^{(n)}(q_0) = \frac{1}{\hbar}\frac{\partial E_q^{(n)}}{\partial q}\bigg|_{q_0}. \tag{2.83}$$

Examples for the Bloch wavefunctions $\phi_q^{(n)}(x)$, which are obtained in a band structure calculation, are shown in Fig. 2.9. Being composed of a discrete sum of plane waves, the Bloch functions are delocalized and extend over the complete, infinitely large lattice. In the lowest band ($n = 1$), the Bloch state in the center of the Brillouin zone ($q = 0$) is symmetric, while at the Brillouin zone edges ($q = \pm\hbar k$) the wavefunction is antisymmetric. The symmetries alternate for higher bands as do the curvatures of the energy bands. The wavefunctions with the highest energy within a certain band are antisymmetric. Furthermore, Fig. 2.9 shows that the wavefunctions of the second band feature an enhanced probability of finding the particle within the potential barrier separating the lattice sites.

The Bloch states of the three-dimensional simple cubic lattice can be constructed as product wavefunctions of the one-dimensional Bloch states for each axis x, y and z, because the Hamiltonian is fully separable. The energy of the product state is given by the sum of the eigenenergies of the three individual Bloch states. In a

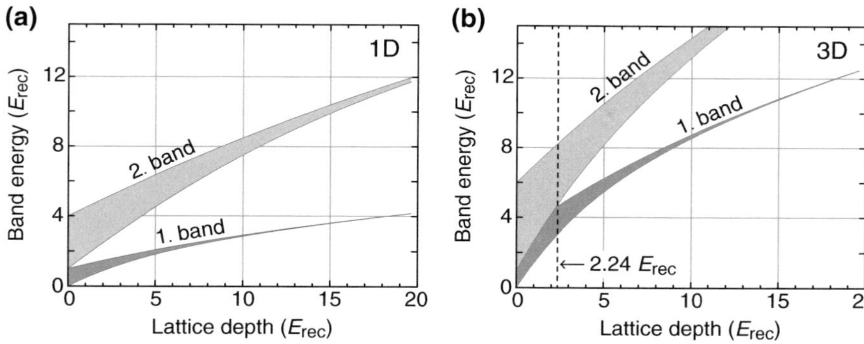

Fig. 2.10 Bandwidths of the first and second Bloch band as a function of lattice depth in one (**a**) and three (**b**) dimensions. While the bands are fully separated in the one-dimensional case, in the three-dimensional case the band gap only starts to open at about 2.24 E_{rec}. The lattice depth is assumed to be identical for each lattice axis

one-dimensional lattice the first and second Bloch band are fully separated for any nonvanishing lattice depth (see Fig. 2.10a). However, in three dimensions the first excited band corresponds to product wavefunctions of two ($n = 1$) and one ($n = 2$) Bloch states. For low lattice depths and certain quasi-momenta $\mathbf{q} = (q_x, q_y, q_z)$, those can have a lower total energy than wavefunctions of the lowest band consisting of a product of three ($n = 1$) Bloch states. For a three-dimensional lattice a band gap between the lowest lattice bands starts to open for depths larger than about 2.24 E_{rec} (see Fig. 2.10b).

All experiments reported in this thesis start by adiabatically loading an ultracold quantum gas or quantum gas mixture into a three-dimensional optical lattice. The lattice depths of all axes are ramped up slowly in order to ensure, that the many-body system remains in its ground state and, in particular, stays in the first lattice band. For the case of a Bose-Einstein condensate this last requirement is not very critical as the atoms dominantly accumulate in the $q = 0$ Bloch state of the first band, which is the lowest energy state and well separated from the second band even for low lattices. However, for fermionic atoms a Fermi sea develops, in which the Bloch states of the first Brillouin zone are filled from bottom up (see Sect. 2.1). Depending on the filling of the Brillouin zone, which is determined by the system size, higher lying Bloch states may be populated, whose energy is degenerate with states of the second band. This situation bears the risk of populating the second band during lattice loading. In Chap. 6 an experimental solution to this problem is presented.

2.2.4 Wannier Basis

Instead of working within the basis of fully delocalized Bloch wavefunctions, it is often more convenient to use a basis set of wavefunctions that are localized at individual lattice sites. Particularly, in the limit of deep lattices the individual sites

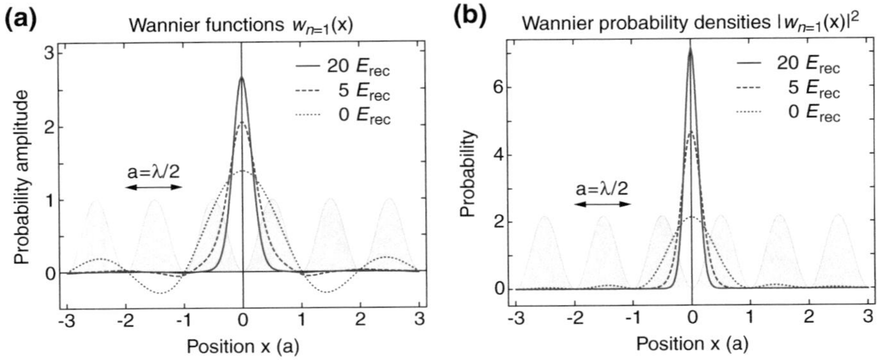

Fig. 2.11 Wannier functions (**a**) and probability densities (**b**) at several lattice depths plotted together with the schematic lattice potential (*gray* shading). For deeper lattices, the Wannier functions are strongly localized. Their overlap with neighboring lattice sites is reduced, which corresponds to a suppression of tunnel coupling

are more and more decoupled from each other and the motion of atoms tends to be restricted to single wells. If the lattice is occupied by many particles, interactions may further enhance localization (see Chap. 3) (Fig. 2.11).

The Wannier functions form a basis set of maximally localized wavefunctions composed by coherent superpositions of Bloch states. The Wannier function of a localized particle at the jth lattice site in the nth Bloch band is defined by [31]

$$w_{(n)}(x - x_j) = \frac{1}{\sqrt{\mathcal{N}}} \sum_q e^{-iqx_j} \phi_q^{(n)}(x). \tag{2.84}$$

Here, the phase factor $\exp(-iqx_j)$ compensates the factor $\exp(iqx)$, which appears in the definition of the Bloch states 2.74, at the coordinate x_j, inducing constructive superposition of the states $\phi_q^{(n)}(x)$ at the corresponding lattice site. The sum runs over the quasi-momenta q within the first Brillouin zone taking discrete values, if the lattice has a finite size. Assuming normalized Bloch states in a system with M lattice sites, there are M different quasi-momenta q and the normalization is given by $\mathcal{N} = M$.

Because both the Bloch and Wannier wavefunctions form a basis, also the reverse transformation is possible

$$\phi_q^{(n)}(x) = \frac{1}{\sqrt{\mathcal{N}}} \sum_j e^{iqx_j} w_{(n)}(x - x_j), \tag{2.85}$$

where the summation is performed over all M lattice sites. This expression shows that the quasi-momentum q determines the phase relation between the localized wavefunctions on the individual lattice sites.

Owing to the decoupling of the spatial directions in a simple cubic lattice, the three-dimensional Wannier function for a particle in the nth band at lattice site $\mathbf{r}_j = (x_j, y_j, z_j)$ is given by the product

$$w_{(n)}(\mathbf{r} - \mathbf{r}_j) = w_{(n)}(x - x_j) \cdot w_{(n)}(y - y_j) \cdot w_{(n)}(z - z_j) \qquad (2.86)$$

of the nth band Wannier functions $w_{(n)}(\alpha - \alpha_j)$ of the individual one-dimensional lattices ($\alpha = x, y, z$). When the first Bloch band is concerned, we usually drop the band index of the Wannier function and simply write $w(\mathbf{r} - \mathbf{r}_j)$.

2.3 Feshbach Resonances

Feshbach resonances allow to control the interactions between atoms by an external magnetic field and have become one of the most important tools in ultracold atom experiments [32]. In this section we will briefly summarize the basics on interatomic interactions and Feshbach resonances as far as they are relevant to this thesis. The discussion also includes the creation of loosely bound Feshbach molecules and the effects of a tight external confinement on the collision physics of two interacting particles.

2.3.1 *Elastic Scattering and Low Energy Collisions*

As in classical mechanics, scattering of two particles in quantum mechanics is described in a coordinate system, in which the center of mass of the two particles is at rest. By doing this the problem reduces to an effective single-particle Hamiltonian in terms of the coordinate $\mathbf{r} = \mathbf{r}_2 - \mathbf{r}_1$, which is the relative distance between the two particles at positions $\mathbf{r}_1$ and $\mathbf{r}_2$, and the reduced mass $\mu = m_1 m_2/(m_1 + m_2)$. In the absence of scattering the solution would simply be a plane wave e^{ikz}, which can be chosen to propagate along the z-direction without loss of generality. k denotes the momentum of the effective particle with the reduced mass μ. When a potential $V(r)$, which we assume to be spherically symmetric for simplicity ($r = |\mathbf{r}|$), is introduced, there will be a finite probability for the particle to be scattered. The resulting asymptotic wavefunction at large distances has the from

$$\psi(\mathbf{r}) \propto e^{ikz} + f(\theta)\frac{e^{ikr}}{r}, \qquad (2.87)$$

consisting of the incoming wave e^{ikz} and an outgoing spherical wave $f(\theta)e^{ikr}/r$, where the scattering amplitude $f(\theta)$ is a function of the scattering angle θ (and generally also of the momentum k). Based on this wavefunction the incoming and the outgoing probability flux can be calculated, which yields for the differential cross section

$$\frac{d\sigma}{d\Omega} = |f(\theta)|^2. \tag{2.88}$$

Since we have assumed a spherically symmetric potential, the wavefunction $\psi(\mathbf{r})$ can be expanded in terms of spherical waves with angular momentum l and vanishing projection on the z-axis ($m = 0$). In this basis the scattering amplitude takes the form [9, 33]

$$f(\theta) = \sum_{l=0}^{\infty}(2l+1)\left(\frac{e^{2i\delta_l}-1}{2ik}\right)P_l(\cos\theta), \tag{2.89}$$

where $P_l(\cos\theta)$ are the Legendre polynomials. The strength of this formulation lies in the fact, that the whole scattering process is captured by the parameters δ_l, which simply denote the spatial phase shift that is added to each spherical wave with angular momentum l due to the scattering process. Using the optical theorem we can directly compute the total scattering cross section [33]

$$\sigma_{\text{tot}} = \frac{4\pi}{k}\mathfrak{Im}[f(\theta=0)] = \frac{4\pi}{k^2}\sum_{l=0}^{\infty}(2l+1)\sin^2\delta_l. \tag{2.90}$$

It can be shown, that the phase shifts scale as a function of the collisional momentum k according to $\delta_l \propto k^{2l+1}$. This means the phase shifts for partial waves $l \neq 0$ essentially vanish for collisions with very low momentum—and so do their contributions to the total scattering cross section. More pictorially, at low momenta, i.e. low collision energy $E_{\text{kin}} = \hbar^2k^2/2\mu$, the effective particle is not able to penetrate the centrifugal barrier [33]. This is the second term of the effective potential $V_{\text{eff}}(r) = V(r) + \hbar^2 l(l+1)/(2mr)$ that arises when the scattering problem is formulated in spherical coordinates. Hence, the scattering amplitude reduces to

$$f(\theta) \approx f_s = \frac{e^{2i\delta_0}-1}{2ik} = \frac{1}{k\cot\delta_0 - ik}, \tag{2.91}$$

which is even independent of the scattering angle θ as $P_0(\cos\theta) = 1$. It is possible to argue that $k\cot\delta_0$ must be an even function of k [4, 33] and correspondingly one can introduce the expansion

$$k\cot\delta_0 \approx -\frac{1}{a_s} + \frac{1}{2}r_{\text{eff}}k^2 \tag{2.92}$$

for low momenta, where a_s denotes the s-wave scattering length and r_{eff} the effective range of the scattering potential. The effective range is typically of the order of the van der Waals length $r_{\rm eff} \approx (\mu C_6/\hbar^2)^{1/4}/2$.[8] For the case $k|a_s| \ll 1$ and $kr_{\rm eff} < 1$ the scattering amplitude becomes independent of the momentum k reading $f_s = -a_s$ and leads to a total cross section for s-wave collisions of

$$\sigma_{\rm tot,s} \approx 4\pi a_s^2, \tag{2.93}$$

which equals the scattering cross section of a hard sphere with a radius a_s [33].

Identical Particles

For now we have derived the theory of collisions based on the assumption of distinguishable particles. However, indistinguishability of quantum particles and their quantum statistics have a profound impact on scattering. On the one hand, the scattering wavefunction $\psi(\mathbf{r})$ must be spatially symmetric for two identical bosons. Therefore, only partial waves with even l enter and they enter twice due to symmetrization. The s-wave scattering cross section therefore amounts to $\sigma_{\rm tot,s} = 8\pi a_s$. On the other hand, the spatial wavefunction for two identical fermions must be antisymmetric as the spin wavefunction is necessarily symmetric. In this case only partial waves with odd l contribute and s-wave collisions of identical fermions are forbidden. Therefore, spin polarized ultracold Fermi gases show essentially no interactions, because scattering of higher partial waves is energetically suppressed.

Pseudo-Potentials

Assuming that the collisional momentum is so low, that the corresponding de Broglie wavelength $\lambda_{\rm dB} = 2\pi/k$ is much larger than the range of the interatomic potential, the details of the potential do not matter for the scattering process. In this case it is convenient to replace the complicated full interatomic potential by a much simpler pseudo-potential that nevertheless reproduces the s-wave scattering correctly. This purpose is served by a the simple contact potential operator

$$V(\mathbf{r}) = \frac{2\pi\hbar^2 a_s}{\mu}\delta(\mathbf{r}). \tag{2.94}$$

Using plane waves it is easily shown, that this expression reproduces the s-wave scattering amplitude $f_s = -a_s$ in first Born approximation [33]. However, for practical use in three dimensions it often must be regularized to avoid $1/r$ divergences of wavefunctions when $r \to 0$. This is typically done by replacing the delta function with $\delta(\mathbf{r})\frac{\partial}{\partial r}r$ [1, 35].

[8] References [32, 34] quote van der Waals lengths of 64.90 a_0 for ^{40}K and 82.58 a_0 for ^{87}Rb, where a_0 denotes the Bohr radius.

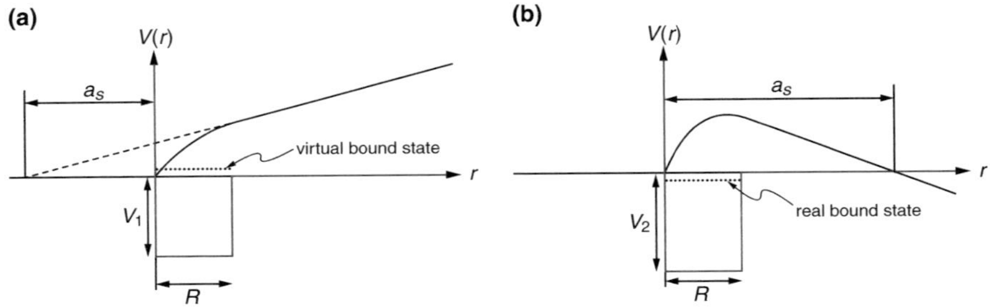

Fig. 2.12 Scattering resonance. We assume two colliding particles with an attractive interparticle box potential $V(r)$ with a finite range R (*gray solid lines*). **a** Due to the attractive potential with a depth V_1 the radial wavefunction $u(r) \propto r - a_s$ has a stronger curvature inside the range R than outside. The intercept a_s of the outside wavefunction with the abscissa corresponds to the scattering length $a_s < 0$. When the potential is deep enough that a bound state almost enters, the scattering length diverges $a_s \rightarrow -\infty$. **b** When the potential is deeper ($V_2 < V_1$) and a real, very loosely bound state has just entered, the bend of the wavefunction becomes stronger and gives rise to a very large positive scattering length a_s. Thus, by tuning the position of a bound state, it can be possible to vary the scattering length over a huge range

2.3.2 Feshbach Resonances

The phenomenon of resonance scattering is treated in many textbooks [9, 33]. Assuming low collision momenta and modeling the attractive scattering potential by a box, it can be shown that the s-wave scattering length diverges and changes sign (from negative to positive) when the potential is deepened and a new bound state enters (see Fig. 2.12). Practically this means, that it would be possible to tune the interactions between particles when there was a knob that allowed to tune a bound state into resonance with the energy of the colliding particles.

For atoms it is not directly possible to tune the depth of the interatomic potential; however, Feshbach resonances offer an ingenious way to bring a bound state into resonance with colliding atoms simply by changing an external magnetic field (see Fig. 2.13a). Let us assume that two (distinguishable) atoms in the hyperfine states $|F_1, m_{F_1}\rangle$ and $|F_2, m_{F_2}\rangle$ collide with a very low relative kinetic energy in the entrance channel (also called: open channel). The atoms can perform s-wave scattering into scattering channels that conserve $M = m_{F_1} + m_{F_2}$ [32]. However, those scattering channels typically have a different magnetic moment than the open channel as they correspond to a different spin configuration. By varying the external magnetic field, it can thus be possible to tune a bound state of one of the scattering channels (also called: closed channel) into resonance with the open channel. If there was no coupling between the open and the closed channel, the atoms would just acquire a phase shift corresponding to the open potential. They would scatter off each other without resonance effects. However, the spin configurations of the open and the closed channels are coupled, because the hyperfine interaction of the two atom system is neither diagonal in the total electronic spin $\hat{S} = \hat{S}_1 + \hat{S}_2$ nor in the total nuclear spin $\hat{I} = \hat{I}_1 + \hat{I}_2$ [4, 32].[9]

This coupling leads to a coherent superposition of the unbound open channel and the bound closed channel wavefunction during the collision. When the admixture of the bound state is relatively small such that the region of the avoided crossing is large (see Fig. 2.13b), the Feshbach resonance is commonly classified as *broad* or *open-channel dominated*. In turn, when the admixture of the bound state is large, the extend of the avoided crossing is small and the resonance is called *narrow* or *closed-channel dominated*.

Generally, the divergent behavior of the s-wave scattering length a_s at a Feshbach resonance can be efficiently parametrized as a function of the magnetic field B by the expression

$$a_s(B) = a_{\mathrm{BG}} \left(1 - \frac{\Delta}{B - B_0}\right). \tag{2.95}$$

Here B_0 denotes the position of the resonance and Δ the width, that is the distance between B_0 and the magnetic field, at which the scattering length vanishes (see Fig. 2.13b). The background scattering length a_{BG} is the scattering length far away from the resonance that is determined by the last bound state or first virtual bound state of the open channel molecular potential.

2.3.3 Creation of Molecules

In the proximity of the Feshbach resonance, on the side of large and positive scattering lengths, a dressed bound state develops with a strong admixture of the open channel wavefunction. Two atoms occupying this bound dressed state can be viewed as a molecule that, in the region of the avoided crossing close to the resonance, has the binding energy

$$E_{\mathrm{b}} \approx \frac{\hbar^2}{2\mu a_s^2}, \tag{2.96}$$

which corresponds to a quadratic dependence on $(B - B_0)$. The molecules in this region are often called Feshbach molecules and are extremely weakly bound. Their molecular wavefunction extends to a very large size on the order of the scattering length a_s. When the magnetic field is tuned further below the resonance the dressed molecule asymptotically turns into the purely bound state of the bare closed channel potential. In this regime the binding energy tunes linearly as a function of the magnetic field following $E_{\mathrm{b}} = \delta\mu \cdot B$, where $\delta\mu$ is the difference in the magnetic moment of the open and the closed channel.

[9] In ultracold atom experiments typically alkali atoms are used in their ground states with vanishing orbital angular momentum $\hat{L}$.

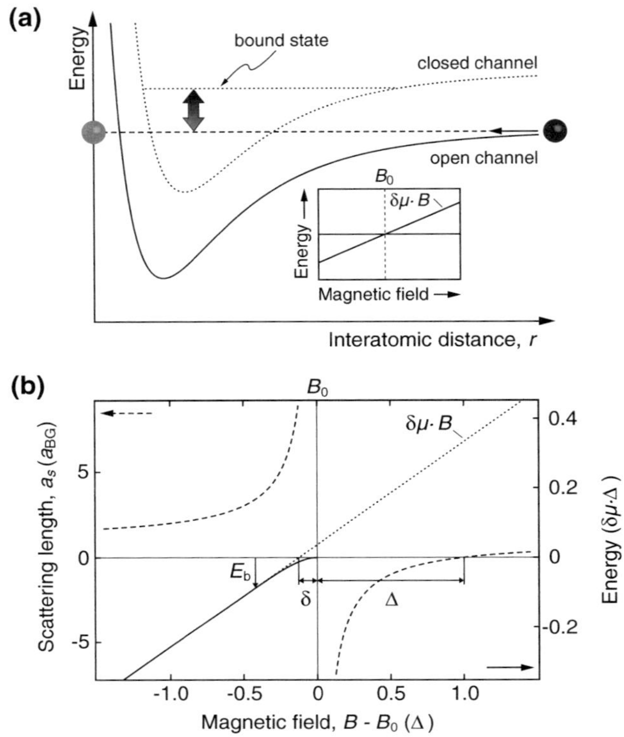

Fig. 2.13 Feshbach resonance. **a** Two atoms that perform a low energy s-wave collision in the open channel (*solid line*) can couple to the bound state of a closed channel potential (*dotted line*). The bound state can be tuned into resonance using an external magnetic field owing to the different magnetic moments of the open and closed channel spin configurations (inset). **b** The Feshbach resonance gives rise to a divergence of the scattering length a_s (*dashed line*). The width Δ of the resonance is the distance between the resonance position and the zero crossing of a_s. The energy of the bare molecular state in the close channel (*dotted line*) depends linearly on the magnetic field and $\delta\mu$ is the difference of the magnetic moments in the open and closed channel. The binding energy E_b of the dressed molecular state (*solid line*) flattens upon approaching the resonance due to mixing of the open and the closed channel in the region of the avoided crossing. The position of the resonance is shifted by δ with respect to the zero crossing of the bare molecular state energy

Feshbach resonances allow to create molecules from free atoms by adiabatically sweeping the magnetic field across the resonance starting from the attractive side [4, 18, 36]. If the Landau-Zener type sweep is performed slowly with respect to the coupling between the open and closed channel and if the temperature of the atomic cloud is low enough, the experimentally achieved conversion efficiencies can reach more than 90 % [37, 38]. Alternatively, the molecular state can be populated using radio-frequency (RF) techniques, which are less efficient, but allow to measure the molecular binding energy [18, 39–41]. In the special case of ^{6}Li a cloud of Feshbach molecules can simply be created by evaporatively cooling the sample on the $a_s > 0$ side of the resonance [38].

When a Feshbach resonance is used to create molecules of two distinct fermionic species, such as two spin states in ^{6}Li or ^{40}K, it has been observed that the lifetime of the molecules is particularly long close to the resonance position. In this regime, the molecule is well approximated by two individual atoms, such that three-body collisions are strongly suppressed by Pauli's principle prohibiting s-wave collisions of two identical fermions [42]. Further away from the resonance, the molecule is more closely bound and it becomes less stable against three-body collisions, because here the bosonic character, which favors collisions, tends to dominate over the fermionic quantum statistics of the constituent atoms. Fermionic molecules can be created by using a heteronuclear mixture of fermionic and bosonic species, for example ^{40}K and ^{87}Rb or ^{6}Li and ^{23}Na [43]. In contrast to bosonic molecules, those are particularly long-lived far away from the resonance, where close binding lets the molecules clearly show fermionic character and collisions among them are prohibited by Pauli's principle. In turn, close to the resonance the lifetime is expected to be very short. The use of an optical lattice can help to increase the lifetime of Feshbach molecules considerably. Isolating atom pairs on individual lattice sites and transferring them into the molecular state, helps to strongly suppress three-body collisions [39].

Feshbach resonances have become a very important tool in experimental ultracold atom physics. In the context of spin mixtures of interacting fermions, they have enabled the fruitful investigation of the so-called BEC-BCS crossover. The ability to tune the interactions between the spins has dramatic consequences: While for vanishing interactions the two spin states populate independent Fermi seas, they pair up in momentum space for attractive interactions forming a superfluid of Cooper pairs [19, 20]. These pairs in momentum space smoothly connect to the molecular state on the repulsive side of the resonance, where a BEC of molecules forms [44–46]. The superfluidity in the crossover region has been demonstrated by the observation of vortices [21]. Another productive branch of experimental efforts uses Feshbach resonances to create ultracold ground state molecules. Here, the weakly bound Feshbach molecules form the starting point, from which the rovibrational ground state is typically addressed using a STIRAP pulse sequence [47–51].

In the experiments presented of this thesis, the formation of molecules by a Feshbach sweep is used to detect double occupation of lattice sites in an interacting fermionic spin mixture of ^{40}K atoms (see Chap. 6). Furthermore, RF spectroscopy on KRb molecules has been performed in our experimental setup. These efforts are reported in the PhD thesis of Thorsten Best [52].

2.3.4 Two Interacting Atoms in a Harmonic Trap

For the creation of molecules in an optical lattice, it is important to understand the physics of two interacting atoms on an individual lattice site. If the on-site potential is approximated by a harmonic oscillator and the atoms are assumed to interact through a regularized contact interaction, the problem can be solved analytically [35]. As shown in Fig. 2.14 the energy levels of the bare 3D harmonic oscillator, which correspond to the noninteracting case ($a_s = 0$), get mixed when interactions set in. The contact interaction between the atoms leads to the emergence of a bound state, whose binding energy diverges towards negative infinity at $a_s = 0$ (see Fig. 2.14a). The application of this level structure to the case of the $B_0 = 202.1$ G Feshbach resonance of ^{40}K is shown in Fig. 2.14b. In contrast to the free space situation, the bound molecular state is not entered exactly on resonance, but already at higher magnetic fields starting at $B_0 + \Delta$, where the scattering length vanishes [36, 53]. The so-called confinement-induced molecules [39] that exist between B_0 and $B_0 + \Delta$ dissociate smoothly when the lattice is adiabatically ramped down.

The energy levels of atom pairs on the sites of an optical lattice have been experimentally investigated using RF spectroscopy [39, 40]. Additionally, Fig. 2.14b shows that the first and the second band are smoothly connected, when the magnetic field is ramped from below to above the resonance, which has also been observed experimentally [54].

2.3.5 Feshbach Resonances for Rubidium and Potassium

The Fermi-Fermi Resonance

In Chap. 6 many-body quantum phases of fermionic spin mixtures in a three-dimensional optical lattice are investigated using ^{40}K in the hyperfine sublevels $|F, m_F\rangle = |9/2, -9/2\rangle$ and $|9/2, -7/2\rangle$. For this combination of spins a broad (open-channel dominated) s-wave Feshbach resonance is available at $B_0 = 202.1$ G to tune interspecies interactions and perform molecule conversion [18, 55–58]. We rely on the parametrization $a_{\mathrm{BG}} = 174\ a_0$ and $\Delta = 7.8 \pm 0.06$ G given in Ref. [19] (see Fig. 2.15). A recent measurement of the zero crossing of the scattering length in our group suggests an updated width of $\Delta = 7.0 \pm 0.2$ G [59]. Below the s-wave resonance, a p-wave resonance among the $|9/2, -7/2\rangle$ spins is located at about 199 G [55, 56]. Unfortunately, the presence of this resonance hinders a close approach of the s-wave resonance on the repulsive side of interactions due to increased losses. The working point at about 175 G corresponds to a scattering length of about $a_{\mathrm{FF}} = 225$ a_0 and is the strongest repulsion that can safely be addressed for our experimental parameters.

An exemplary molecule conversion and dissociation measurement performed in a harmonically trapped sample is shown in Fig. 2.16. Such measurements allow for a precise determination of the resonance position. Additionally, they can serve as a thermometer as the molecule conversion efficiency has been shown to be a sensitive function of the initial dimensionless temperature T/T_{F} of the spin mixture. The conservative value of 80 % conversion efficiency suggests $T/T_{\mathrm{F}} < 0.15$ according to reference [37].

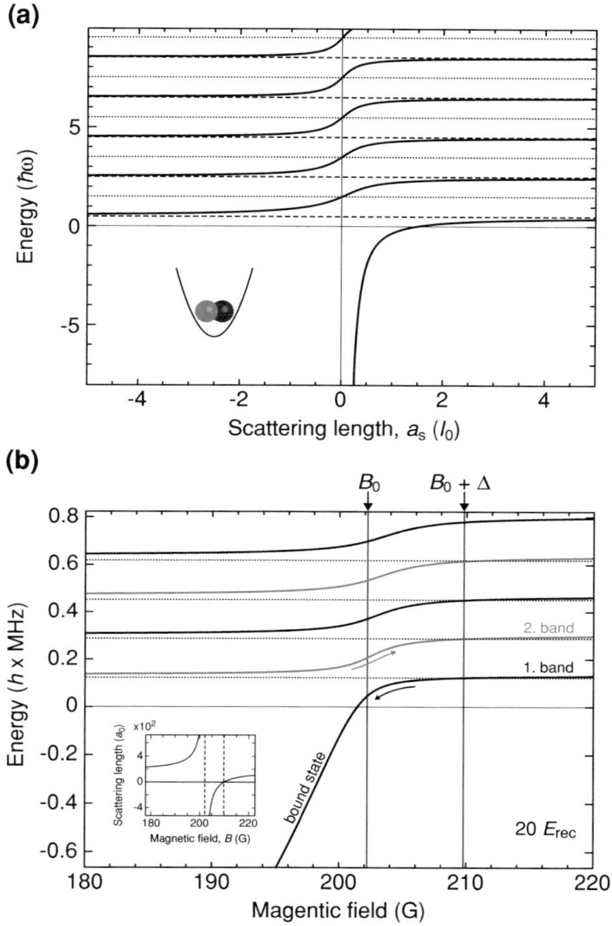

Fig. 2.14 Two interacting atoms in a harmonic oscillator potential. **a** The general structure of the energy levels obtained in an analytical calculation according to reference [35] is shown as a function of the scattering length a_s in units of the harmonic oscillator length $\ell_0 = \sqrt{\hbar/\mu\omega}$. For vanishing interactions ($a_s = 0$) the energy is given by $E = \hbar\omega(2n+3/2)$, while for infinitely strong attraction or repulsion ($a_s \rightarrow \pm\infty$) the energies are shifted by $\hbar\omega$ yielding $E = \hbar\omega(2n+1/2)$. **b** Application of the analytical result to a ^{40}K $|9/2, -9/2\rangle + |9/2, -7/2\rangle$ spin mixture in a 20 E_{rec} lattice in the vicinity of the Feshbach resonance at $B_0 = 202.1$ G (see Sect. 2.3.5)

The Bose-Fermi Resonance

In Chaps. 8 and 9 experiments with interacting Bose-Fermi mixtures are reported. In those investigations we tune the interspecies scattering length with the s-wave Feshbach resonance at $B_0 = 546.75(6)$ G between ^{40}K and ^{87}Rb in the absolute

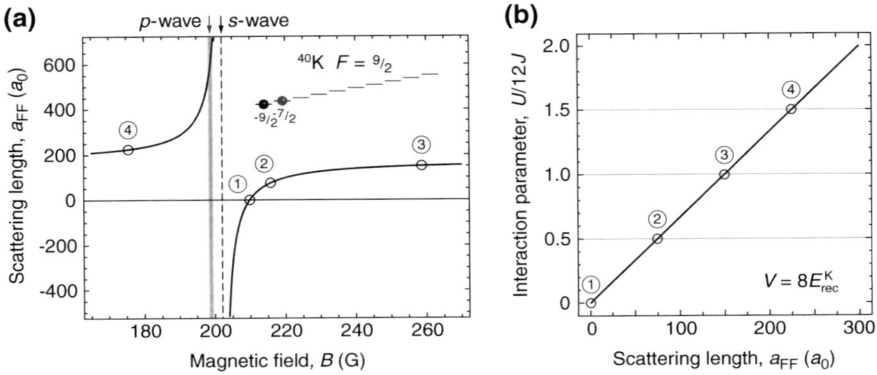

Fig. 2.15 Tunable fermionic ^{40}K spin mixture. **a** An s-wave Feshbach resonance between the hyperfine levels $|9/2, -9/2\rangle$ and $|9/2, -7/2\rangle$ is located at $B_0 = 202.1$ G. Slightly below at 198.8(5) G, there is a p-wave resonance (*gray shaded*) of atoms in the $|9/2, -7/2\rangle$ state [55, 56]. **b** In Chap. 6 we investigate repulsively interacting Fermi-Fermi mixtures in an optical lattice ($\lambda = 738$ nm) at discrete values of the interaction parameter $U/12J = 0, 0.5, 1.0$ and 1.5. For the fixed lattice depth of $V = 8\ E_{rec}$, these working points correspond to the magnetic fields ① 209.9 G, ② 215.8 G, ③ 258.7 G and ④ 175.5 G

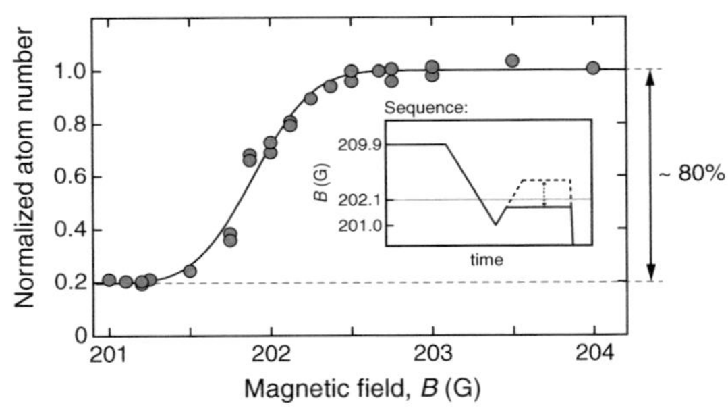

Fig. 2.16 In a first step, molecules consisting of a $|9/2, -9/2\rangle$ and a $|9/2, -7/2\rangle$ atom are created by slowly ramping the magnetic field to $B = 201$ G below the Feshbach resonance. During time-of-flight expansion the magnetic field is again ramped up and the molecules dissociate when the resonance is crossed (see inset). The number of atoms (normalized to the maximal number of recovered atoms) is shown as a function of the final magnetic field. The *solid line* is a fit to the data using an error function. The difference of the highest and lowest normalized atom number yields a lower bound to the molecule conversion efficiency, here about 80 %

ground state hyperfine sublevels $|9/2, -9/2\rangle$ and $|1, +1\rangle$, respectively (see Fig. 2.17). The resonance is the most useable of many, mostly extremely narrow ones between ^{40}K and ^{87}Rb. It has been characterized in several references [60–64], but the most accurate parametrization has recently been reported by Simoni et al. [65] quoting a background scattering length of $a_{BG} = -189\ a_0$ and a width of $\Delta = -3.1$ G.

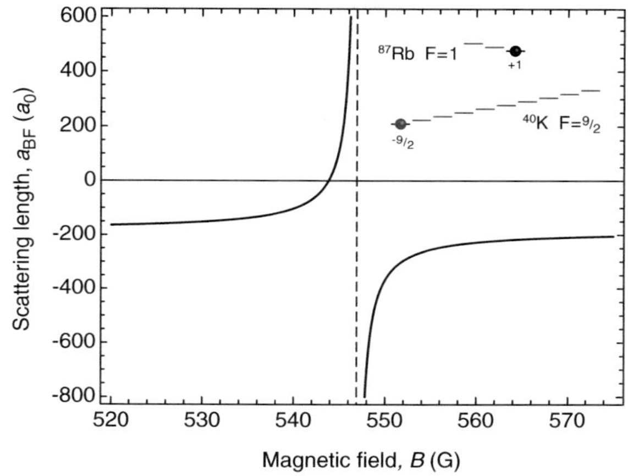

Fig. 2.17 Interspecies Feshbach resonance between the absolute ground state hyperfine sublevels $|9/2, -9/2\rangle$ and $|1, +1\rangle$ of ^{40}K and ^{87}Rb, respectively

Based on the experimental data in Chap. 9, we have been able to precisely extract the magnetic field $B_0 + \Delta$, at which the interspecies interactions vanish, confirming the parametrization on a 0.1 G level.

References

1. K. Huang, *Statistical Mechanics*, 2nd edn. (Wiley, New York, 1987)
2. A. Fetter, J. Walecka, *Quantum Theory of Many-Particle Systems* (Dover Publications, New York, 2003)
3. F. Dalfovo, S. Giorgini, L.P. Pitaevskii, S. Stringari, Theory of Bose-Einstein condensation in trapped gases. Rev. Mod. Phys. **71**, 463 (1999)
4. W. Ketterle, M.W. Zwierlein, in *Making, Probing and Understanding Ultracold Fermi Gases*, ed. by M. Inguscio, W. Ketterle, C. Salomon Ultracold Fermi Gases, Proceedings of the International School of Physics "Enrico Fermi", Course CLXIV, Varenna, 20–30 June 2006, (IOS Press, Amsterdam, 2008)
5. S. Giorgini, L.P. Pitaevskii, S. Stringari, Theory of ultracold atomic Fermi gases. Rev. Mod. Phys. **80**, 1215 (2008)
6. W. Zwerger, Mott-Hubbard transition of cold atoms in optical lattices. J. Opt. B **5**, S9 (2003)
7. I. Bloch, J. Dalibard, W. Zwerger, Many-body physics with ultracold gases. Rev. Mod. Phys. **80**, 885 (2008)
8. W. Pauli, The connection between spin and statistics. Phys. Rev. **58**, 716 (1940)
9. L.D. Landau, E.M. Lifshitz, *Quantum Mechanics: Non-relativistic Theory* (Pergamon Press, New York, 1977)
10. L.D. Carr, G.V. Shlyapnikov, Y. Castin, Achieving a BCS transition in an atomic Fermi gas. Phys. Rev. Lett. **92**, 150404 (2004)
11. M. Köhl, Thermometry of fermionic atoms in an optical lattice. Phys. Rev. A **73**, 031601 (2006)
12. K. Mølmer, Bose condensates and Fermi gases at zero temperature. Phys. Rev. Lett. **80**, 1804 (1998)
13. R. Roth, H. Feldmeier, Mean-field instability of trapped dilute boson-fermion mixtures. Phys. Rev. A **65**, 021603 (2002)

14. R. Roth, Structure and stability of trapped atomic boson-fermion mixtures. Phys. Rev. A **66**, 013614 (2002)
15. C. Ospelkaus, S. Ospelkaus, K. Sengstock, K. Bongs, Interaction-driven dynamics of ^{40}K ^{87}Rb fermion-boson gas mixtures in the large-particle-number limit. Phys. Rev. Lett. **96**, 020401 (2006)
16. D. Jaksch, C. Bruder, J.I. Cirac, C.W. Gardiner, P. Zoller, Cold bosonic atoms in optical lattices. Phys. Rev. Lett. **81**, 3108 (1998)
17. M. Greiner, O. Mandel, T. Esslinger, T. Hänsch, I. Bloch, Quantum phase transition from a superfluid to a Mott insulator in a gas of ultracold atoms. Nature **415**, 39 (2002)
18. C.A. Regal, C. Ticknor, J.L. Bohn, D.S. Jin, Creation of ultracold molecules from a Fermi gas of atoms. Nature **424**, 47 (2003)
19. C.A. Regal, M. Greiner, D.S. Jin, Observation of resonance condensation of fermionic atom pairs. Phys. Rev. Lett. **92**, 040403 (2004)
20. M.W. Zwierlein, C.A. Stan, C.H. Schunck, S.M.F. Raupach, A.J. Kerman, W. Ketterle, Condensation of pairs of fermionic atoms near a Feshbach resonance. Phys. Rev. Lett. **92**, 120403 (2004)
21. M. Zwierlein, J. Abo-Shaeer, A. Schirotzek, C. Schunck, W. Ketterle, Vortices and superfluidity in a strongly interacting Fermi gas. Nature **435**, 1047 (2005)
22. W.D. Phillips, in *Laser Cooling and Trapping of Neutral Atoms*, ed. by W.D. Phillips, E. Arimondo, F. Strumia. Laser Manipulation of Atom and Ions, Proceesdings of the International School of Physics "Enrico Fermi" Course CXVIII, (North-Holland, Amsterdam, 1992)
23. O. Morsch, M. Oberthaler, Dynamics of Bose-Einstein condensates in optical lattices. Rev. Mod. Phys. **78**, 179 (2006)
24. H. Metcalf, P. van der Straten, *Laser Cooling and Trapping* (Springer, Berlin, 1999)
25. R. Grimm, M. Weidemüller, Y. Ovchinnikov, Optical dipole traps for neutral atoms. Adv. At. Mol. Opt. Phys. **42**, 95 (2000)
26. L. Allen, J.H. Eberly, *Optical Resonance and Two-Level Atoms* (Dover Publications, Inc., New York, 1987)
27. C. Cohen-Tannoudji, J. Dupont-Roc, G. Grynberg, *Atom-Photon Interactions* (Wiley, New York, 1998)
28. M. Scully, S. Zubairy, *Quantum Optics* (Cambridge University Press, Cambridge, 1997)
29. R. Loudon, *The Quantum Theory of Light* (*Oxford Science Publications*), 3rd edn. (OUP, Oxford, 2000)
30. B. Saleh, M. Teich, *Fundamentals of Photonics* (Wiley-Interscience, New York, 1991)
31. N.W. Ashcroft, N.D. Mermin, *Solid State Physics* (Saunders College Publishing, Fort Worth, 1976)
32. C. Chin, R. Grimm, P. Julienne, E. Tiesinga, Feshbach resonances in ultracold gases. Rev. Mod. Phys. **82**, 1225 (2010)
33. J.J. Sakurai, *Modern Quantum Mechanics* (Revised edn). (Addison Wesley, Reading, 1994)
34. E.G.M. van Kempen, S.J.J.M.F. Kokkelmans, D.J. Heinzen, B.J. Verhaar, Interisotope determination of ultracold rubidium interactions from three high-precision experiments. Phys. Rev. Lett. **88**, 093201 (2002)
35. T. Busch, B.-G. Englert, K. Rzażewski, M. Wilkens, Two cold atoms in a harmonic trap. Found. Phys. **28**, 549 (1998)
36. T. Köhler, K. Góral, P.S. Julienne, Production of cold molecules via magnetically tunable Feshbach resonances. Rev. Mod. Phys. **78**, 1311 (2006)
37. C. Regal, *Experimental realization of BCS-BEC crossover physics wiht a Fermi gas of atoms*. Ph.D. Thesis, University of Colorado, 2005
38. M.W. Zwierlein, *High-temperature superfluidity in an ultracold Fermi gas*. Ph.D. Thesis, Massachusetts Institute of Technology, 2006
39. C. Ospelkaus, S. Ospelkaus, L. Humbert, P. Ernst, K. Sengstock, K. Bongs, Ultracold heteronuclear molecules in a 3D optical lattice. Phys. Rev. Lett. **97**, 120402 (2006)
40. T. Stöferle, H. Moritz, K. Günter, M. Köhl, T. Esslinger, Molecules of fermionic atoms in an optical lattice. Phys. Rev. Lett. **96**, 030401 (2006)

41. C. Klempt, T. Henninger, O. Topic, M. Scherer, L. Kattner, E. Tiemann, W. Ertmer, J.J. Arlt, Radio-frequency association of heteronuclear Feshbach molecules. Phys. Rev. A **78**, 061602 (2008)
42. D.S. Petrov, C. Salomon, G.V. Shlyapnikov, Weakly bound dimers of fermionic atoms. Phys. Rev. Lett. **93**, 090404 (2004)
43. C.A. Stan, M.W. Zwierlein, C.H. Schunck, S.M.F. Raupach, W. Ketterle, Observation of Feshbach resonances between two different atomic species. Phys. Rev. Lett. **93**, 143001 (2004)
44. S. Jochim, M. Bartenstein, A. Altmeyer, G. Hendl, S. Riedl, C. Chin, J. Hecker Denschlag, R. Grimm, Bose-Einstein condensation of molecules. Science **302**, 2101 (2003)
45. M. Greiner, C. Regal, D. Jin, Emergence of a molecular Bose-Einstein condensate from a Fermi gas. Nature **426**, 537 (2003)
46. M.W. Zwierlein, C.A. Stan, C.H. Schunck, S.M.F. Raupach, S. Gupta, Z. Hadzibabic, W. Ketterle, Observation of Bose-Einstein condensation of molecules. Phys. Rev. Lett. **91**, 250401 (2003)
47. S. Ospelkaus, A. Pe'er, K. Ni, J. Zirbel, B. Neyenhuis, S. Kotochigova, P. Julienne, J. Ye, D. Jin, Efficient state transfer in an ultracold dense gas of heteronuclear molecules. Nat. Phys. **4**, 622 (2008)
48. K.-K. Ni, S. Ospelkaus, M.H.G. de Miranda, A. Pe'er, B. Neyenhuis, J.J. Zirbel, S. Kotochigova, P.S. Julienne, D.S. Jin, J. Ye, A high phase-space-density gas of polar molecules. Science **322**, 231 (2008)
49. J. Danzl, E. Haller, M. Gustavsson, M. Mark, R. Hart, N. Bouloufa, O. Dulieu, H. Ritsch, H.-C. Nägerl, Quantum gas of deeply bound ground state molecules. Science **321**, 1062 (2008)
50. J. Danzl, M. Mark, E. Haller, M. Gustavsson, R. Hart, J. Aldegunde, J. Hutson, H.-C. Nägerl, An ultracold high-density sample of rovibronic ground-state molecules in an optical lattice. Nat. Phys. **6**, 265 (2010)
51. S. Ospelkaus, K.-K. Ni, D. Wang, M.H.G. de Miranda, B. Neyenhuis, G. Quemener, P.S. Julienne, J.L. Bohn, D.S. Jin, J. Ye, Quantum-state controlled chemical reactions of ultracold potassium-rubidium molecules. Science **327**, 853 (2010)
52. T. Best, *Interacting Bose-Fermi mixtures in optical lattices*. Ph.D. Thesis, Universität Mainz, 2011
53. D.B.M. Dickerscheid, U. Al Khawaja, D. van Oosten, H.T.C. Stoof, Feshbach resonances in an optical lattice. Phys. Rev. A **71**, 043604 (2005)
54. M. Köhl, H. Moritz, T. Stöferle, K. Günter, T. Esslinger, Fermionic atoms in a three dimensional optical lattice: observing Fermi surfaces, dynamics, and interactions. Phys. Rev. Lett. **94**, 080403 (2005)
55. C.A. Regal, C. Ticknor, J.L. Bohn, D.S. Jin, Tuning p-wave interactions in an ultracold Fermi gas of atoms. Phys. Rev. Lett. **90**, 053201 (2003)
56. C. Ticknor, C.A. Regal, D.S. Jin, J.L. Bohn, Multiplet structure of Feshbach resonances in nonzero partial waves. Phys. Rev. A **69**, 042712 (2004)
57. T. Loftus, C.A. Regal, C. Ticknor, J.L. Bohn, D.S. Jin, Resonant control of elastic collisions in an optically trapped Fermi gas of atoms. Phys. Rev. Lett. **88**, 173201 (2002)
58. C.A. Regal, D.S. Jin, Measurement of positive and negative scattering lengths in a Fermi gas of atoms. Phys. Rev. Lett. **90**, 230404 (2003)
59. U. Schneider, L. Hackermüller, J.P. Ronzheimer, S. Will, S. Braun, T. Best, I. Bloch, E. Demler, S. Mandt, D. Rasch, A. Rosch, Fermionic transport and out-of-equilibrium dynamics in a homogeneous Hubbard model with ultracold atoms. Nat. Phys. **8**, 213 (2012)
60. K. Günter, T. Stöferle, H. Moritz, M. Köhl, T. Esslinger, Bose-Fermi mixtures in a three-dimensional optical lattice. Phys. Rev. Lett. **96**, 180402 (2006)
61. S. Ospelkaus, C. Ospelkaus, O. Wille, M. Succo, P. Ernst, K. Sengstock, K. Bongs, Localization of bosonic atoms by fermionic impurities in a three-dimensional optical lattice. Phys. Rev. Lett. **96**, 180403 (2006)
62. M. Zaccanti, C. D'Errico, F. Ferlaino, G. Roati, M. Inguscio, G. Modugno, Control of the interaction in a Fermi-Bose mixture. Phys. Rev. A **74**, 041605 (2006)

63. F. Ferlaino, C. D'Errico, G. Roati, M. Zaccanti, M. Inguscio, G. Modugno, A. Simoni, Feshbach spectroscopy of a K–Rb atomic mixture. Phys. Rev. A **73**, 040702 (2006)
64. C. Klempt, T. Henninger, O. Topic, J. Will, W. Ertmer, E. Tiemann, J. Arlt, ^{40}K-^{87}Rb Feshbach resonances: modeling the interatomic potential. Phys. Rev. A **76**, 020701 (2007)
65. A. Simoni, M. Zaccanti, C. D'Errico, M. Fattori, G. Roati, M. Inguscio, G. Modugno, Near-threshold model for ultracold KRb dimers from interisotope Feshbach spectroscopy. Phys. Rev. A **77**, 052705 (2008)

Chapter 3
Hubbard Models for Bosons and Fermions

The concepts of the Hubbard model have largely shaped the way we think about atoms in optical lattices. Originally, the Hubbard model has been developed in theoretical condensed matter physics [1]. It constitutes the simplest model system that describes the competition between the kinetic energy and interactions of electrons in the lattice potential of an ionic crystal [2]. This competition gives rise to a quantum phase transition from a metallic state of delocalized atoms to an insulating state—the famous Mott insulator transition. Nevertheless, many ground state properties of the Hubbard model are yet unknown even after decades of theoretical research.

In 1998 Jaksch et al. [3] have proposed the implementation of the Bose-Hubbard model [4] with ultracold bosonic atoms in an optical lattice. The proposal has been realized in a seminal experiment by Greiner et al. [5] that lead to the observation of the superfluid to Mott insulator transition. Since then ultracold atoms in optical lattices are regarded as an near ideal experimental implementation of the Hubbard model, much cleaner, much more controllable and much more tractable than any real strongly correlated solid state system. The remarkable correspondence between a theoretical model system and its experimental realization has fueled the idea to use ultracold atoms in optical lattices as a quantum simulator [6], operating to find the ground states of complex theoretical model systems.

In the first section I derive the Hubbard Hamiltonian departing from the full many-body Hamiltonian of interacting particles in a periodic potential. The derivation is presented in a general form that is valid both for bosonic and fermionic atoms. Then, the implementation of the Bose-Hubbard model using interacting bosonic atoms in a simple cubic optical lattice is presented, including a brief discussion of the superfluid to Mott insulator transition [3, 5, 7]. The third section introduces the Fermi-Hubbard model that has originally been developed in condensed matter physics to describe electrons in real materials, which form a many-body system of spin one-half fermions.[1] It can be experimentally realized with two-component spin

[1] Originally, this model is called *the* Hubbard model. In this thesis we will mostly refer to it as the Fermi-Hubbard model for better distinction from the bosonic version, the Bose-Hubbard model.

S. Will, *From Atom Optics to Quantum Simulation*, Springer Theses,
DOI: 10.1007/978-3-642-33633-1_3,

mixtures of fermionic atoms. The phases of the Fermi-Hubbard model at half-filling are discussed with a special emphasis on the ground state phases at strong repulsion. Here the Fermi-Hubbard model reduces to a quantum Heisenberg model, which favors antiferromagnetic ordering of spins. The chapter concludes with a discussion of the band structure and single-particle wavefunctions in a harmonically confined lattice system, as it is typical in experimental realizations. Based on this geometry, the formation of a band insulator of noninteracting fermions is explained.

3.1 Derivation of the Hubbard Model

In order to assess under which conditions ultracold atoms in optical lattices form a good quantum simulator of the Hubbard model, it is crucial to understand the approximations that are applied in its derivation. In fact, care must be taken to operate experimental optical lattice setups within the bounds of the Hubbard model. We will discuss in Chaps. 7–9 that effects beyond the Hubbard model can play an important role in optical lattice experiments.

The derivation of the single-band Hubbard model [1] starts with the general many-body Hamiltonian of a d-dimensional lattice system of bosonic or fermionic atoms. The spin state of the atoms is indicated by the index σ, which, for example, represents the spin up and spin down states in a fermionic spin mixture. The atoms are exposed to a periodic lattice potential $V_{\text{lat}}(\mathbf{r})$ with an additional underlying trapping potential $V_{\text{trap}}(\mathbf{r})$. Within this section no specific assumptions about the dimensionality, the exact lattice geometry or the underlying potential are made. The interactions between atom pairs are parametrized by a contact interaction as shown in Eq. 2.94.

Using the field operator formalism introduced in Sect. 2.1.1, the full Hamiltonian $\hat{H} = \hat{H}_0 + \hat{H}_{\text{int}}$ reads

$$\hat{H}_0 = \sum_{\sigma} \int d^3r\, \hat{\psi}_{\sigma}^{\dagger}(\mathbf{r}) \left[-\frac{\hbar^2}{2m}\Delta + V_{\text{lat}}(\mathbf{r}) + V_{\text{trap}}(\mathbf{r}) \right] \hat{\psi}_{\sigma}(\mathbf{r}), \tag{3.1}$$

$$\hat{H}_{\text{int}} = \frac{g}{2} \sum_{\sigma\sigma'} \int d^3r\, \hat{\psi}_{\sigma}^{\dagger}(\mathbf{r}) \hat{\psi}_{\sigma'}^{\dagger}(\mathbf{r}) \hat{\psi}_{\sigma'}(\mathbf{r}) \hat{\psi}_{\sigma}(\mathbf{r}), \tag{3.2}$$

where the interaction strength is defined by $g = 4\pi\hbar^2 a_s/m$ with the s-wave scattering length a_s. The field operators $\hat{\psi}_{\sigma}(\mathbf{r})$ and $\hat{\psi}_{\sigma}^{\dagger}(\mathbf{r})$ describe the annihilation and creation of an atom with spin σ at position $\mathbf{r}$, respectively. In quantum systems of noninteracting particles there is an intrinsic tendency for delocalization to minimize the kinetic energy. However, interactions favor the localization of particles to reduce the interaction energy. This competition between kinetic and interaction energy is the physical core of the Hubbard model. Therefore, it is natural to expand the field operators in terms of the localized Wannier functions (see Sect. 2.2.4). The Wannier function at a lattice site j in the nth Bloch band is given by

$$w_{(n)}(\mathbf{r}-\mathbf{r}_j) = \frac{1}{\sqrt{M}} \sum_{\mathbf{q}} e^{-i\mathbf{q}\mathbf{r}_j} \phi_{\mathbf{q}}^{(n)}(\mathbf{r}), \tag{3.3}$$

where $\mathbf{q}$ denotes the d-dimensional quasi-momentum vector, running over the first Brillouin zone, and M is the number lattice of sites in the system. The operator that creates a Wannier state at site j is defined by

$$\hat{c}^{\dagger}_{(n)j\sigma} = \int d^3r\, w_{(n)}(\mathbf{r}-\mathbf{r}_j)\hat{\psi}^{\dagger}_{\sigma}(\mathbf{r}). \tag{3.4}$$

For brevity, the fermionic operators $\hat{c}_{j\sigma}$ and $\hat{c}^{\dagger}_{j\sigma}$ are used in this section, but can be replaced by bosonic operators at any point of the derivation. The expansion of the field operator reads

$$\hat{\psi}^{\dagger}_{\sigma}(\mathbf{r}) = \sum_{n=1}^{\infty}\sum_{j=1}^{M} w^{*}_{(n)}(\mathbf{r}-\mathbf{r}_j)\,\hat{c}^{\dagger}_{(n)j\sigma} \xrightarrow{\text{single-band}} \sum_{j=1}^{M} w^{*}(\mathbf{r}-\mathbf{r}_j)\,\hat{c}^{\dagger}_{j\sigma}. \tag{3.5}$$

In the second step, we have introduced the first major simplification of the Hubbard model: The restriction to the first Bloch band.[2] Since higher bands are excluded, all interband couplings are suppressed that otherwise inherently arise as a consequence of interactions. The experimental observation of interaction-induced coupling between bands is a central topic of Chap. 7.

With the single-band field operators of Eq. 3.5 the Hamiltonian $\hat{H}$ takes the form

$$\hat{H} = -\sum_{\sigma}\sum_{ij} J_{ij}\,\hat{c}^{\dagger}_{i\sigma}\hat{c}_{j\sigma} + \frac{1}{2}\sum_{\sigma\sigma'}\sum_{ijkl} U_{ijkl}\,\hat{c}^{\dagger}_{i\sigma}\hat{c}^{\dagger}_{j\sigma'}\hat{c}_{k\sigma'}\hat{c}_{l\sigma} + \sum_{\sigma}\sum_{i} \epsilon_i\,\hat{n}_{i\sigma}, \tag{3.6}$$

where $\hat{n}_{i\sigma} = \hat{c}^{\dagger}_{i\sigma}\hat{c}_{i\sigma}$ counts the number of particles in spin state σ at lattice site i.

The first term denotes the kinetic energy, which is expressed by the tunneling matrix elements J_{ij}. They denote the gain in energy when a particle hops from site i to site j and read

$$\begin{aligned} J_{ij} &= -\int d^3r\, w^{*}(\mathbf{r}-\mathbf{r}_i)\left[-\frac{\hbar^2}{2m}\Delta + V_{\text{lat}}(\mathbf{r})\right] w(\mathbf{r}-\mathbf{r}_j) \\ &= -\frac{1}{M}\sum_{\mathbf{q}} e^{-i(\mathbf{r}_j-\mathbf{r}_i)\mathbf{q}}\, E_{\mathbf{q}}, \end{aligned} \tag{3.7}$$

---

[2] This approximation is justified when the Fermi energy of the system lies within the first band. The interaction energy of a particle pair should be much smaller than the band gap between the first and second band. Therefore, in experimental realizations care must be taken to chose atom number, lattice depth and system size in a way that only the first band is populated. Furthermore, population of the second band during lattice loading must be avoided.

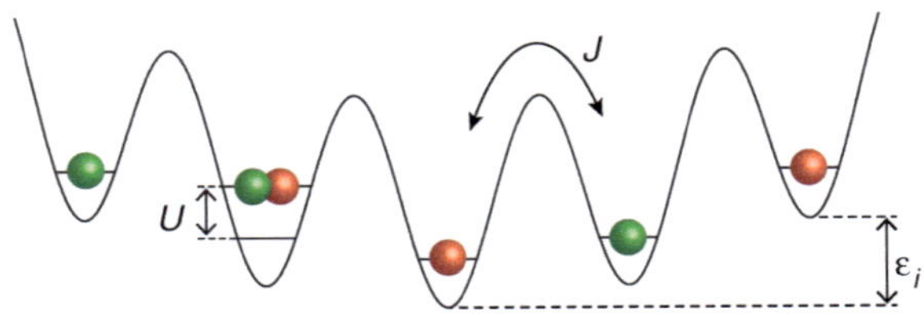

Fig. 3.1 Illustration of the matrix elements of the Hubbard model (see text)

where $\mathbf{q}$ runs over the first Brillouin zone. In the second step the definition of the Wannier function is inserted and the property of the Bloch states to be the eigenfunctions of the enclosed operator is used (see Sect. 2.2.3). The equation establishes an important relation between the tunneling of particles and the dispersion relation $E_{\mathbf{q}}$. Using $\sum_i e^{-i(\mathbf{q}-\mathbf{q}')\mathbf{r}_i} = M\delta_{\mathbf{q},\mathbf{q}'}$ for $M \gg 1$, it can be cast in the instructive form

$$E_{\mathbf{q}} = -\sum_i J_{ij}\, e^{i(\mathbf{r}_j - \mathbf{r}_i)\mathbf{q}}, \tag{3.8}$$

where $\mathbf{r}_j$ denotes a fixed, but arbitrary lattice site (Fig. 3.1).

The second term describes all possible forms of interactions between the different lattice sites within the first Bloch band. The interaction matrix elements are given by

$$U_{ijkl} = g \int d^3r\, w^*(\mathbf{r}-\mathbf{r}_i)w^*(\mathbf{r}-\mathbf{r}_j)w(\mathbf{r}-\mathbf{r}_k)w(\mathbf{r}-\mathbf{r}_l) \tag{3.9}$$

and yield notable contributions when the Wannier functions at lattice sites i, j, k and l have a significant overlap.

The third term accounts for the external trapping potential

$$\epsilon_i = \int d^3r\, |w(\mathbf{r}-\mathbf{r}_i)|^2 V_{\text{trap}}(\mathbf{r}) \approx V_{\text{trap}}(\mathbf{r}_i). \tag{3.10}$$

Here, we have used the orthonormality of the Wannier function and the fact, that the trapping potential typically varies very slowly over the extent of the on-site Wannier function.

Based on the general single-band Hamiltonian (see Eq. 3.6), the Hubbard model is obtained by omitting almost all of the intersite couplings. Only the strongest terms are kept:

- Tunneling between adjacent lattice sites i and j, for which the notation $\langle i, j\rangle$ is used in the following. This approximation is often called the tight-binding limit [1, 8, 9].
- On-site interactions. U_{iiii} is the dominating on-site interaction matrix element, particularly, when a contact interaction between the particles is assumed. All intersite couplings like U_{ijji} or the exchange terms U_{ijij} are ignored.

Using the notations $J = J_{ij}$ for the nearest-neighbor tunneling matrix element between sites i and j and $U = U_{iiii}$ for the on-site interaction matrix element, the Hubbard model takes the form

$$\hat{H} = -J \sum_{\sigma} \sum_{\langle i,j \rangle} \hat{c}^{\dagger}_{i\sigma} \hat{c}_{j\sigma} + \frac{U}{2} \sum_{\sigma\sigma'} \sum_{i} \hat{c}^{\dagger}_{i\sigma} \hat{c}^{\dagger}_{i\sigma'} \hat{c}_{i\sigma'} \hat{c}_{i\sigma} + \sum_{\sigma} \sum_{i} \epsilon_i \, \hat{n}_{i\sigma}. \quad (3.11)$$

Compared to the full Hamiltonian, the Hubbard model comprises extensive simplifications. Nevertheless, it lucidly features the competition between tunneling and interaction, which gives rise to highly correlated ground states.

3.1.1 *Matrix Elements of the Hubbard Model*

The experiments of this thesis use a three-dimensional ($d = 3$) optical lattice with an underlying harmonic confinement. The lattice has simple cubic symmetry and the depth is identical on all three axes ($\alpha = x, y, z$). We summarize the most important relations for the matrix elements in this experimental setting.

Tunneling Matrix Element

In a simple cubic lattice the movement of particles can be considered independently for each lattice axis. With the lattice constant $a = \lambda/2$, which corresponds to the distance between adjacent lattice sites, the tunneling matrix element reads

$$J = -\int d\alpha \, w^*(\alpha \pm a) \left(-\frac{\hbar^2}{2m} \frac{\partial^2}{\partial \alpha^2} + V_{\text{lat}}(\alpha) \right) w(\alpha). \quad (3.12)$$

This energy scale defines the tunneling time scale by $\tau_J = \hbar/(zJ)$, where z is the coordination number. The coordination number is the number of nearest neighbors of an individual lattice site, such that $z = 6$ for a three-dimensional simple cubic lattice.

Applying the tight-binding approximation to Eq. 3.8 the one-dimensional dispersion relation takes a sinusoidal form and has a band width of $4J$,

$$E_{q_\alpha} = -2J \cos(q_\alpha a). \quad (3.13)$$

Therefore, the tunneling matrix element can be inferred from a band structure calculation via [7, 10]

$$J = \frac{1}{4} \left(E_{\pm\hbar k} - E_0 \right), \quad (3.14)$$

provided the lattice is deep enough to fulfill the tight-binding approximation.[3] Because the three-dimensional dispersion relation corresponds to $E_{\mathbf{q}} = -2J\sum_\alpha \cos(q_\alpha a)$, the band width of the three-dimensional lattice is given by $12J$.

In the limit of very deep lattices, an analytic expression for the dependence of the tunneling matrix element on the lattice depth can be derived by solving a one-dimensional Mathieu equation [10, 12] yielding

$$J \approx \frac{4E_{\mathrm{rec}}}{\sqrt{\pi}} s_\alpha^{3/4} e^{-2\sqrt{s_\alpha}}. \tag{3.15}$$

This expression shows that the tunnel coupling between nearest neighbors decreases exponentially for increasing lattice depth $s_\alpha = V_\alpha / E_{\mathrm{rec}}$. However, for experimentally relevant lattice depths ($s_\alpha < 50\,E_{\mathrm{rec}}$), the accuracy of Eq. 3.15 is mostly not sufficient and J must be calculated numerically based on a band structure calculation and Eq. 3.12 (see Fig. 3.2a).

Interaction Matrix Element

Applying definition 3.9 to the case $i = j = k = l$, the on-site interaction matrix element is given by

$$U = \frac{4\pi\hbar^2 a_s}{m}\int d^3r\,|w(\mathbf{r})|^4 = \frac{4\pi\hbar^2 a_s}{m}\left(\int_{-\infty}^{\infty} d\alpha\,|w(\alpha)|^4\right)^3, \tag{3.16}$$

where the integral corresponds to the density-density overlap integral of the on-site wavefunction. The second step is possible, because the three-dimensional Wannier function factorizes and the one-dimensional Wannier functions $w(\alpha)$ are identical for each direction due to identical lattice depths. The matrix element U represents the on-site interaction energy of an atom pair within the single-band approximation of the Hubbard model. However, interactions between atoms promote population to higher bands, which effectively modifies the on-site interaction matrix element. The details of this process are discussed in Chap. 7.

For very deep lattices, when single lattice sites can be approximated by an harmonic oscillator potential, the Wannier function may be replaced by the Gaussian ground state wavefunction of the harmonic oscillator. In this limit the interaction energy can be expressed as a function of the dimensionless lattice depth s_α,

$$U \approx \sqrt{\frac{8}{\pi}} k a_s s_\alpha^{3/4} E_{\mathrm{rec}}. \tag{3.17}$$

---

[3] This is typically the case for lattice depth larger than 5 E_{rec}. Under these conditions the higher-order tunneling processes are suppressed with respect to nearest-neighbor tunneling by at least an order of magnitude [7, 11].

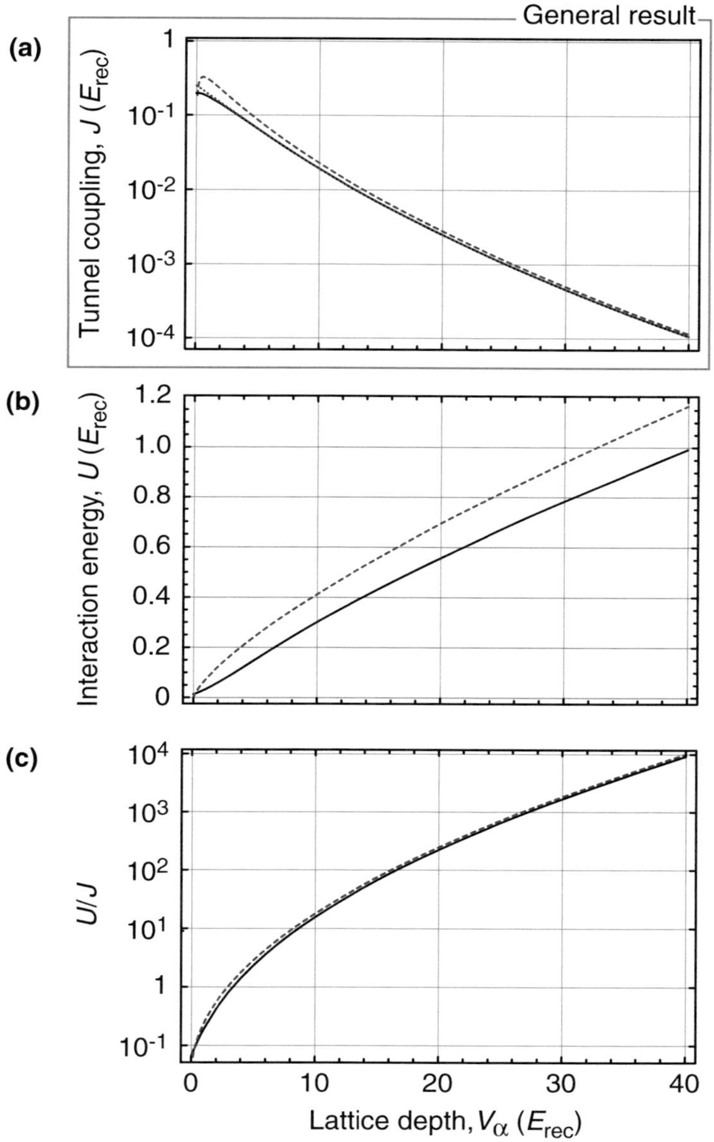

Fig. 3.2 Tunneling and interaction matrix elements. **a** The tunneling matrix element is calculated numerically (*solid*) and using the approximate formulas 3.14 (*dashed*) and 3.15 (*dotted*). The results are independent of the atomic species or lattice constant ($a = \lambda/2$), because V_α ($\alpha = x, y, z$) and J are given in units of E_{rec}. **b** On-site interaction energy U for ^{87}Rb in the $|F, m_F\rangle = |1, +1\rangle$ state ($a_s = 102\, a_0$) as a function of the depth of a three-dimensional 738 nm-lattice. The exact numerical result and the harmonic approximation 3.17 are shown as *solid* and *dashed* lines, respectively. **c** The relative strength of the interaction U/J can be tuned over several orders of magnitude. The *solid curve* shows the exact numerical result, the *dashed curve* uses the approximations in 3.15 and 3.17

For practically relevant lattice depths ($s_\alpha < 50\ E_{\text{rec}}$) and typical interactions ($a_s \approx 100\ a_0$) the relative error compared to a calculation with Wannier functions amounts to at least 20 % as shown in Fig. 3.2b. Therefore, the harmonic approximation must be applied carefully and should only be used as a rough estimate of the interaction energy (see variational model in Sect. 9.2.3).

According to Eqs. 3.15 and 3.17 the tunnel coupling decreases exponentially and the interaction matrix element features a power law increase when the lattice depth is raised. Therefore the ratio U/J can be tuned over several orders of magnitude simply by changing the lattice depth as shown in Fig. 3.2c for the case of ^{87}Rb atoms. In this way, optical lattices are used to enter the strongly interacting regime. With the additional application of a Feshbach resonance to control the s-wave scattering length a_s (see Sect. 2.3), the parameters J and U can even be tuned independently from each other. This approach can be advantageous, because it allows to have strong interactions and fast tunneling at the same time, ensuring a fast redistribution of atoms. Due to the limited lifetime of ultracold atom samples, fast adiabatic parameter changes are experimentally crucial as discussed in Chap. 6.

3.2 Bose-Hubbard Model

The Bose-Hubbard model [3, 4, 7, 13] describes a single bosonic species in a periodic potential. It is obtained from the general Hubbard Hamiltonian 3.11 by omitting the spin degree of freedom and inserting the bosonic operators $\hat{a}_i$ and $\hat{a}_i^\dagger$ that obey the commutation relations $[\hat{a}_i, \hat{a}_j^\dagger] = \delta_{ij}$. The Bose-Hubbard Hamiltonian reads

$$\hat{H}_{\text{BH}} = -J \sum_{\langle i,j \rangle} \hat{a}_i^\dagger \hat{a}_j + \frac{U}{2} \sum_i \hat{n}_i(\hat{n}_i - 1) + \sum_i (\epsilon_i - \mu)\hat{n}_i, \tag{3.18}$$

where $\hat{n}_i = \hat{a}_i^\dagger \hat{a}_i$ counts the number of bosons at lattice site i.

The first term indicates that an energy corresponding to the tunneling matrix element J is gained, when a boson is delocalized between neighboring lattice sites. When the tunneling matrix element J is large in comparison to the interaction energy U, many-body quantum states with strong number fluctuations are energetically favored.

The second term captures the total interaction energy of n_i atoms at lattice site i. U corresponds to the interaction energy per atom pair. It can take negative and positive values depending on whether the interaction is attractive or repulsive. Taking into account the indistinguishability of quantum particles, there are $n_i(n_i - 1)/2$ atom pairs that perform collisions. When the interaction energy U is large and positive, the localization of atoms is favored. This is because delocalized many-body states also feature high local occupations that are energetically costly due to a quadratic dependence of the interaction energy on n_i.

The third term accounts for an external trapping potential by assigning an energy offset $\epsilon_i = V_{\text{trap}}(\mathbf{r}_i)$ to each lattice site i. Typically, the external confinement is a harmonic trap. For a homogeneous system ϵ_i is zero. When the system is treated in terms of a grand canonical ensemble, the chemical potential μ fixes the mean total atom number of the system.

3.2.1 Superfluid and Mott Insulating Ground State

Two Bosons in a Double Well System

The basic physics of the Bose-Hubbard model can be nicely illustrated by considering a double well system with two interacting bosons, which can easily be solved. It reveals how the ground state evolves from delocalized to localized atoms, when the ratio of the interaction energy U to the tunneling matrix element J is increased. The full Hamiltonian reads

$$\hat{H}_{\text{BH}}^{(2)} = -J\left(\hat{a}_1^\dagger \hat{a}_2 + \hat{a}_2^\dagger \hat{a}_1\right) + \frac{U}{2}\left[\hat{n}_1(\hat{n}_1 - 1) + \hat{n}_2(\hat{n}_2 - 1)\right], \tag{3.19}$$

where a possible energy offset $\Delta = \epsilon_2 - \epsilon_1$ between the sites is neglected. Using the basis set $\{|2,0\rangle, |1,1\rangle, |0,2\rangle\}$ we obtain the matrix form

$$\hat{H}_{\text{BH}}^{(2)} = \begin{pmatrix} U & -\sqrt{2}J & 0 \\ -\sqrt{2}J & 0 & -\sqrt{2}J \\ 0 & -\sqrt{2}J & U \end{pmatrix}. \tag{3.20}$$

The ground state of the model has the eigenenergy

$$E_g = \frac{U}{2} - \sqrt{4J^2 + \frac{U^2}{4}} \tag{3.21}$$

and is generally formed by the superposition of all basis states

$$|\psi_g\rangle \propto |2,0\rangle + |0,2\rangle + \left(\frac{U}{2\sqrt{2}J} + \sqrt{2 + \left(\frac{U}{2\sqrt{2}J}\right)^2}\right)|1,1\rangle, \tag{3.22}$$

where normalization is neglected for better readability. In the limit of strong attractive interactions ($U/J \rightarrow -\infty$), double occupation is energetically favored and the state approaches $|\psi_g\rangle \rightarrow |20\rangle + |02\rangle$. At vanishing interactions ($U/J = 0$) the atoms are maximally delocalized expressed by the binomial state $|\psi_g\rangle \rightarrow |2,0\rangle + |0,2\rangle + \sqrt{2}|1,1\rangle$. This state has the strongest possible number fluctuations and corresponds to the superfluid state of this miniature Bose-Hubbard system. Finally, strongly repulsive interactions ($U/J \to \infty$) suppress double occupation and number fluctuations. The energetic cost of delocalization becomes too high as an energy U must be expended for double occupancies. The particles are localized at the individual lattice sites $|\psi_g\rangle \to |1,1\rangle$, which is the miniature version of a Mott insulator.

Many Bosons in a Homogeneous Lattice

The limiting cases of the miniature Bose-Hubbard model can readily be extended to N bosons in a system of M lattice sites [10]. When the tunneling matrix element J is much larger than the interaction energy U, the tunneling term dominates and favors the maximal delocalization of atoms. Maximal delocalization is achieved when all atoms occupy the same single-particle Bloch state $\phi_{\mathbf{q}=0}(\mathbf{r})$. The system is in a superfluid state that in the Wannier basis takes the form

$$|\Psi_{\mathrm{SF}}\rangle_{U/J\approx 0} \propto \left(\frac{1}{\sqrt{M}}\sum_{i=1}^{M}\hat{a}_i^\dagger\right)^N |0\rangle. \tag{3.23}$$

In this notation the maximal number fluctuations are directly visible: There is a finite probability to find any atom number between zero and N at a given lattice site i.

In the limit of dominating interactions ($U/J \to \infty$), the atoms are localized to individual lattice sites and number fluctuations are fully suppressed. The many-body wavefunction of the Mott insulating state is given by a product of local Fock states with commensurate integer filling,

$$|\Psi_{\mathrm{MI}}\rangle_{U/J\to\infty} \propto \prod_{i=1}^{M}(\hat{a}_i^\dagger)^n|0\rangle. \tag{3.24}$$

The Mott insulator state with fixed local atom numbers minimizes the total interaction energy.

3.2.2 Quantum Phase Transition

The Bose-Hubbard model contains a second-order quantum phase transition between the superfluid and the Mott insulating state, which is controlled by the ratio U/J. The phase boundaries at zero temperature can be calculated within a grand-canonical ensemble using a mean-field treatment similar to a Bogoliubov approach [14]. In second-order perturbation theory the phase boundary between the superfluid and Mott insulating state is given by

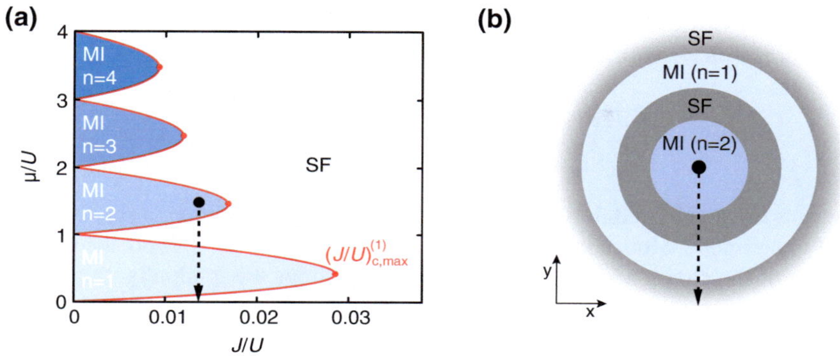

Fig. 3.3 Phase diagram of the Bose-Hubbard model at $T = 0$. **a** As the ratio J/U decreases the system undergoes a phase transition from a superfluid (SF) to a Mott insulator (MI). The phase boundaries (Eq. 3.25) are shown as *red lines*. The Mott lobes (*shaded in blue*) contain the incompressible Mott phases at integer filling n. **b** In a trapped geometry, the phase diagram (**a**) gives rise to a shell structure: moving from the trap center outwards corresponds to sampling a continuously decreasing chemical potential. The Mott insulator shells are separated by superfluid layers, which become thinner for decreasing J/U

$$\left(\frac{J}{U}\right)_{\mathrm{c}} = \frac{1}{z}\frac{(n-\mu/U)\,(1-n+\mu/U)}{1+\mu/U}, \tag{3.25}$$

where z is the coordination number. The filling factors n of the Mott phases are integer numbers that depend on the chemical potential via $n-1 \leq \mu/U \leq n$. The resulting phase diagram is shown in Fig. 3.3a.

The optical lattice in experimental realizations of the Bose-Hubbard model is inhomogeneous, featuring an underlying confining potential $V_{\mathrm{trap}}(\mathbf{r})$. However, in a local-density approximation the system can be locally viewed as being homogeneous.[4] This allows to define a local chemical potential according to

$$\mu_i = \mu - \epsilon_i = \mu - V_{\mathrm{trap}}(\mathbf{r}_i), \tag{3.26}$$

where $\mathbf{r}_i$ is the position of lattice site i. Therefore, at a fixed lattice depth, corresponding to a fixed value of J/U, the phases in a trapped system correspond to a cut through the phase diagram along the μ/U-axis. Starting with the maximal chemical potential μ in the center of the trap, the local chemical potential μ_i decreases towards zero when moving away from the trap center. Accordingly, the system forms Mott insulating shells with fixed filling factors n that are separated by intermediate superfluid layers (see Fig. 3.3b).

When the lattice depth is raised the parameter J/U decreases and the system enters the Mott insulating phases at the tip of each Mott lobe with filling n. Taking

[4] Note that the tight-binding limit with its restriction to nearest-neighbor tunneling processes favors the applicability of a local-density approximation.

the derivative of Eq. 3.25 with respect to the chemical potential yields the critical values for the phase transition of each Mott lobe,

$$\left(\frac{J}{U}\right)^{(n)}_{\mathrm{c,max}} = \frac{1}{z}\left(1 + 2n - 2\sqrt{n^2 + n}\right), \tag{3.27}$$

which are marked by red dots in Fig. 3.3a. When the optical lattice is slowly raised in the experiment, Mott domains emerge starting with a $n = 1$ Mott insulator at the edge of the cloud. Mott domains with higher fillings are gradually reached for deeper lattices in the center of the trap. The mean-field calculation presented here predicts a critical value for the $n = 1$ Mott domain of $(J/U)^{(1)}_{\mathrm{c,max}} = 0.0286$, while a quantum Monte Carlo simulation of the Bose-Hubbard model has yielded the exact value $(J/U)^{(1)}_{\mathrm{c,max}} = 0.0341$ [15].

3.3 Fermi-Hubbard Model

The Fermi-Hubbard model describes an interacting spin mixture (spin up $|\uparrow\rangle$ and down $|\downarrow\rangle$) of fermionic particles in a lattice potential. In solid state physics the Hubbard model resembles electrons that interact via Coulomb interactions and move in a crystal of positively charged ions. Equivalent physics is expected, when an ultracold mixture of two hyperfine states of a fermionic species is loaded into an optical lattice. Using the fermionic anticommutation relations $[\hat{c}_{i\sigma}, \hat{c}^\dagger_{j\sigma'}]_+ = \delta_{ij}\delta_{\sigma\sigma'}$ the Fermi-Hubbard model is readily derived from Eq. 3.11 reading

$$\hat{H}_{\mathrm{FH}} = -J\sum_{\sigma}\sum_{\langle i,j\rangle} \hat{c}^\dagger_{i\sigma}\hat{c}_{j\sigma} + U\sum_i \hat{n}_{i\uparrow}\hat{n}_{i\downarrow} + \sum_i \epsilon_i(\hat{n}_{i\uparrow} + \hat{n}_{i\downarrow}), \tag{3.28}$$

where σ runs over the spin states $|\uparrow\rangle$, $|\downarrow\rangle$. Analogous to the Bose-Hubbard model the parameter J denotes the tunneling energy that is gained when a particle is delocalized over neighboring lattice sites. The tunnel coupling is identical for both spins σ given the lattice potential is spin-independent. The interaction matrix element U denotes the energy cost when a spin up and a spin down atom occupy the same lattice site. Pauli's principle prohibits occupations of more than one atom per spin state and lattice site.

The Fermi-Hubbard model gives rise to complex phase diagrams, which result from an intricate interplay of interaction, delocalization, filling, spin ordering and temperature. The ground states of the Hubbard model are the topic of vast theoretical investigations. Due to the lack of analytical solutions in the two- and three-dimensional case, generally, approximate numerical methods are employed. Exact quantum Monte Carlo methods are often troubled by the notorious sign problem that originates from the antisymmetry of fermionic many-body wavefunctions. One of the most important questions under debate is, whether the two-dimensional

Fermi-Hubbard model can explain the properties of cuprate high-temperature superconductors [16]. The complexity of the theoretical investigations makes a clean experimental realization of the Fermi-Hubbard highly desirable with the prospect to measure the phase diagrams in a perfectly controlled way.

In this section, we briefly discuss limiting cases of the Fermi-Hubbard model, which are relevant to the experiments of this thesis (see Chap. 6), aiming at an intuitive understanding. For thorough discussions of the phases of the Fermi-Hubbard model, we refer the reader to many textbooks and reviews [1, 16–20].

3.3.1 Spins in a Double Well System

As the smallest nontrivial instance of the Fermi-Hubbard model, we consider a double well system filled with a spin up and a spin down fermion. Already in this miniature model basic properties are revealed, illustrating the notions of band width, double occupancy and antiferromagnetism. The full two-site Hamiltonian reads

$$\hat{H}_{\mathrm{FH}}^{(2)} = -J\left(\hat{c}_{1\uparrow}^{\dagger}\hat{c}_{2\uparrow} + \hat{c}_{1\downarrow}^{\dagger}\hat{c}_{2\downarrow} + \hat{c}_{2\uparrow}^{\dagger}\hat{c}_{1\uparrow} + \hat{c}_{2\downarrow}^{\dagger}\hat{c}_{1\downarrow}\right) + U\left(\hat{n}_{1\downarrow}\hat{n}_{1\uparrow} + \hat{n}_{2\downarrow}\hat{n}_{2\uparrow}\right). \tag{3.29}$$

Because the Hamiltonian conserves the atom number, we can restrict the discussion to the Fock space with one spin up and one spin down atom $\{|\uparrow\downarrow, 0\rangle, |\uparrow, \downarrow\rangle, |\downarrow, \uparrow\rangle, |0, \uparrow\downarrow\rangle\}$, where the first entry corresponds to site 1 and the second entry to site 2. Using this basis the Hamiltonian can be cast in the matrix form

$$\hat{H}_{\mathrm{FH}}^{(2)} = \begin{pmatrix} U & -J & -J & 0 \\ -J & 0 & 0 & -J \\ -J & 0 & 0 & -J \\ 0 & -J & -J & U \end{pmatrix}. \tag{3.30}$$

A straightforward analytical calculation yields the eigenenergies E_i and eigenvectors $|\psi_i\rangle$ (normalization constants are omitted for better readability):

- $E_1 = 0$ for the singlet eigenstate $|\psi_1\rangle \propto |\uparrow, \downarrow\rangle - |\downarrow, \uparrow\rangle$,
- $E_2 = U$ for the eigenstate $|\psi_2\rangle \propto |\uparrow\downarrow, 0\rangle - |0, \downarrow\uparrow\rangle$ featuring a superposition of doubly occupied sites,
- $E_{3/4} = \frac{U}{2} \pm \sqrt{4J^2 + \frac{U^2}{4}}$ for the eigenstates that are formed by a superposition of all basis states $|\psi_{3/4}\rangle \propto |\uparrow\downarrow, 0\rangle + |0, \downarrow\uparrow\rangle + \left(\frac{U}{4J} \mp \sqrt{1 + \left(\frac{U}{4J}\right)^2}\right) (|\uparrow, \downarrow\rangle + |\downarrow, \uparrow\rangle)$, where the upper sign applies to $|\psi_3\rangle$ and the lower one to $|\psi_4\rangle$.

The spectrum and the eigenstates (see Fig. 3.4) have important features that are similarly observed in the many-body case. At vanishing interaction $U/4J = 0$ the energy difference between the ground and highest excited state, $|\psi_3\rangle$ and $|\psi_4\rangle$,

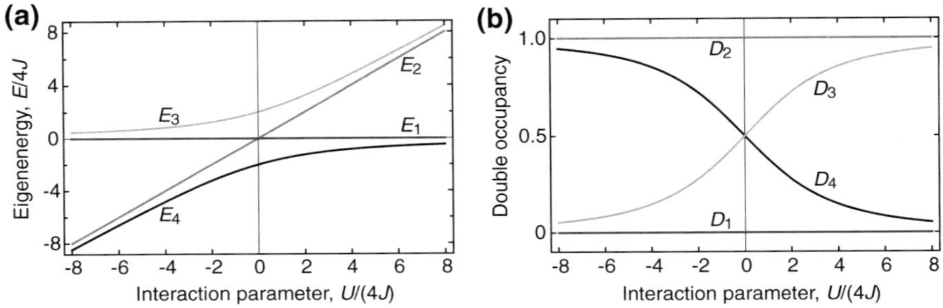

Fig. 3.4 Properties of the double-well Fermi-Hubbard model with a spin up and a spin down particle. The eigenenergies (**a**) and the double occupancy (**b**) of the four eigenstates are shown as a function of the normalized interaction $U/(4J)$

amounts to $4J$. This corresponds to the band width of the first Bloch band in an infinite homogeneous lattice within the tight-binding approximation (see Eq. 3.14). Accordingly, the states $|\psi_3\rangle$ and $|\psi_4\rangle$ may be viewed as the Bloch states $\phi_{q=\hbar k}$ and $\phi_{q=0}$ with maximal and minimal quasi-momentum, respectively. In those states the fermions are delocalized over the system and there is a 50 % probability for double occupation in both cases. For non-vanishing interactions the probability to find doubly occupied lattice sites is given by

$$D_{3/4} = \frac{1}{1 + \left[\frac{U}{4J} \mp \sqrt{1 + \left(\frac{U}{4J}\right)^2}\right]^2}, \tag{3.31}$$

where the upper (lower) sign applies to state $|\psi_3\rangle$ ($|\psi_4\rangle$) (see Fig. 3.4). The eigenstates $|\psi_1\rangle$ and $|\psi_2\rangle$ do not depend on the interaction parameter and the probability for double occupation remains fixed at $D_1 = 0$ and $D_2 = 1$, respectively.

The eigenstate $|\psi_4\rangle$ has the lowest eigenenergy and forms the ground state. In the limit of strongly attractive interactions the atoms pair up in each well and the double occupancy reaches a value of one. The eigenenergy can be approximated by $U - 4J^2/|U|$ and the correction term $4J^2/|U|$ corresponds to the energy that is gained when the atom pair is delocalized over the double well [21]. In the opposite limit of strong repulsion the ground state shows antiferromagnetic ordering of the spins. The corresponding eigenenergy is given by the Heisenberg exchange energy $-4J^2/U$ that is gained when neighboring spins exchange places via intermediate double occupation [22]. This is in close analogy to the Fermi-Hubbard model in an infinite homogeneous lattice system, which maps to a quantum Heisenberg model in the limit of strong repulsion and half-filling (see below). The zero temperature phase of the Heisenberg model is antiferromagnetically ordered and the lowest lying excitations have an energy of $J_{ex} = 4J^2/U$ (see Fig. 3.6) [1, 23, 24].

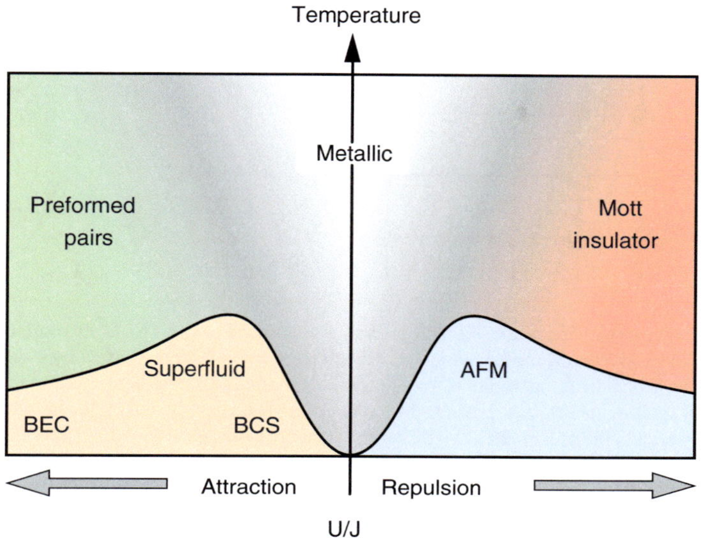

Fig. 3.5 Sketch of the phase diagram for the Fermi-Hubbard model at half-filling in a three-dimensional optical lattice (adapted from reference [20]). The following abbreviations have been used: molecular Bose-Einstein condensate (BEC), Bardeen-Cooper-Shrieffer (BCS) and anti-ferromagnetic phase (AFM)

3.3.2 Schematic Phase Diagram

A schematic phase diagram of the three-dimensional Hubbard model at half-filling is shown in Fig. 3.5 summarizing results for repulsive [23–25] and attractive [17, 26–28] interactions. Half-filling corresponds to a *mean* local filling factor per spin state of $\bar{n}_\uparrow = \bar{n}_\downarrow = 0.5$.

For the case of attractive interactions the on-site interaction energy between the spin up and down fermions is negative. At sufficiently low temperatures this gives rise to a superfluid regime with a BEC-BCS crossover. For weak interactions the particles form BCS pairs and the critical temperature increases monotonically with $|U|/J$. When the interactions get stronger, the atom pairs get more strongly bound until they eventually form hardcore bosons that undergo Bose-Einstein condensation. In the BEC regime the critical temperature decreases as J^2/U. This is also the energy scale for nearest-neighbor tunneling and *nearest-neighbor repulsion* between the hardcore bosons [28]. In the competition of energy scales, a charge density wave emerges, in which the density is reduced at every other lattice site [1]. Even above the critical temperature for superfluidity, a pseudo-gap regime of preformed, uncondensed pairs is predicted. The experimental observation of those pairs and the thermodynamic consequences of their presence are briefly discussed in Sect. 6.3 [29].

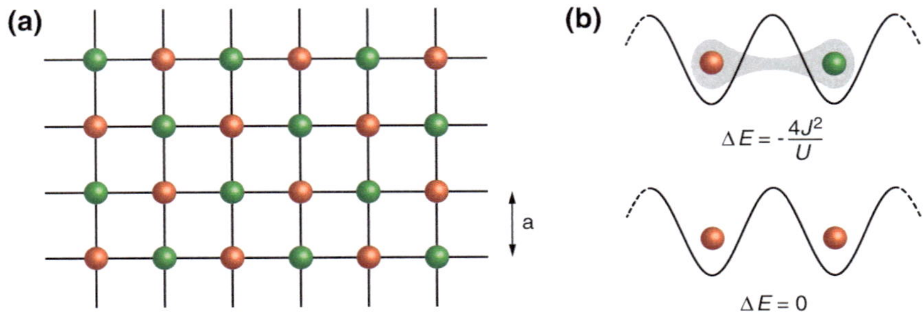

Fig. 3.6 Antiferromagnetic ordering and exchange interactions. **(a)** The Mott insulating ground state of the Fermi-Hubbard model, which is characterized by localized spins, shows antiferromagnetic ordering when the temperature of the system is low enough. **(b)** When opposite spins occupy neighboring lattice sites (*upper panel*), they reduce their energy by $-4J^2/U$ through a virtual exchange process. This process has two possible exchange paths, during which either the left or the right well are virtually doubly occupied. The intermediate state is detuned by the interaction energy U. For identical spins (*lower panel*) the exchange process is not possible, because Pauli's principle forbids the intermediate doubly occupied state [1]

At half-filling the phases of the attractive Fermi-Hubbard model can be mapped to the repulsive Fermi-Hubbard model using a particle-hole transformation [28]. In the regime of higher temperatures the phase of disordered preformed pairs maps to a paramagnetic Mott insulator without spin ordering. In the metallic Fermi liquid phase at weak repulsion the particles are delocalized to minimize their kinetic energy. When the repulsion is increased, the interactions of particles with opposite spin smoothly lead to localization [23]. Deep in the Mott insulating regime double occupation of lattice sites is suppressed due to the repulsion (and Pauli's principle). Owing to its paramagnetic character, the Mott insulating phase can accommodate a large amount of spin entropy.

Upon lowering of the temperature less spin entropy needs to be stored and the system can display magnetic ordering (see Fig. 3.6a). Antiferromagnetic spin order sets in below the Néel temperature after crossing a second-order phase transition. When the repulsive interactions are small, a weak spin-density wave modulation develops. In this regime the Néel temperature is exponentially suppressed as a function of U/J [24]. For increasing repulsion, the system undergoes a crossover towards an antiferromagnetically ordered Mott insulator. At very strong repulsion the Hubbard Hamiltonian at half-filling can be reduced to a quantum Heisenberg model

$$\hat{H}_{\text{HB}} = J_{ex} \sum_{\langle i,j \rangle} \hat{\mathbf{S}}_i \cdot \hat{\mathbf{S}}_j, \tag{3.32}$$

where $J_{ex} = 4J^2/U$ is the Heisenberg exchange energy. The spin operator at lattice site i is given by

$$\hat{\mathbf{S}}_i = \frac{1}{2} \sum_{\sigma\sigma'} \hat{c}^\dagger_{i\sigma} \boldsymbol{\sigma} \, \hat{c}_{i\sigma'}, \tag{3.33}$$

where $\boldsymbol{\sigma} = (\sigma_x, \sigma_y, \sigma_z)$ is the vector of Pauli matrices [1]. The exchange energy $J_{ex} = 4J^2/U$ expresses the energy that is gained in a virtual tunneling process between lattice sites that are singly occupied by atoms of opposite spin (see Fig. 3.6b). In this process an atom tunnels to the neighboring lattice site, where it interacts at an energy U before it tunnels back. When the neighboring lattice sites are occupied by identical spin, this superexchange interaction is forbidden by the Pauli principle. Therefore, antiferromagnetic ordering is energetically favored (see Sect. 3.3.1). For a three-dimensional simple cubic lattice, quantum Monte Carlo simulations have shown that the Néel temperature is essentially given by the exchange energy, $T_N \approx 0.96 J_{ex}$ [23, 24].

The case of half-filling gives a flavor of the rich physics that is contained in the Fermi-Hubbard model. In the experiments of this thesis the atoms are trapped in an optical lattice with underlying harmonic confinement, which gives rise to a strong variation of the local fillings. One expects that several phases coexist in the trap, ranging from metallic over Mott insulating to band insulating phases. This intriguing situation is elucidated in Chap. 6 both theoretically and experimentally. In addition to the parameters considered so far, the impact of doping on the phases of the Fermi-Hubbard model is an extremely active field of research [16]. It is believed that superconducting ground states with d-wave symmetry can arise within the two-dimensional repulsive Fermi-Hubbard model when the system is doped [30]. This might be an explanation for the high-temperature superconductivity in cuprates, but the exact mechanism is not well understood. Therefore, experimental realizations of the two-dimensional Fermi-Hubbard model are highly sought after [25, 31–33].

3.3.3 Band Insulator of Noninteracting Fermions

In order to investigate the role of the underlying harmonic confinement in optical lattices, it is instructive to consider a gas of noninteracting fermions. It can experimentally be created using an ultracold cloud of spin polarized fermionic atoms. In such a sample s-wave collisions do not occur owing to Pauli's principle and collisions at higher partial waves are suppressed by low temperature (see Sect. 2.1.3). Alternatively, vanishing interactions can be achieved in a mixture of spin up and spin down fermions by means of a Feshbach resonance. This approach is chosen in the measurements of Chap. 6.

For the proper preparation and interpretation of quantum phases, it is crucial to understand the influence of the underlying potential on the band structure of the

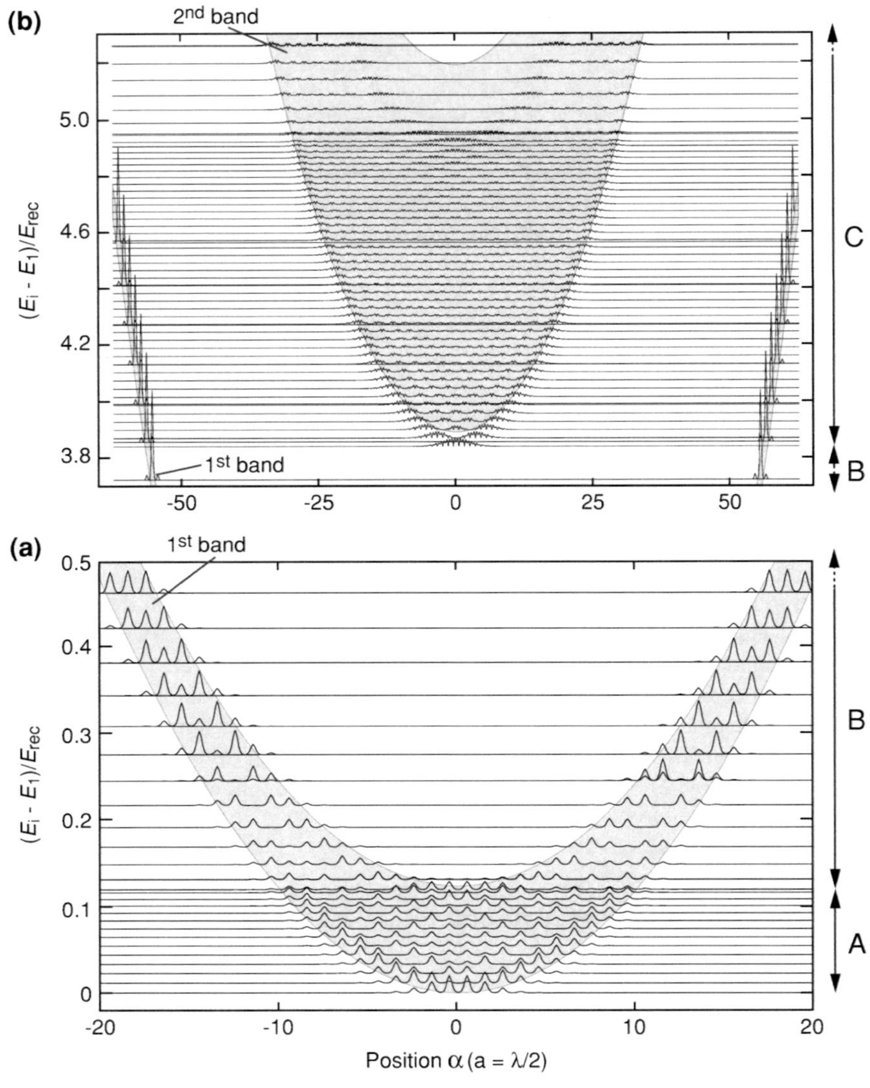

Fig. 3.7 Numerically calculated single-particle wavefunctions $\phi_i(\alpha)$ and eigenenergies E_i in an one-dimensional optical lattice ($\lambda = 738\,\mathrm{nm}$) at 8 E_{rec} with an underlying harmonic potential ($\omega_\alpha = 2\pi \times 200\,\mathrm{Hz}$ for ^{40}K). *Solid lines* indicate the probability density $|\phi_i(x)|^2$ that is vertically positioned at the respective eigenenergy E_i. Panel **(a)** shows the lowest energy states of the first band, while panel **(b)** displays the onset of the second band. Mind the different scaling of the position axis and the gap on the energy axis. *Gray areas* show the widths of the first and second band derived from a band structure calculation (see Sect. 2.2.3). The band widths are locally offset by the underlying harmonic potential

lattice.[5] Owing to the harmonic confinement, translational symmetry is broken and the Bloch states of the homogeneous lattice are no longer appropriate (see Sect. 2.2.3). The new single-particle states are limited to a finite spatial extend. They can even be localized to a single lattice site when the lattice is deep and external confinement is tight, as we will see below.

The single-particle eigenstates are derived via numerical diagonalization of the Schrödinger equation. Because the physics of a simple cubic lattice is separable into its individual axes, it suffices to consider only one dimension. The Schrödinger equation for one axis ($\alpha = x, y, z$) reads [34, 35]

$$\left(-\frac{\hbar^2}{2m}\frac{\partial^2}{\partial\alpha^2} + V_\alpha \cos^2(k\alpha) + \frac{1}{2}m\omega_\alpha^2\alpha^2\right)\phi_i(\alpha) = E_i\phi_i(\alpha), \tag{3.34}$$

where ω_α denotes the trapping frequency of the harmonic potential, $k = 2\pi/\lambda$ the wave vector and λ the wavelength of the lattice laser. The position coordinate α is discretized into typically ten intervals per lattice period $a = \lambda/2$, such that the Schrödinger equation can be cast in matrix form. Diagonalization of the matrix yields the eigenvalues E_i along with the eigenstates $\phi_i(\alpha)$.

In Fig. 3.7 the result of a calculation for typical experimental parameters is displayed. It shows that the areas of finite probability density $|\phi_i(\alpha)|^2 > 0$ can also approximately be deduced from the band structure of the homogeneous lattice: The gray shaded areas display the sum of the energy bands in the homogeneous lattice and the local energy offset from the harmonic confinement. For low energies this is a good approximation, because the energy offset between adjacent sites is small and the lattice appears almost homogeneous. However, for higher energies deviations can become significant.

It is important to note that the usual meaning of bands is somewhat obscured by the presence of the confining potential. The spectrum of a homogeneous lattice has energetically forbidden regions, in which no eigenstates exist. Those regions separate the bands containing the eigenstates (see Figs. 2.8 and 2.10). In the inhomogeneous case, however, a true gap between the bands only exists locally. The eigenenergies in lower bands can be larger than those of higher lying bands when one departs from the center of the lattice. Generally, the spatial overlap between states of comparable energy in different bands is low, which implies poor interband coupling. Therefore, equilibration across different lattice bands is very slow compared to typical experimental time scales.

Figure 3.8a shows the single-particle spectrum of the eigenstates in Fig. 3.7, distinguishing three regions A, B and C. For low energies, E_i increases linearly with the state index, similar to the eigenenergies of a purely harmonic potential. Region B starts when the eigenstates are localized either to the left or right side of the lattice,

[5] It is worth noting that most optical lattice setups have an approximately harmonic confining potential. This is a natural consequence of the use of Gaussian laser beams for optical lattices. However, techniques are currently being investigated that aim at the projection of optical lattice potentials. Thereby, it may become possible to create engineered potential landscapes, for example, box potentials.

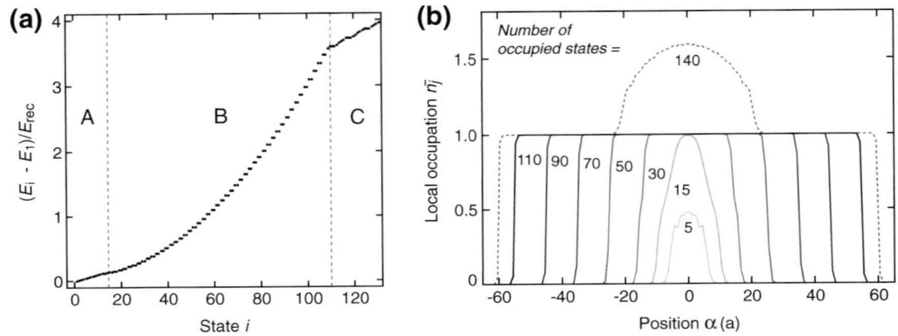

Fig. 3.8 (**a**) Single-particle spectrum as a function of the quantum number i for a one-dimensional $8\ E_{rec}$ lattice with a harmonic confinement of $\omega_\alpha = 2\pi \times 200\,\text{Hz}$ (same parameters as in Fig. 3.7). (**b**) Formation of a band insulator. Noninteracting fermions at $T = 0$ fill the eigenstates from bottom up. The local occupation $\bar{n}_j$ is shown for an increasing number of fermions N (*solid lines* with increasing darkness). The band insulating state is reached at about $N \approx 110$. When more single-particle states are occupied, population enters the second band shown as *gray dashed line*

while the probability density vanishes in the center. Those states are pairwise degenerate and become more and more localized to single lattice sites for increasing i. For quantum numbers i beyond a certain threshold (here about 110), delocalized states in the center become available again, belonging to the second band. The probability densities of the eigenstates in the second band (see Fig. 3.7b) approach zero in the center of each lattice site. This is similar to the second band Bloch functions in the homogeneous lattice as shown in Fig. 2.9 and indicates the dominant admixture of plane waves with momenta in the second Brillouin zone $[\hbar k, 2\hbar k]$ and $[-\hbar k, -2\hbar k]$.

Based on the above results we can analyze the formation of a band insulator in an optical lattice with underlying harmonic confinement. For a system of N noninteracting fermions at zero temperature, the N lowest single-particle eigenstates $\phi_i(\alpha)$ are filled from bottom up. The local filling $\bar{n}_j$ is calculated by integrating all density distributions $|\phi_i(\alpha)|^2$ for $i \leq N$ over the extend a of lattice site j and subsequent summation. The result is shown for several fermion numbers N in Fig. 3.8b. Unity filling is reached in the center of the trap as soon as all states of region A are filled at $N \approx 15$. Departing from there, the central density does not increase further, but the distribution broadens and forms a wide flat-top profile until all states of region B are added. At $N \approx 110$ the band insulating state with maximal extent is reached. The populated eigenstates correspond to Bloch states with quasi-momenta in the first Brillouin zone. Experimental Brillouin zone mapping in time-of-flight expansion would show a square-shaped atomic cloud (see Sect. 4.2.4).[6]

[6] However, it should be noted that Brillouin zone mapping in time-of-flight yields an almost square-shaped distribution with a flat top for any filling inside region B, only the atomic density differs [36]. Thus, for the given parameters the qualitative differences in experimental images between $N = 15$ and $N = 110$ would only be marginal, but they could, in principle, be distinguished by a measurement of the absolute atom numbers.

The band insulator is an incompressible many-body state. In a homogeneous lattice it corresponds to a completely filled first Bloch band. An increase of the local filling is only possible when the compression is strong enough to overcome the gap between the first and second band. In the harmonically confined lattice such a gap does not exist in a strict sense. Nevertheless, the incompressibility can be experimentally probed in a very direct manner: For a fixed number of fermions inside the band insulating region B, an increase of the trap frequency ω_α does initially not change the size of the cloud. The system is incompressible. Upon further increasing ω_α, the highest populated single-particle state in the first band eventually reaches the energy of the lowest eigenstate in the second band. It becomes energetically favorable to populate states of the second band leading to double occupation of the central lattice sites. At this point the system starts to shrink in size, which implies finite compressibility.

In the typical experimental preparation scheme, an ultracold atomic cloud is initially held in a harmonic trap and the optical lattice is smoothly ramped up. Usually the aim is to exclusively populate the first band. This imposes an upper bound on the atom number depending on the system parameters, as can be conjectured from Fig. 3.7. Is there a simple criterion to predict whether the atomic cloud will solely populate the first band? By inspecting the single-particle spectrum as a function of the lattice depth (at fixed harmonic confinement) one can show that states with an energy $E_i < E_{\text{rec}}$ in the purely harmonic potential evolve into the first band (regions A and B), while states with $E_i > E_{\text{rec}}$ either lie in the first, second or higher bands (region C and upwards) [34]. Therefore, exclusive population of the first lattice band is achieved when the Fermi energy of the harmonically trapped cloud is smaller than the recoil energy $\epsilon_F < E_{\text{rec}}$. Using Eq. 2.24 this results in an upper bound for the number of fermions

$$N_{\text{max}} \leq \frac{1}{6}\left(\frac{E_{\text{rec}}}{\hbar\bar{\omega}}\right)^3, \tag{3.35}$$

where $\bar{\omega} = (\omega_x\omega_y\omega_z)^{1/3}$ is the geometric mean of the trap frequencies in the harmonic potential.

Many-Body Hamiltonian

It is instructive to consider the many-body system of noninteracting fermions in an optical lattice, because it is an important reference point to the changes that arise when interactions come into play. The single-particle Schrödinger Eq. 3.34 gives rise to the corresponding many-body Hamiltonian

$$\hat{H} = -J\sum_{\langle i,j\rangle}\hat{c}_i^\dagger\hat{c}_j + \sum_j \epsilon_j\hat{n}_j, \tag{3.36}$$

where the tight-binding approximation is introduced by restricting the tunneling to nearest neighbors $\langle i, j\rangle$ and $\hat{n}_j = \hat{c}_j^\dagger \hat{c}_j$ counts the number of fermions at site j. Alternatively, the Hamiltonian is obtained from the Fermi-Hubbard model 3.28 by setting $U = 0$ and omitting σ, because only a single spin state is considered.

For a three-dimensional optical lattice with $a = \lambda/2$ and spherically symmetric harmonic confinement characterized by the trap frequency ω, the local energy offset is given by

$$\epsilon_{\mathbf{j}} = \frac{m\omega^2\lambda^2}{8}\mathbf{j}^2 = V_t \mathbf{j}^2. \tag{3.37}$$

Here, $\mathbf{j} = (j_x, j_y, j_z)$ is the site index and $V_t = m\omega^2\lambda^2/8$ is the energy offset between adjacent lattice sites at the trap center.

The exemplary calculation in Fig. 3.7 shows that higher lying single-particle states of the first band are already localized to single lattice sites at moderate lattice depths. Only the lowest states close to the trap center are delocalized over several lattice sites. Neglecting the effect of those central states, we can use Eq. 3.37 to derive an analytical expression for the density of states of the inhomogeneous three-dimensional lattice, reading

$$\rho_{3\mathrm{D}}(\epsilon) = 2\pi\, V_t^{-3/2}\epsilon^{1/2}. \tag{3.38}$$

With the relation $N = \int_0^{\epsilon_F} d\epsilon\, \rho_{3\mathrm{D}}(\epsilon)$ we obtain the Fermi energy as a function of the total atom number N

$$\epsilon_F(N) = V_t\left(\frac{3N}{4\pi}\right)^{2/3}, \tag{3.39}$$

which plays an important role for the definition of rescaled units in Chap. 6. The density of states 3.38 allows to calculate the total entropy of the system at a given temperature using the general prescription 2.19. To the lowest order in temperature, the Sommerfeld approximation yields [37, 38]

$$\frac{S}{k_B} = \frac{\pi^2}{2}N\frac{T}{T_F} + \mathcal{O}\left[\left(\frac{T}{T_F}\right)^2\right]. \tag{3.40}$$

It is interesting to compare this result to the entropy of a harmonically trapped Fermi gas without a lattice, which has been derived in Eq. 2.25. The expressions differ exactly by a factor of two for a fixed dimensionless temperature T/T_F. This has an important implication for the loading of ultracold fermions into an optical lattice: when a harmonically trapped noninteracting Fermi gas is adiabatically ($S = \text{const.}$) transferred into a deep lattice, the dimensionless temperature is doubled. If the loading is nonadiabatic, the temperature increase will even be larger.

References

1. A. Auerbach, *Interacting Electrons and Quantum Magnetism* (Springer, New York, 1994)
2. G. Kotliar, D. Vollhardt, Strongly correlated materials: insights from dynamical mean-field theory. Phys. Today **57**, 53 (2004)
3. D. Jaksch, C. Bruder, J.I. Cirac, C.W. Gardiner, P. Zoller, Cold bosonic atoms in optical lattices. Phys. Rev. Lett. **81**, 3108 (1998)
4. M.P.A. Fisher, P.B. Weichman, G. Grinstein, D.S. Fisher, Boson localization and the superfluid-insulator transition. Phys. Rev. B **40**, 546 (1989)
5. M. Greiner, O. Mandel, T. Esslinger, T. Hänsch, I. Bloch, Quantum phase transition from a superfluid to a Mott insulator in a gas of ultracold atoms. Nature **415**, 39 (2002)
6. R. Feynman, Simulating physics with computers. Int. J. Theor. Phys. **21**, 467 (1982)
7. D. Jaksch, *Bose-Einstein Condensation and Applications.* Ph.D. Thesis, Leopold-Franzens-Universität Innsburck, (1999)
8. N.W. Ashcroft, N.D. Mermin, *Solid State Physics* (Saunders College Publishing, Fort Worth, 1976)
9. A. Altland, B. Simons, *Condensed Matter Field Theory* (Cambridge University Press, Cambridge, 2006)
10. W. Zwerger, Mott-Hubbard transition of cold atoms in optical lattices. J. Opt. B **5**, S9 (2003)
11. M. Greiner, *Ultracold quantum gases in three-dimensional optical lattice potentials.* Ph.D. Thesis, LMU München, (2003)
12. I. Bloch, J. Dalibard, W. Zwerger, Many-body physics with ultracold gases. Rev. Mod. Phys. **80**, 885 (2008)
13. D. Jaksch, P. Zoller, The cold atom Hubbard toolbox. Ann. Phys. **315**, 52 (2005)
14. D. van Oosten, P. van der Straten, H.T.C. Stoof, Quantum phases in an optical lattice. Phys. Rev. A **63**, 053601 (2001)
15. B. Capogrosso-Sansone, N.V. Prokof'ev, B.V. Svistunov, Phase diagram and thermodynamics of the three-dimensional Bose-Hubbard model. Phys. Rev. B **75**, 134302 (2007)
16. P.A. Lee, N. Nagaosa, X.-G. Wen, Doping a Mott insulator: physics of high-temperature superconductivity. Rev. Mod. Phys. **78**, 17 (2006)
17. R. Micnas, J. Ranninger, S. Robaszkiewicz, Superconductivity in narrow-band systems with local nonretarded attractive interactions. Rev. Mod. Phys. **62**, 113 (1990)
18. A. Georges, G. Kotliar, W. Krauth, M.J. Rozenberg, Dynamical mean-field theory of strongly correlated fermion systems and the limit of infinite dimensions. Rev. Mod. Phys. **68**, 13 (1996)
19. M. Imada, A. Fujimori, Y. Tokura, Metal-insulator transitions. Rev. Mod. Phys. **70**, 1039 (1998)
20. T. Esslinger, Fermi-Hubbard physics with atoms in an optical lattice. Annu. Rev. Condens. Matter Phys. **1**, 129 (2010)
21. S. Fölling, S. Trotzky, P. Cheinet, M. Feld, R. Saers, A. Widera, T. Müller, I. Bloch, Direct observation of second-order atom tunnelling. Nature **448**, 1029 (2007)
22. S. Trotzky, P. Cheinet, S. Fölling, M. Feld, U. Schnorrberger, A.M. Rey, A. Polkovnikov, E.A. Demler, M.D. Lukin, I. Bloch, Time-resolved observation and control of superexchange interactions with ultracold atoms in optical lattices. Science **319**, 295 (2008)
23. R. Staudt, M. Dzierzawa, A. Muramatsu, Phase diagram of the three-dimensional Hubbard model at half filling. Eur. Phys. J. B **17**, 411 (2000)
24. F. Werner, O. Parcollet, A. Georges, S.R. Hassan, Interaction-induced adiabatic cooling and antiferromagnetism of cold fermions in optical lattices. Phys. Rev. Lett. **95**, 056401 (2005)
25. W. Hofstetter, J.I. Cirac, P. Zoller, E. Demler, M.D. Lukin, High-temperature superfluidity of fermionic atoms in optical lattices. Phys. Rev. Lett. **89**, 220407 (2002)
26. A. Toschi, P. Barone, M. Capone, C. Castellani, Pairing and superconductivity from weak to strong coupling in the attractive Hubbard model. New J. Phys. **7**, 7 (2005)
27. C.-C. Chien, Q. Chen, K. Levin, Fermions with attractive interactions on optical lattices and implications for correlated systems. Phys. Rev. A **78**, 043612 (2008)
28. A.F. Ho, M.A. Cazalilla, T. Giamarchi, Quantum simulation of the Hubbard model: the attractive route. Phys. Rev. A **79**, 033620 (2009)

29. L. Hackermüller, U. Schneider, M. Moreno-Cardoner, T. Kitagawa, T. Best, S. Will, E. Demler, E. Altman, I. Bloch, B. Paredes, Anomalous expansion of attractively interacting fermionic atoms in an optical lattice. Science **327**, 1621 (2010)
30. P.W. Anderson, P.A. Lee, M. Randeria, T.M. Rice, N. Trivedi, F.C. Zhang, The physics behind high-temperature superconducting cuprates: the 'plain vanilla' version of RVB. J. Phys.: Condens. Matter **16**, R755 (2004)
31. S. Trebst, U. Schollwöck, M. Troyer, P. Zoller, *d*-Wave resonating valence bond states of fermionic atoms in optical lattices. Phys. Rev. Lett. **96**, 250402 (2006)
32. A. Cho, The mad dash to make light crystals. Science **320**, 312 (2008)
33. A.M. Rey, R. Sensarma, S. Fölling, M. Greiner, E. Demler, M.D. Lukin, Controlled preparation and detection of d-wave superfluidity in two-dimensional optical superlattices. Europhys. Lett. **87**, 60001 (2009)
34. L. Viverit, C. Menotti, T. Calarco, A. Smerzi, Efficient and robust initialization of a qubit register with fermionic atoms. Phys. Rev. Lett. **93**, 110401 (2004)
35. H. Ott, E. de Mirandes, F. Ferlaino, G. Roati, V. Türck, G. Modugno, M. Inguscio, Radio frequency selective addressing of localized atoms in a periodic potential. Phys. Rev. Lett. **93**, 120407 (2004)
36. T. Rom, *Bosonische und fermionische Quantengase in dreidimensionalen optischen Gittern.* Ph.D. Thesis, Ludwig-Maximilians-Universität München, (2009)
37. L.D. Carr, G.V. Shlyapnikov, Y. Castin, Achieving a BCS transition in an atomic fermi gas. Phys. Rev. Lett. **92**, 150404 (2004)
38. M. Köhl, Thermometry of fermionic atoms in an optical lattice. Phys. Rev. A **73**, 031601 (2006)

Chapter 4
Detection and Observables

Powerful detection methods and reliable observables are the cornerstones in the analysis of many-body quantum phases of ultracold atoms. By far the most detection techniques aim at direct imaging of atomic densities either in-situ or after time-of-flight expansion. On the basis of experimental images, observables are defined that characterize the quantum many-body state. The quality of these observables determines how much can be learned from a sample of ultracold atoms.

In the first part of this chapter, the theoretical foundations of absorption and phase-contrast imaging are described [1]. In the second part we summarize the fundamental experimental observables used in this work, including the in-situ density distribution, the momentum and the quasi-momentum distribution as well as measures of first- and second-order correlations.

4.1 Detecting Ultracold Atoms

All information that we gain about clouds of ultracold atoms is acquired either by imaging the in-situ density distribution in the trap or the momentum distribution obtained after time-of-flight expansion. The vast majority of the existing experimental setups relies on optical detection techniques.[1] In the experiments of this thesis we employ absorption and phase-contrast imaging.

4.1.1 Interaction Between Atoms and Light

When a light field passes through a medium generally both its amplitude and phase are modified. The changes in the amplitude and the phase originate from a complex

[1] Currently the only notable exception being the electron microscope realized in Herwig Ott's group, which is based on the ionization of ultracold atoms and subsequent detection with a channeltron [2].

S. Will, *From Atom Optics to Quantum Simulation*, Springer Theses,
DOI: 10.1007/978-3-642-33633-1_4,

susceptibility of the medium that gives rise to a complex index of refraction n_{ref} [3, 4]. A cloud of ultracold atoms can be idealized as an ensemble of two-level systems with a spatially dependent density $n(\mathbf{r})$. During imaging the cloud is illuminated by a monochromatic laser at wavelength λ. If the laser intensity I_0 is low compared to the saturation intensity I_{sat}, the complex index of refraction is given within the rotation wave approximation by [1, 4]

$$n_{\text{ref}}(\mathbf{r}) = 1 + n(\mathbf{r})\frac{\lambda\sigma_0}{4\pi}\left(\frac{i}{1+4\Delta^2/\Gamma^2} - \frac{2\Delta/\Gamma}{1+4\Delta^2/\Gamma^2}\right). \tag{4.1}$$

Here, $\sigma_0 = 3\lambda^2/2\pi$ denotes the resonant scattering cross section of an atomic two-level system, $\Delta = \omega - \omega_0$ the detuning between the laser frequency ω and the atomic resonance ω_0, and Γ the natural line width of the transition.

We let the light propagate along the z-direction and apply the thin-lens approximation, which means that the light is assumed to enter and exit the atom cloud at the same x- and y-coordinates. The propagation through the cloud transforms a laser field $E_0 \propto e^{ikz}$ into

$$E = \exp\left(ik\int dz\,[n_{\text{ref}}(\mathbf{r}) - 1]\right)E_0 = t(x, y)\,e^{i\phi(x,y)}E_0, \tag{4.2}$$

where the spatially dependent transmission coefficient and phase shift are given by

$$t(x, y) = e^{-D(x,y)/2} \quad \text{and} \quad \phi(x, y) = \frac{\Delta}{\Gamma}D(x, y). \tag{4.3}$$

The optical density $D(x, y)$ of the medium is given by

$$D(x, y) = \frac{\sigma_0}{1+4\Delta^2/\Gamma^2}n_{\text{col}}(x, y), \tag{4.4}$$

where the column density $n_{\text{col}}(x, y) = \int dz\,n(\mathbf{r})$ is defined as the spatial density integrated along the z-axis. The integration indicates that no spatially resolved information is obtained along the line-of-sight. The above equations form the basis for the theoretical treatment of absorption imaging and phase-contrast imaging.

4.1.2 Absorption Imaging

Absorption imaging is our standard method to measure the density distributions of ultracold potassium and rubidium atoms after time-of-flight expansion [1]. The atom clouds are illuminated by a resonant laser beam and partly absorb the light. This imprints a shadow cast on the intensity profile of the laser beam that is imaged on a charge coupled device (CCD) camera. The absorption process is followed by spontaneous emission that leads to the transfer of recoil momentum and strong heating of the cloud. Assuming that 150 photons are scattered during 50 μs of illumination, momentum diffusion blurs the image by about 2 μm (4 μm) for rubidium (potassium) [1]. Therefore, the duration of illumination should not be too long. Obviously, absorption imaging is destructive.

In order to extract the atomic density distribution $n(\mathbf{r})$, three individual images are recorded: First, the shadow cast of the atom cloud $I_{\mathrm{w}}(x, y)$, second, the intensity profile of the imaging beam without atoms $I_{\mathrm{wo}}(x, y)$, and, third, the stray light distribution $I_s(x, y)$ without atoms and without imaging beam. The stray light is subtracted both from the image with atoms $I(x, y) \equiv I_{\mathrm{w}}(x, y) - I_s(x, y)$ and the image without atoms $I_0(x, y) \equiv I_{\mathrm{wo}}(x, y) - I_s(x, y)$. Noting that $I \propto |E|^2$ and using Eqs. 4.2 and 4.3, a relation between the two images is easy to derive:

$$I(x, y) = e^{-D(x,y)}\, I_0(x, y). \tag{4.5}$$

According to the approximations introduced in Sect. 4.1.1, the relation is valid in the limit of low laser intensity $I_0(x, y) \ll I_{\mathrm{sat}}$. This is a favorable regime, because the optical density is solely determined by the relative intensity ratio $D(x, y) = -\ln(I/I_0)$ and no absolute intensities are required.

The applicability of absorption imaging is technically limited by the dynamic range of the CCD camera that typically corresponds to $D(x, y) \leq 4$. The optical density of a trapped degenerate atom cloud is generally much larger, up to $D = 100$ and even more. Therefore, the optical density must be reduced by about two orders of magnitude. This can be achieved by releasing the atom cloud from the trap and allowing for free expansion in time-of-flight. In principle, the optical density could also be reduced by choosing a large detuning Δ of the imaging beam (see Eq. 4.4). However, this increases the real part of the refractive index and gives rise to a finite phase shift $\phi(x, y)$. The thin-lens approximation breaks down and the experimental images show lensing and distortions. While large detuning is delicate in absorption imaging, it is beneficially employed in the phase-contrast method (see next section).

An ideal implementation of absorption imaging is often difficult due to technical constraints. In our setup, we are bound to use linearly polarized light. Therefore, the atoms cannot be treated as two-level systems and several degenerate transitions must be taken into account (at vanishing magnetic field). Averaging of the respective squared Clebsch-Gordon coefficients yields a reduced scattering cross section σ_0. The theoretical correction factors for ^{87}Rb and ^{40}K are $\alpha_{\mathrm{Rb}} = 7/15$ and $\alpha_{\mathrm{K}} = 2/5$, respectively [5]. These values have been validated by independent atom number calibrations based on the in-situ size of Bose-Einstein condensates and noninteracting degenerate Fermi clouds (see Sect. 6.2.3).

4.1.3 Phase-Contrast Imaging

The implementation of phase-contrast imaging (PCI) has been a pivotal step towards a dependable measurement of in-situ cloud sizes of interacting fermionic spin mixtures,

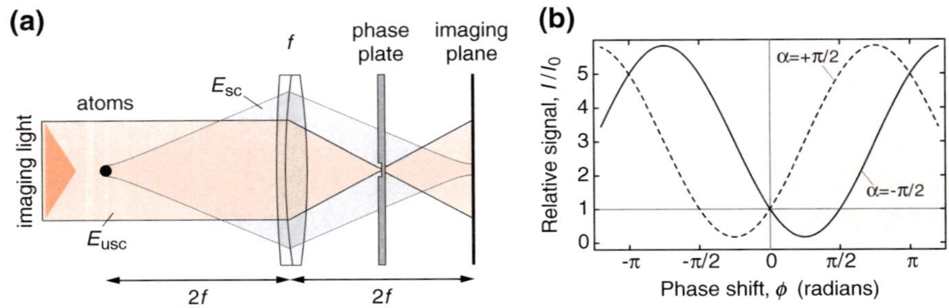

Fig. 4.1 Theory of phase-contrast imaging. **a** Elementary phase-contrast setup. Far-detuned imaging light is coherently scattered by the atoms (*gray shaded area*). A phase plate in the Fourier plane shifts the relative phase between scattered (E_{sc}) and unscattered (E_{usc}) light. In the imaging plane, the phase modulation $\phi(x, y)$, which the atoms imprint on the imaging beam, can be converted into an intensity modulation $I(x, y)$, if the phase shift α of the phase plate is chosen appropriately (see main text). **b** Relative signal I/I_0 as a function of the phase shift ϕ for advancing ($\alpha = +\pi/2$, *dashed*) and retarding ($\alpha = -\pi/2$, *solid*) phase plates. For simplicity a transparent atom cloud is assumed ($t = 1$) corresponding to the limit of far detuning. The shading indicates that the phase shift ϕ is positive (negative) for *red* (*blue*) detuned imaging light (see Eq. 4.3). The method is usually used in the linear regime $I/I_0 \propto \phi$. Only here the phase-contrast signal is proportional to the column density of the atom cloud, because $\phi \propto n_{\text{col}}(\mathbf{r})$

which led to the observation of a fermionic Mott insulator (see Sect. 5.5 for technical details and Chap. 6 for the experiments).

Phase-contrast imaging has two remarkable features: First, it is an almost nondestructive technique, because it relies on elastic scattering of far-detuned laser light. Therefore, multiple images of the same atomic sample can be recorded. Second, in-situ density distributions can be detected, which usually have an optical density that is too high for absorption imaging (see previous section) [1, 6–9]. These properties have led to several intriguing applications of the phase-contrast method, for example, in the detection of local magnetization and in-situ spin textures of spinor condensates [10, 11] or in studies of in-trap distributions of imbalanced strongly interacting Fermi gases [12, 13].

Phase-contrast imaging was invented by Frits Zernike in 1934 and earned him the 1953 Nobel prize in physics. The invention was originally driven by the aim to visualize transparent objects, such as biological cells [14]. The phase-contrast method relies on converting the phase shift that is imprinted on a light beam passing through a refractive object into intensity information that can be observed in a microscope or recorded by a camera. This conversion is achieved by interfering light that is coherently scattered by the object with unscattered light.

The interference effect of the phase-contrast method is achieved by an elegant distinction between scattered and unscattered light: When light passes through a refractive object it acquires an angular spread that is not present for light that does not pass through the object. Therefore, scattered and unscattered light propagate along different paths allowing for a spatial distinction as shown in Fig. 4.1a. In an appropriately designed imaging system, the unscattered light has an intermediate focus at a plane, where the spatial extent of the scattered light is much larger than the focussed unscattered beam. This plane is called the Fourier plane [3]. Here, a differential phase shift between scattered and unscattered light can be applied by placing a special phase plate into the beam path. At the center of the phase plate, where the focussed unscattered light propagates, the optical path length is different from the outer regions. After passing the plate the scattered and phase-shifted unscattered light propagate further and, finally, recombine in the imaging plane (see Fig. 4.1a). At this point, they complete a homodyne detection scheme, in which the unscattered light acts as a local oscillator and interferes destructively or constructively with the coherently scattered light.

Theoretical Background

In order to properly account for interference and phase sensitivity, the phase-contrast method is theoretically treated on the level of the electric field. The laser field (see Eq. 4.2) is decomposed into a scattered and an unscattered part after passing through the atom cloud

$$E = t\, e^{i\phi} E_0 = E_{\mathrm{sc}} + E_{\mathrm{usc}}, \tag{4.6}$$

where the definitions $E_{\mathrm{sc}} = (t\, e^{i\phi} - 1)E_0$ and $E_{\mathrm{usc}} = E_0$ are introduced. The transmission coefficient $t = t(x, y)$ and the phase shift $\phi = \phi(x, y)$ are understood to be functions of the spatial coordinates x and y.

The phase plate in the Fourier plane (see Fig. 4.1) imprints a phase shift α on the unscattered light with respect to the scattered light. Positive (negative) α is generated by a phase plate that advances (retards) the unscattered light by propagation through a bump (dimple). Using $E = E_{\mathrm{sc}} + e^{i\alpha} E_{\mathrm{usc}}$ the intensity at the imaging plane $I_{\mathrm{PCI}} \propto |E|^2$ is given by

$$\begin{aligned} I_{\mathrm{PCI}} &= I_0\, |t\, e^{i\phi} - 1 + e^{i\alpha}|^2 \\ &= I_0 \left(t^2 + 2 - 2\cos(\alpha) - 2t\sqrt{2 - 2\cos(\alpha)}\, \cos\left(\phi + \frac{\alpha}{2}\right)\right). \end{aligned} \tag{4.7}$$

Phase shifts of $\alpha = \pm\pi/2$ are most favorable, because in these cases the dependence between I_{PCI}/I_0 and ϕ is almost linear. Furthermore, a large dynamic range of up to $I_{\mathrm{PCI}}/I_0 \lesssim 6$ is covered before the signal rolls over (see Fig. 4.1b).[2] For these values of α the phase contrast signal is given by

$$I_{\mathrm{PCI}} = I_0 \left(t^2 + 2 - 2\sqrt{2} t \cos\left(\phi \pm \frac{\pi}{4}\right)\right). \tag{4.8}$$

---

[2] Reference [9] points out that a phase shift of $\alpha = \pm\pi/3$ ensures even better linearity of I_{PCI}/I_0 in particular for small ϕ. However, the cost is a limitation of the dynamic range to $I_{\mathrm{PCI}}/I_0 \lesssim 4$.

This expression can be further simplified in the limit of small optical densities (corresponding to small phase shifts ϕ) and full transmission ($t \rightarrow 1$) to

$$I_{\mathrm{PCI}} = I_0 \left(t^2 - 2t + 2 \pm 2t\phi\right) \rightarrow I_0\,(1 \pm 2\phi)\,. \tag{4.9}$$

Small optical density can experimentally be enforced by choosing a sufficiently large detuning Δ. Far-detuned phase-contrast imaging is linear in ϕ and, consequently, linear in the column density $n_{\mathrm{col}}(x, y)$ (see Eqs. 4.3 and 4.4). Furthermore, Eq. 4.9 indicates that experimental images have a background corresponding to I_0, which must be subtracted in the evaluation process.

Outside of the linear regime the phase-contrast signal I/I_0 is a periodic function of the optical density (see Eq. 4.8), whereas absorption imaging saturates for large optical densities (see Eq. 4.5). Nonlinear phase-contrast imaging has been successfully applied for thermometry in Bose-Einstein condensates [9] .

Linear phase-contrast imaging with a retarding phase plate $\alpha = -\pi/2$ is used in Chap. 6 to detect the in-situ density distribution of potassium atoms. Blue-detuning of the imaging laser is induced by shifting the respective atomic transition with a strong magnetic field (see Appendix B). Details on the experimental setup are provided in Sect. 5.5.

4.2 Observables

Based on the recorded images, experimental observables are defined that yield information about the many-body quantum state. In this section, we focus on observables that characterize many-body states of ultracold atoms in simple cubic optical lattice potentials. We discuss the in-situ density distribution as well as the distributions of the momenta and the quasi-momenta. On the basis of the momentum distribution, we introduce and analyze measures of bosonic coherence—the visibility and the fraction of coherent atoms—that are used in Chap. 7–9. While these quantify first-order correlations, second-order correlations can also be extracted from the quantum noise in images of the momentum distribution, which, for example, provides information on the spatial ordering of atoms.

To simplify the discussion we consider a homogeneous simple cubic lattice, in which all sites are equivalent. Furthermore, we make the idealizing assumption that time-of-flight imaging transforms the in-situ real-space density distribution into the momentum distribution. In the experimental realization, this would require a purely ballistic expansion of the atom cloud after the release from the trap and a very long time of flight. Both requirements are rather hard to fulfill [1, 15], but the visibility and the fraction of coherent atoms are rather insensitive to the resulting deviations [15]. More care must be taken for observables that are sensitive to the fine details of the momentum distribution, for example, when the absolute degree of coherence in a bosonic system is to be determined [16].

4.2.1 In-Situ Density Distribution

We start by considering the general in-situ density distribution of a quantum gas in an optical lattice. Due to the periodic potential the single-particle states can be expressed in terms of the Wannier functions $w_{(n)}(\mathbf{r} - \mathbf{r}_j)$ (where $\mathbf{r}_j = 2\pi/a\,\mathbf{n}_j$ with lattice constant a and $\mathbf{n}_j$ being a three-dimensional vector of integers labeling lattice site j) (see Sect. 2.2.4). If all atomic population is contained in the first lattice band, the index n can be dropped and the field operator can be expanded according to $\hat{\psi}(\mathbf{r}) = \sum_j w(\mathbf{r} - \mathbf{r}_j)\hat{a}_j$. The annihilation operator $\hat{a}_j$ destroys a particle in the Wannier state $w(\mathbf{r} - \mathbf{r}_j)$, while the field operator $\hat{\psi}(\mathbf{r})$ as a whole annihilates a particle at position $\mathbf{r}$. Using the definition of the in-situ real space density distribution we obtain

$$n(\mathbf{r}) = \langle \hat{\psi}^\dagger(\mathbf{r})\hat{\psi}(\mathbf{r})\rangle = \sum_{ij} w^*(\mathbf{r} - \mathbf{r}_i)\, w(\mathbf{r} - \mathbf{r}_j)\langle \hat{a}_i^\dagger \hat{a}_j\rangle. \tag{4.10}$$

Here, the expectation value $\langle \hat{a}_i^\dagger \hat{a}_j\rangle$ is understood to be taken within the many-body quantum state $|\Psi\rangle$ in second quantized form, which can generally also feature a time dependence. When the off-diagonal ($i \neq j$) correlations $\langle \hat{a}_i^\dagger \hat{a}_j\rangle$ vanish, corresponding to the absence of long-range phase coherence, the density distribution turns into a simple sum $\sum_j \bar{n}_j |w(\mathbf{r} - \mathbf{r}_j)|^2$ of Wannier functions weighted by the local filling $\bar{n}_j = \langle \hat{n}_j\rangle$.

The locally resolved detection of the in-situ density is not possible with common imaging techniques due to the small spacing between lattice sites ($a \approx 400$ μm). Only recently quantum gas microscopes have been demonstrated that employ fluorescence imaging and permit single-site and and single-atom resolved detection [17–19]. However, even without ultra high resolution the detection of the real space density can yield valuable information [6, 7, 20, 21]. For example, the full three-dimensional density distributions $n(\mathbf{r})$ of imbalanced Fermi mixtures have been reconstructed from the recorded column density $n_{\text{col}}(x, y)$ by exploiting the symmetries of the harmonic trap in an inverse Abel transformation [12]. Furthermore, the first direct images of the shell structure in a bosonic Mott insulator have been obtained via in-situ absorption imaging [22, 23] (compare Sect. 3.2.2).

In Chap. 6, in-situ phase-contrast images of fermionic spin mixtures in an optical lattice are recorded (see Fig. 6.5b). We measure the radius of the column density as a function of the external trapping potential and use these data to extract the global compressibility of the many-body quantum state. The compressibility permits a direct distinction between metallic (compressible) and insulating (incompressible) phases and reveals a fermionic Mott insulator for strong interactions.

4.2.2 Momentum Distribution

The measurement of momentum distributions has a very long tradition in the field of ultracold atoms. The time-of-flight technique, in which all trapping potentials are abruptly switched off before atoms undergo ballistic expansion, enabled the first unambiguous detection of Bose-Einstein condensates [24, 25]. It was equally important in the first detection of the superfluid to Mott insulator transition. Here, the loss of contrast in a lattice Bose gas revealed the suppression of long-range coherence as an indicator for the localization of atoms [26] (see Sect. 3.2). Within this thesis the momentum distribution is mostly used for the analysis of bosonic many-body quantum states. Therefore, we use bosonic operators $\hat{a}$ and $\hat{a}^\dagger$ in the following derivation, although the relations also account for fermions.

In order to derive the atomic density in momentum space, we first Fourier transform the field operator $\hat{\psi}(\mathbf{r})$ and obtain

$$\hat{\psi}(\mathbf{k}) = \sum_j \hat{a}_j \int d^3r\, e^{-i\mathbf{k}\cdot\mathbf{r}}\, w(\mathbf{r}-\mathbf{r}_j) = \left(\sum_j \hat{a}_j e^{-i\mathbf{k}\cdot\mathbf{r}_j}\right) w(\mathbf{k}) \equiv \hat{a}_\mathbf{k} w(\mathbf{k}), \tag{4.11}$$

where $w(\mathbf{k})$ is the Fourier transform of the Wannier function. Furthermore, the annihilation operator for an atom at momentum $\mathbf{k}$ is given by $\hat{a}_\mathbf{k} = \sum_j \hat{a}_j e^{-i\mathbf{k}\cdot\mathbf{r}_j}$. With the above relations the momentum distribution is given by its definition

$$n(\mathbf{k}) \equiv \langle \hat{\psi}(\mathbf{k})^\dagger \hat{\psi}(\mathbf{k})\rangle = |w(\mathbf{k})|^2 \sum_{ij} e^{-i\mathbf{k}\cdot(\mathbf{r}_i-\mathbf{r}_j)} \langle \hat{a}_i^\dagger \hat{a}_j\rangle. \tag{4.12}$$

To further elucidate Eq. 4.12, we divide it into diagonal ($i = j$) and off-diagonal ($i \neq j$) terms according to

$$\begin{aligned} n(\mathbf{k}) &= |w(\mathbf{k})|^2 \left(\sum_j \langle \hat{n}_j\rangle + \sum_{i\neq j} e^{-i\mathbf{k}\cdot(\mathbf{r}_i-\mathbf{r}_j)} \langle \hat{a}_i^\dagger \hat{a}_j\rangle\right) \\ &\rightarrow |w(\mathbf{k})|^2 \left(\sum_j \langle \hat{n}_j\rangle + |\langle \hat{a}\rangle|^2 \sum_{i\neq j} e^{-i\mathbf{k}\cdot(\mathbf{r}_i-\mathbf{r}_j)}\right). \end{aligned} \tag{4.13}$$

The squared modulus of $w(\mathbf{k})$ gives rise to an envelope of the momentum distribution, the so-called Wannier background. In the absence of phase coherence ($\langle \hat{a}_i^\dagger \hat{a}_j\rangle = 0$ for all $i \neq j$) the momentum distribution is given by this background, whose width is inversely proportional to the width of the on-site Wannier function $w(\mathbf{r})$ owing to the Fourier relation. If phase coherence is perfectly established, the single-particle density matrix $\langle \hat{a}_i^\dagger \hat{a}_j\rangle$ will take finite values even for lattice sites i and j that are far away from each other. In this case, the width of the individual interference peaks is inversely proportional to the system size. For an infinitely large theoretical system

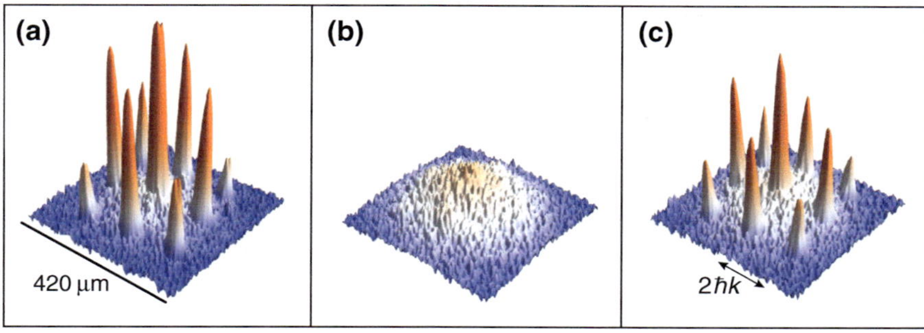

Fig. 4.2 Momentum distribution of ^{87}Rb atoms during collapse and revival dynamics in a 41 E_{rec} deep lattice. The data show three points of the evolution at hold times of **a** 40 μs (start), **b** 120 μs (collapse) and **c** 320 μs (partial revival). The data are taken from the set in Fig. 7.8. While the overall atom number is almost constant, the off-diagonal correlations vary in time and result in the emergence or suppression of interference

δ-peaks are obtained. For partial coherence the width of the interference peaks is inversely proportional to the coherence length.

In order to make the replacement $\langle \hat{a}_i^\dagger \hat{a}_j \rangle \rightarrow |\langle \hat{a} \rangle|^2$ in the second step of the above equation, we follow a Gutzwiller ansatz, in which the many-body quantum state $|\Psi\rangle$ factorizes into equivalent on-site wavefunctions (compare Sect. 7.2.1). In this case, all information on the phase coherence of the system is contained in the squared modulus of the field amplitude $|\langle \hat{a} \rangle|^2$, which determines how strongly the interference term $\sum_{i \neq j} e^{-i\mathbf{k}\cdot(\mathbf{r}_i - \mathbf{r}_j)}$ contributes.

Measurements of the bosonic phase coherence play a central role in Chaps. 7–9. Specifically, the collapse and revival dynamics of coherent bosonic states require a robust observable for the dynamical evolution of the coherence (see Fig. 4.2). As a consequence of Eq. 4.13, the contrast of the interference pattern forms a measure for $|\langle \hat{a} \rangle|^2$. We make use of two experimental observables, the visibility (Chap. 7) and the fraction of coherent atoms (Chap. 9), that yield a maximal value when all atoms are found in the interference peaks and vanish when all atoms are part of the featureless background.

4.2.2.1 Visibility

In analogy to typical definitions of contrast, the visibility $\mathcal{V}$ of a bosonic interference pattern is defined by [27]

$$\mathcal{V} = \frac{N_{\mathrm{max}} - N_{\mathrm{min}}}{N_{\mathrm{max}} + N_{\mathrm{min}}}. \tag{4.14}$$

N_{max} denotes the total number of atoms inside four boxes around the first order interference peaks and N_{min} the total number of atoms in a set of identical boxes that are rotated by 45° around the central peak (see inset of Fig. 4.3a).

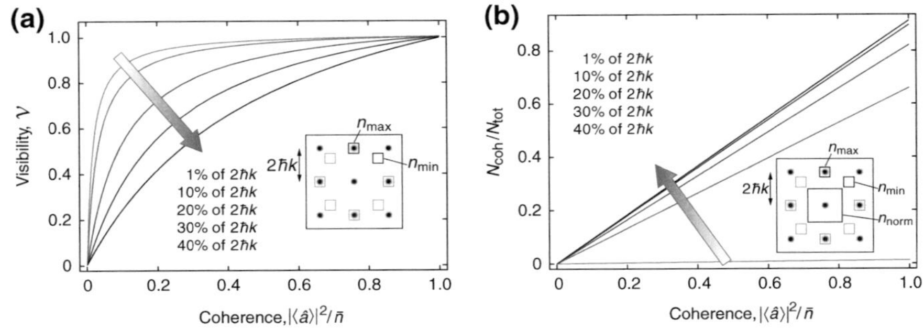

Fig. 4.3 Comparison of visibility and fraction of coherent atoms as measures for bosonic coherence. Based on simulated momentum distributions of a two-dimensional lattice system (11 × 11 sites) with a known value of $|\langle\hat{a}\rangle|^2/\bar{n}$ ($\bar{n}$ is the mean on-site filling) the experimental observables are extracted: **a** The visibility is defined by $\mathcal{V} = (n_{\max} - n_{\min})/(n_{\max} + n_{\min})$ and **b** the fraction of coherent atoms follows from $(N_{\max} - n_{\min})/n_{\text{norm}}$. Both quantities are evaluated for several sizes of the $n_{\max}$- and $n_{\min}$-counting boxes given in units of the Brillouin zone width $2\hbar k$

4.2.2.2 Fraction of Coherent Atoms

The fraction of coherent atoms is obtained from an experimental image as follows: First, the Wannier background is fitted by a Gaussian and subtracted from the image. In the resulting image, the atom numbers in boxes that contain the central, first- and second-order coherence peaks are counted and summed up to obtain N_{coh}. The total atom number N_{tot} is the sum of N_{coh} and the atoms contained in the Gaussian background. The ratio $N_{\text{coh}}/N_{\text{tot}}$ yields the fraction of coherent atoms [28].

Figure 4.3 shows a simulation that elucidates the exact relation of the visibility and the fraction of coherent atoms to the theoretical coherence $|\langle\hat{a}\rangle|^2$. For this purpose, the momentum distribution of a small two-dimensional lattice is calculated as a function of $|\langle\hat{a}\rangle|^2$. From the resulting numerical momentum distributions the visibility $\mathcal{V}$ and $N_{\text{coh}}/N_{\text{tot}}$ are derived, as if they were experimental images (see insets of Fig. 4.3). Furthermore, the dependence of the observables on the size of the evaluation boxes is investigated.

As expected, both observables have a strictly monotonic relation to the coherence. The visibility shows a nonlinear behavior with a strong enhancement for weak coherences (see Fig. 4.3a). This can be helpful, if a high sensitivity to small values of $|\langle\hat{a}\rangle|^2$ is required. In fact, the nonlinearity has helped us to identify the beat signal in the quantum phase revival dynamics in Chap. 7 (see Fig. 4.2 for exemplary images). Nevertheless, it bears the risk of frequency mixing in the evaluation of the experimental collapse and revival time traces. The nonlinearity is strongest for small evaluation boxes and the relation becomes increasingly linear for larger ones.

The fraction of coherent atoms is an almost perfectly linear function of $|\langle\hat{a}\rangle|^2$ (see Fig. 4.3b). While the absolute values of $N_{\text{coh}}/N_{\text{tot}}$ show a strong variation as a function of the box sizes, the linearity is always preserved. The linear behavior makes it harder to detect small values of the coherence, but it is conceptionally favorable for the evaluation of quantum phase revival measurements: The experimental time traces and the time-dependent oscillating coherence $|\langle\hat{a}\rangle(t)|^2$ only differ by a fixed factor. Therefore, the Fourier transform of the experimental trace precisely yields the spectral content of the quantum phase revival dynamics contained in $|\langle\hat{a}\rangle(t)|^2$. The fraction of coherent atoms is extensively used in the analyses of Chap. 9.

When fermionic atoms are released from optical lattices, they also show an interference pattern (see Sect. 9.3). As in the case of bosonic atoms, the width of the interference peaks contains information on the delocalization of atoms and, consequently, their coherence length [29] (see Fig. 9.11a, b). Necessarily, the interference peaks are broad, because fermions cannot macroscopically occupy delocalized Bloch states due to Pauli's principle. As shown in Fig. 3.7 only few of the lowest lying single-particle states show a notable delocalization.

4.2.3 Noise Correlations

In strongly correlated quantum phases, such as a bosonic Mott insulator, the atoms are localized at individual lattice sites: The off-diagonal elements of the single-particle density matrix $\langle\hat{a}_i^\dagger\hat{a}_j\rangle$ vanish, which corresponds to the absence of long-range phase coherence. When the potential is switched-off to obtain a time-of-flight image, all atoms leave the lattice individually and no interference occurs. The resulting momentum distribution merely shows the practically featureless Wannier background (see above). Many of the intriguing strongly correlated phases that are predicted to emerge within ultracold atom systems suffer from this problem, for example, antiferromagnetically ordered states as well as spin- and charge density waves.

While the momentum distribution measures first-order correlations $\langle\hat{\psi}(\mathbf{k})^\dagger\hat{\psi}(\mathbf{k})\rangle$, it is also possible to extract second-order density-density correlations from experimental images [30]. The density-density correlations in momentum space are theoretically defined by

$$C(\mathbf{k},\mathbf{k}') \equiv \langle\hat{n}(\mathbf{k})^\dagger\hat{n}(\mathbf{k}')\rangle = \langle\hat{\psi}(\mathbf{k})^\dagger\hat{\psi}(\mathbf{k})\hat{\psi}(\mathbf{k}')^\dagger\hat{\psi}(\mathbf{k}')\rangle, \tag{4.15}$$

which is evaluated within the many-body state of interest $|\Psi\rangle$ in second quantized form. Altman and collaborators [30] have shown that such correlations can be extracted from the atomic shot-noise in images of the momentum distribution using the prescription [31]

$$C(\mathbf{d}) = \frac{\int d^2\mathbf{r}\,\langle n_{\mathrm{col}}(\mathbf{r}-\mathbf{d}/2)n_{\mathrm{col}}(\mathbf{r}+\mathbf{d}/2)\rangle}{\int d^2\mathbf{r}\,\langle n_{\mathrm{col}}(\mathbf{r}-\mathbf{d}/2)\rangle\langle n_{\mathrm{col}}(\mathbf{r}+\mathbf{d}/2)\rangle}, \tag{4.16}$$

where $n_{\mathrm{col}}(\mathbf{r})$ denotes the column density recorded in experimental images. The coordinate $\mathbf{r} = (x, y)$ in the image corresponds to a momentum $\mathbf{k} = m\mathbf{r}/(\hbar t)$, where t is the time-of-flight and m the atomic mass. In the prescription 4.16 the

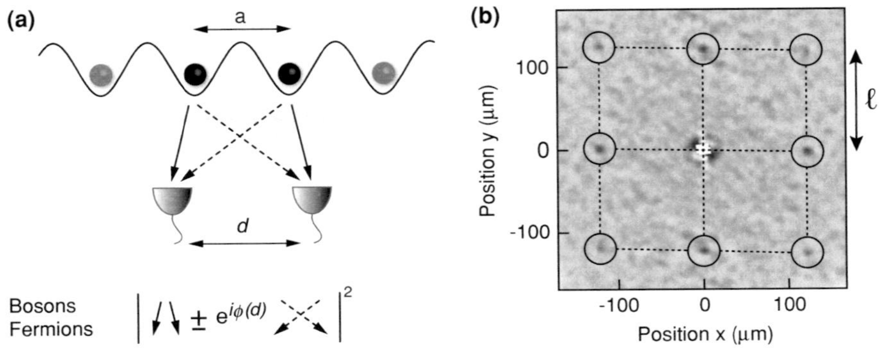

Fig. 4.4 Quantum noise correlations for bosonic and fermionic atoms. **a** When two identical atoms are released from neighboring lattice sites, they can reach the independent detectors via two, fundamentally indistinguishable paths. The respective probability amplitudes interfere constructively (destructively) for bosons (fermions) corresponding to the *upper* (*lower*) sign. The correlation between the detectors is sinusoidally modulated as a function of the distance d. For more than two atoms the correlations evolve from a sinusoidal modulation to narrow peaks, in analogy to optical diffraction from a grating with an increasing number of slits. **b** Experimental noise correlations of a ^{87}Rb Mott insulator released from a three-dimensional optical lattice ($V_{\text{lat}} = 32\ E_{\text{rec}}$). About 30 images have been averaged. The position of the correlation peaks with a periodicity $\ell = ht/(ma)$ resembles the in-situ order with a lattice constant a

angled brackets $\langle \ldots \rangle$ do not denote a quantum mechanical expectation value, but averaging over an ensemble of independently acquired images.

The emergence of correlation peaks for certain distances **d** can reveal the hidden order in strongly correlated many-body quantum states as well as the quantum statistics of the underlying particles. For example, a Mott insulator of bosons and a band insulator of spin-polarized fermions have identical in-situ density and momentum distributions. In both cases, correlations appear at locations **d** that correspond to reciprocal lattice vectors with the periodicity

$$\ell = \frac{ht}{ma} = \frac{t}{m} 2\hbar k, \tag{4.17}$$

where $k = 2\pi/\lambda$ is the wave vector of the lattice light with wavelength λ. While bosons show a positive signal indicating bunching [31, 32], the antibunching of fermions is revealed by a negative signal (see Fig. 4.4) [5, 33].

Noise correlation analysis has been proposed as a helpful tool for the identification of antiferromagnetically ordered phases, where doubling of the unit cell (see Fig. 3.6) will lead to additional correlation peaks at positions with the periodicity $\ell/2$ [34]. Similar signatures are expected for the detection of density waves [30] and supersolids [35]. The proof of principle that noise correlations can identify modulated in-situ densities has already been provided by the detection of patterned loading of an optical superlattice [36].

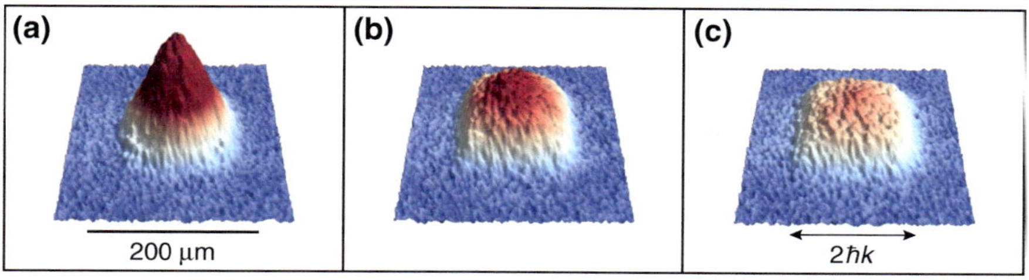

Fig. 4.5 Brillouin zone mapping reveals the quasi-momentum distribution for a noninteracting spin mixture of ^{40}K in a 8 E_{rec} lattice. While the overall atom number is approximately constant, the filling of the Brillouin zone varies as a function of the external harmonic confinement corresponding to **a** $\omega_\perp = 2\pi \times 46$ Hz, **b** $2\pi \times 90$ Hz and **c** and $2\pi \times 157$ Hz. For the strongest harmonic confinement the first Brillouin zone is homogeneously filled and population in higher Brillouin zones is negligible. This observation is compatible with the formation of a band insulator in the center of the trap, which is independently confirmed by a cloud size measurement. (Images are selected from the black data in Fig. 6.6)

Based on second-order correlations, we devise a novel detection scheme for the Schrödinger cat state that is expected to arise in bosonic collapse and revival dynamics (see Sect. 7.4) [28, 37, 38]. At the time of the collapse (see for example Fig. 4.2b) the formation of an on-site superposition of coherent states is expected, which we show to lead to characteristic **k**/-**k** quantum correlations in the momentum distribution. The analysis of **k**/-**k** correlations has also been proposed to reveal the Cooper-pairing in momentum space of the fermionic BCS state [30]. Correlations between opposite momenta, albeit of classical nature, have already been observed after the dissociation of weakly bound Feshbach molecules [39].

4.2.4 Quasi-Momentum Distribution

While the lattice is abruptly switched off for the measurement of the momentum distribution, the quasi-momentum distribution can be revealed when the lattice is ramped down on a time scale that is adiabatic with respect to the band gap, but fast with respect to tunneling. In this case the band gaps become continuously smaller until the free particle dispersion is reached (see Sect. 2.2.3). The Bloch states are adiabatically transformed into the corresponding free-space plane waves, whose momenta p are mapped in time-of-flight expansion. Population of the first band is expected to lie within the first Brillouin zone of the reciprocal lattice [40, 41] and population of higher bands n analogously maps to higher Brillouin zones $(n-1)\hbar k \leq |p| \leq n\hbar k$, where $k = 2\pi/a$.

At first sight, Brillouin zone mapping seems to be an ideal tool to distinguish metallic and band insulating phases of noninteracting fermions [29] (see Fig. 4.5). A box-shaped Fermi surface is indeed expected when the first band is fully occupied and forms a band insulator (see Sect. 3.3.3). However, in practice, the Fermi surface is smeared out, because the spatial variation of the filling in the underlying harmonic trap corresponds to a local variation of the band filling. Additionally, Brillouin zone mapping only reveals the relative population of Bloch states. Therefore, experimental observation of a homogeneously filled Brillouin zone is compatible with a band insulator. But it is not a sufficient condition, because a sample with an irregular arrangement of localized atoms or a Mott insulator have the same quasi-momentum distribution as a zero-temperature band insulator. Nevertheless, in combination with further observables (for example the in-situ cloud size) Brillouin zone mapping can yield useful information on the many-body quantum state.

In Chap. 6 we map the quasi-momentum distributions of a noninteracting spin mixture of fermions. Therefore, we use the fermionic operators $\hat{c}_{j\sigma}$ and $\hat{c}^{\dagger}_{j\sigma}$ with $\sigma \in \{\uparrow, \downarrow\}$ and the fermionic form of the field operator $\hat{\psi}(\mathbf{r}) = \sum_{j\sigma} w(\mathbf{r} - \mathbf{r}_j)\hat{c}_{j\sigma}$ in the following derivation. In analogy to Eq. 4.11, the field operator can be formally expressed in quasi-momentum space as $\hat{\psi}(\mathbf{q}) = \hat{c}_{\mathbf{q}\sigma} w(\mathbf{q})$, where the annihilation operator for an atom in Bloch state $\mathbf{q}$ and spin state σ is given by

$$\hat{c}_{\mathbf{q}\sigma} = \sum_{j\sigma} \hat{c}_{j\sigma} e^{i\mathbf{q}\cdot\mathbf{r}_j}. \tag{4.18}$$

By definition, the quasi-momentum distribution for a spin state σ is given by the occupation of Bloch states

$$\langle \hat{n}_{\mathbf{q}\sigma} \rangle = \langle \hat{c}^{\dagger}_{\mathbf{q}\sigma} \hat{c}_{\mathbf{q}\sigma} \rangle = \sum_{ij} e^{-i\mathbf{q}(\mathbf{r}_i - \mathbf{r}_j)} \langle \hat{c}^{\dagger}_{i\sigma} \hat{c}_{j\sigma} \rangle. \tag{4.19}$$

As our imaging techniques do not distinguish the spin states, the observed distribution corresponds to $\langle \hat{n}_{\mathbf{q}} \rangle = \langle \hat{n}_{\mathbf{q}\uparrow} \rangle + \langle \hat{n}_{\mathbf{q}\downarrow} \rangle$. Under restriction to the first Brillouin zone ($|\mathbf{q}| \leq \hbar k$) this yields a simple relation between the momentum distribution (Eq. 4.12) and quasi-momentum distribution (Eq. 4.19)

$$n(\mathbf{q}) = |w(\mathbf{q})|^2 \langle \hat{n}_{\mathbf{q}} \rangle. \tag{4.20}$$

The spatial position of the atomic density in the experimental image relates to the quasi-momentum via $\mathbf{r} = \mathbf{q}\hbar t/m$, where t is the time-of-flight and m the atomic mass. Relation 4.20 is utilized in Sect. 9.3, where the momentum distribution of spin-polarized fermions is used to extract information on the quasi-momentum distribution.

References

1. W. Ketterle, D. Durfee, D.M. Stamper-Kurn, Making, probing and understanding Bose-Einstein condensates, in *Bose-Einstein Condensation in Atomic Gases*, ed. by S. Stringari, M. Inguscio, C.E. Wieman (IOS Press, Amsterdam, 1999), pp. 67–176
2. T. Gericke, P. Würtz, D. Reitz, T. Langen, H. Ott, High-resolution scanning electron microscopy of an ultracold quantum gas. Nat. Phys. **4**, 949 (2008)

3. B. Saleh, M. Teich, *Fundamentals of Photonics* (Wiley, New Jersey, 1991)
4. M. Scully, S. Zubairy, *Quantum Optics* (Cambridge University Press, Cambridge 1997)
5. T. Rom, *Bosonische und fermionische Quantengase in dreidimensionalen optischen Gittern.* Ph.D. Thesis, Ludwig-Maximilians-Universität München, 2009
6. M.R. Andrews, M.O. Mewes, N.J. van Druten, D.S. Durfee, D.M. Kurn, W. Ketterle, Direct, nondestructive observation of a Bose condensate. Science **273**, 84 (1996)
7. M.R. Andrews, D.M. Kurn, H.-J. Miesner, D.S. Durfee, C.G. Townsend, S. Inouye, W. Ketterle, Propagation of sound in a Bose-Einstein condensate. Phys. Rev. Lett. **79**, 553 (1997)
8. C.C. Bradley, C.A. Sackett, R.G. Hulet, Bose-Einstein condensation of lithium: observation of limited condensate number. Phys. Rev. Lett. **78**, 985 (1997)
9. R. Meppelink, R.A. Rozendaal, S.B. Koller, J.M. Vogels, P. van der Straten, Thermodynamics of Bose-Einstein-condensed clouds using phase-contrast imaging. Phys. Rev. A **81**, 053632 (2010)
10. J.M. Higbie, L.E. Sadler, S. Inouye, A.P. Chikkatur, S.R. Leslie, K.L. Moore, V. Savalli, D.M. Stamper-Kurn, Direct nondestructive imaging of magnetization in a spin-1 Bose-Einstein gas. Phys. Rev. Lett. **95**, 050401 (2005)
11. L.E. Sadler, J.M. Higbie, S.R. Leslie, M. Vengalattore, D.M. Stamper-Kurn, Spontaneous symmetry breaking in a quenched ferromagnetic spinor Bose-Einstein condensate. Nature **443**, 312 (2006)
12. Y. Shin, M.W. Zwierlein, C.H. Schunck, A. Schirotzek, W. Ketterle, Observation of phase separation in a strongly interacting imbalanced Fermi gas. Phys. Rev. Lett. **97**, 030401 (2006)
13. Y. Shin, C.H. Schunck, A. Schirotzek, W. Ketterle, Tomographic rf spectroscopy of a trapped Fermi gas at unitarity. Phys. Rev. Lett. **99**, 090403 (2007)
14. E. Hecht, *Optics*, 4th edn. (Addison Wesley, San Francisco, 2002)
15. F. Gerbier, S. Trotzky, S. Fölling, U. Schnorrberger, J.D. Thompson, A. Widera, I. Bloch, L. Pollet, M. Troyer, B. Capogrosso-Sansone, N.V. Prokof'ev, B.V. Svistunov, Expansion of a quantum gas released from an optical lattice. Phys. Rev. Lett. **101**, 155303 (2008)
16. S. Trotzky, L. Pollet, F. Gerbier, U. Schnorrberger, I. Bloch, N. Prokof'ev, B. Svistunov, M. Troyer, Suppression of the critical temperature for superfluidity near the Mott transition. Nat. Phys. **6**, 998 (2010)
17. W. Bakr, J. Gillen, A. Peng, S. Fölling, M. Greiner, A quantum gas microscope for detecting single atoms in a Hubbard-regime optical lattice. Nature **462**, 74 (2009)
18. W.S. Bakr, A. Peng, M.E. Tai, R. Ma, J. Simon, J.I. Gillen, S. Fölling, L. Pollet, M. Greiner, Probing the superfluid-to-Mott insulator transition at the single-atom level. Science **329**, 547 (2010)
19. J. Sherson, C. Weitenberg, M. Endres, M. Cheneau, I. Bloch, S. Kuhr, Single-atom-resolved fluorescence imaging of an atomic Mott insulator. Nature **467**, 68 (2010)
20. G.B. Partridge, W. Li, R.I. Kamar, Y.-a. Liao, R.G. Hulet, Pairing and phase separation in a polarized Fermi gas. Science, **311**, 503 (2006)
21. Y. Liao, A. Rittner, T. Paprotta, W. Li, G. Partridge, R. Hulet, S. Baur, E. Mueller, Spin-imbalance in a one-dimensional Fermi gas. Nature **467**, 567 (2010)
22. G.K. Campbell, J. Mun, M. Boyd, P. Medley, A.E. Leanhardt, L.G. Marcassa, D.E. Pritchard, W. Ketterle, Imaging the Mott insulator shells by using atomic clock shifts. Science **313**, 649 (2006)
23. N. Gemelke, X. Zhang, C.-L. Hung, C. Chin, In situ observation of incompressible Mott-insulating domains in ultracold atomic gases. Nature **460**, 995 (2009)
24. M.H. Anderson, J.R. Ensher, M.R. Matthews, C.E. Wieman, E.A. Cornell, Observation of Bose-Einstein condensation in a dilute atomic vapor. Science **269**, 198 (1995)
25. K.B. Davis, M.O. Mewes, M.R. Andrews, N.J. van Druten, D.S. Durfee, D.M. Kurn, W. Ketterle, Bose-Einstein condensation in a gas of sodium atoms. Phys. Rev. Lett. **75**, 3969 (1995)
26. M. Greiner, O. Mandel, T. Esslinger, T. Hänsch, I. Bloch, quantum phase transition from a superfluid to a Mott insulator in a gas of ultracold atoms. Nature **415**, 39 (2002)
27. F. Gerbier, A. Widera, S. Fölling, O. Mandel, T. Gericke, I. Bloch, Phase coherence of an atomic Mott insulator. Phys. Rev. Lett. **95**, 050404 (2005)

28. M. Greiner, O. Mandel, T. Hänsch, I. Bloch, Collapse and revival of the matter wave field of a Bose-Einstein condensate. Nature **419**, 51 (2002)
29. M. Köhl, H. Moritz, T. Stöferle, K. Günter, T. Esslinger, Fermionic atoms in a three dimensional optical lattice: observing Fermi surfaces, dynamics, and interactions. Phys. Rev. Lett. **94**, 080403 (2005)
30. E. Altman, E. Demler, M.D. Lukin, Probing many-body states of ultracold atoms via noise correlations. Phys. Rev. A **70**, 013603 (2004)
31. S. Fölling, F. Gerbier, A. Widera, O. Mandel, T. Gericke, I. Bloch, Spatial quantum noise interferometry in expanding ultracold atom clouds. Nature, 434, 481 (2005)
32. I.B. Spielman, W.D. Phillips, J.V. Porto, Mott-insulator transition in a two-dimensional atomic Bose gas. Phys. Rev. Lett. **98**, 080404 (2007)
33. T. Rom, T. Best, D. van Oosten, U. Schneider, S. Fölling, B. Paredes, I. Bloch, Free fermion antibunching in a degenerate atomic Fermi gas released from an optical lattice. Nature **444**, 733 (2006)
34. G.M. Bruun, O.F. Syljuåsen, K.G.L. Pedersen, B.M. Andersen, E. Demler, A.S. Sørensen, Antiferromagnetic noise correlations in optical lattices. Phys. Rev. A **80**, 033622 (2009)
35. V.W. Scarola, E. Demler, S. Das Sarma, Searching for a supersolid in cold-atom optical lattices. Phys. Rev. A **73**, 051601 (2006)
36. S. Fölling. *Probing Strongly Correlated States of Ultracold Atoms in Optical Lattices*. Ph.D. Thesis, Johannes Gutenberg-Universität Mainz, 2008
37. S. Haroche, J.-M. Raimond, *Exploring the Quantum: Atoms, Cavities, and Photons (Oxford Graduate Texts)*, 1st edn. (Oxford University Press, New York, 2006)
38. K. Pawłowski, K. Rzążewski, Background atoms and decoherence in optical lattices. Phys. Rev. A **81**, 013620 (2010)
39. M. Greiner, C.A. Regal, J.T. Stewart, D.S. Jin, Probing pair-correlated fermionic atoms through correlations in atom shot noise. Phys. Rev. Lett. **94**, 110401 (2005)
40. N.W. Ashcroft, N.D. Mermin, *Solid State Physics* (Saunders College Publishing, Fort Worth, 1976)
41. M. Greiner, I. Bloch, O. Mandel, T.W. Hänsch, T. Esslinger, Exploring phase coherence in a 2D lattice of Bose-Einstein condensates. Phys. Rev. Lett. **87**, 160405 (2001)

Chapter 5
Experimental Apparatus

This chapter summarizes the experimental sequence and the techniques used to produce, manipulate and probe ultracold bosonic and fermionic quantum gases in our setup. Many of the applied cooling and trapping techniques are fairly standard and are only described in passing [1]. More emphasis is put on peculiarities of our setup, such as the optically plugged magnetic quadrupole trap, the combined trapping in a red-detuned dipole trap and a blue-detuned optical lattice as well as phase-contrast imaging. Further details on the experimental setup can be found in the Ph.D. theses of Tim Rom [2] and Thorsten Best [3].

5.1 Overview of the Setup and Experimental Sequence

In the following we briefly outline the experimental sequence, which is used to create quantum degenerate clouds of bosonic ^{87}Rb and fermionic ^{40}K. The relevant energy levels and transitions are summarized in Fig. 5.1 and a cross-sectional view of the apparatus is displayed in Fig. 5.2.

- **Two-species magneto-optical trap**: About 3×10^9 ^{87}Rb atoms and 2×10^7 ^{40}K are captured and laser cooled in a vapor cell magneto-optical trap (MOT) [4]. Commercial rubidium- and homemade potassium-dispensers are used as atom sources. The latter ones contain enriched potassium with 7 % of ^{40}K. The dispensers operate in a pulsed mode: Atoms are only released during the MOT phase of the experimental cycle, which efficiently minimizes the background pressure in the MOT chamber. The duration of the MOT phase is dominated by a loading time of up to 14 s for ^{40}K, while ^{87}Rb is rapidly captured within the last 4 s or less. The largest atom numbers in the two-species MOT are obtained by displacing the rubidium and potassium cloud with respect to each other using slightly imbalanced laser intensities. This reduces the overlap and suppresses heteronuclear atom loss via light-induced collisions.

DOI: 10.1007/978-3-642-33633-1_5,

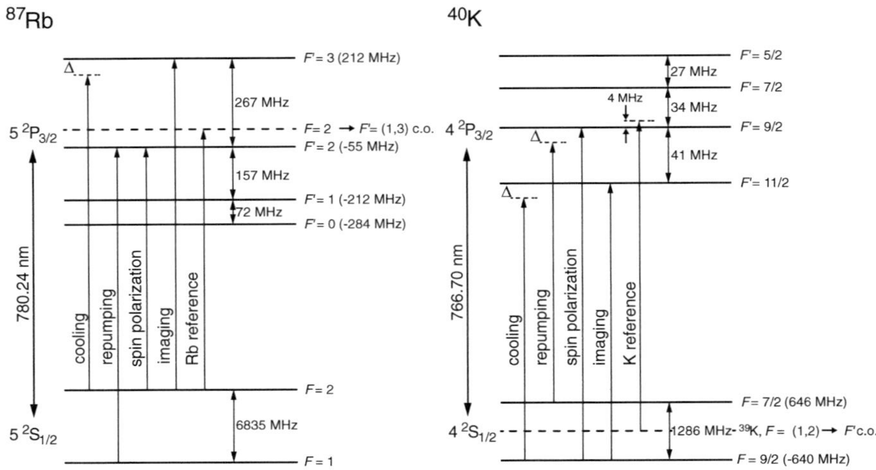

Fig. 5.1 Level scheme for the D_2 lines of ^{87}Rb and ^{40}K and corresponding optical transitions used in the experimental sequence. *Cooling* and *repumping* denote the transitions used for laser cooling. *Spin polarization* is used to transfer the atoms to hyperfine sublevels that are magnetically trappable. Absorption and phase-contrast images are recorded on the *imaging*-transition. The Δ indicates that the potassium cooling and repumping transitions are detuned, but the exact detunings are different. Strong *reference* transitions are stabilized and used to lock the other laser frequencies by means of offset-locks. The label *c.o.* indicates crossover resonances

- **Molasses**: The MOT is followed by a short molasses phase (10 ms), during which the magnetic field is switched off and the laser detuning is shifted further to the red. This mostly affects the rubidium cloud, reaching sub-Doppler temperatures of about 50 μK, while potassium essentially remains at the Doppler temperature of about 150 μK [2].
- **Optical pumping**: Using σ^+-polarized light at a weak homogeneous magnetic field, the atoms are actively transferred into weak-field seeking states that are magnetically trappable. Within 400 μs rubidium is pumped to the $|F, m_F\rangle = |2, +2\rangle$ state by applying light on the $|F = 2\rangle \rightarrow |F' = 2\rangle$ and $|F = 1\rangle \rightarrow |F' = 2\rangle$ transitions, while potassium is transferred to the $|F, m_F\rangle = |9/2, +9/2\rangle$ state by simultaneously driving the $|F = 9/2\rangle \rightarrow |F' = 9/2\rangle$ and $|F = 7/2\rangle \rightarrow |F' = 9/2\rangle$ transitions. At this point rubidium and potassium are prepared in stretched hyperfine states having the same magnetic moment μ_B. This ensures good overlap in the magnetic trap that is crucial for proper thermalization during sympathetic cooling. Furthermore, spin-changing collisions that can lead to atom loss and heating are efficiently suppressed.
- **Magnetic transport**: After optical pumping the atom clouds are transferred into a mode-matched magnetic quadrupole trap [2, 5] and adiabatically compressed within 200 ms to a vertical trap gradient of 100 G/cm. Then, the rubidium and

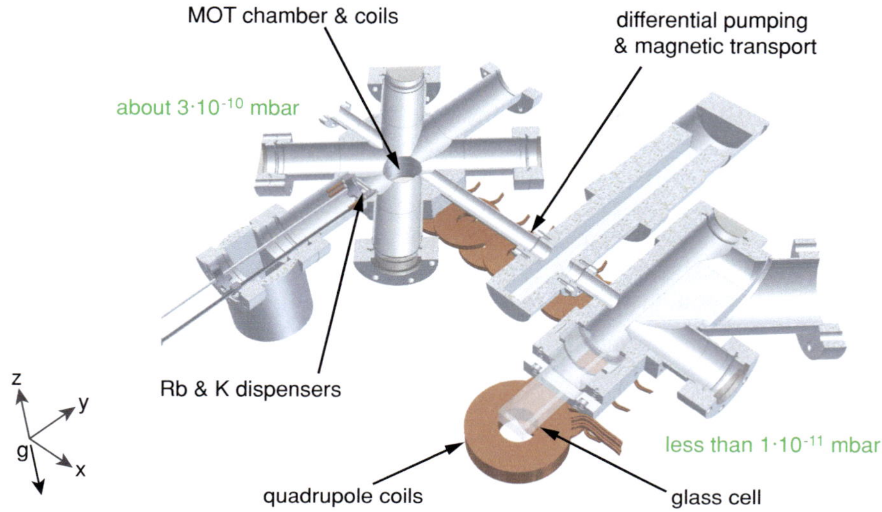

Fig. 5.2 Cross-sectional view of the experimental apparatus showing the vacuum chamber and the most important magnetic field coils. *Green* annotations show the pressure in the corresponding part of the vacuum chamber

potassium clouds are transferred from the MOT chamber to an ultra-high vacuum (UHV) glass cell over a distance of almost 40 cm (see Fig. 5.2). The transport is performed within 2 s by using a chain of quadrupole coil pairs, whose currents are cleverly ramped to smoothly move the clouds with a constant quadrupole field through the vacuum chamber [2, 6]. The chamber design allows for a large pressure gradient between the MOT and the UHV glass cell, where good optical access from all six directions is achieved.

- **Sympathetic cooling in a plugged quadrupole trap**: At the end of the transport the atoms are captured in a powerful quadrupole trap (see Fig. 5.2). In order to perform efficient cooling without Majorana spin flips [7, 8], a focussed blue-detuned laser along the z-axis prevents the atoms from reaching the magnetic field zero in the center of the quadrupole field (see Sect. 5.2). Rubidium is evaporatively cooled by applying a slow radio-frequency (RF) sweep, which continuously flips the most energetic atoms into untrapped Zeeman levels. Potassium is almost not affected by the RF sweep and sympathetically cooled by thermalization with the rubidium atoms. At the end of typically 8 s of RF evaporation 10×10^6 ^{87}Rb and up to 2×10^6 ^{40}K atoms are in thermal equilibrium at a temperature of about 2 μK.
- **Cooling in a crossed dipole trap**: Within 200 ms the two pre-cooled clouds are smoothly transferred into a pancake-shaped crossed dipole trap (see Sect. 5.3). The quadrupole field is continuously transformed into a homogeneous offset field at 13.6 G. Using Landau-Zener microwave and RF sweeps rubidium and

potassium are transferred into the absolute ground states $|F, m_F\rangle = |1, +1\rangle$ and $|F, m_F\rangle = |9/2, -9/2\rangle$, respectively. A short pulse of imaging light is used to clean out the few remaining rubidium atoms in the $F = 2$ manifold. By continuously reducing the dipole trap depth, quantum degeneracy is achieved for both species within about 4 s. Typically, a quasi-pure Bose-Einstein condensate (about 3×10^5 atoms) coexists with a spin-polarized Fermi cloud (about 3×10^5 atoms) at a temperature of slightly below $T/T_F = 0.2$. The balance between bosons and fermions can be controlled over a wide range by adjusting the initial MOT parameters. This cooling sequence is used in the experiments with Bose-Fermi mixtures (see Chaps. 8 and 9).

- **Cooling towards a fermionic spin mixture**: When quantum degenerate spin mixtures of fermions are needed (see Chap. 6), early during evaporation in the dipole trap 50 % of the potassium atoms in the $|F, m_F\rangle = |9/2, -9/2\rangle$ state are transferred into the $|9/2, -7/2\rangle$ state. This is done by means of a Landau-Zener sweep at high magnetic fields, where the Zeeman levels are well separated, either at 165 or 220 G, which is below or above the Feshbach resonance (see Sect. 2.3.5). While the Landau-Zener sweep actually creates a coherent superposition of the two spin states, collisions and inhomogeneities of the external fields lead to rapid decoherence that generates a statistical mixture of the spin states [9, 10]. During evaporation all rubidium atoms are used as a coolant until the trap bottom is crossed (see Sect. 5.3). From this point on, the potassium mixture is cooled via further evaporation and internal thermalization to temperatures down to about $T/T_F = 0.1$. This cooling sequence typically takes about 6 s.[1]

The details of the experimental sequence after preparation of the quantum degenerate gases depend on the intended experiments and are described in each of the Chaps. 6–9 individually. Typically, the dipole trap is adjusted to provide the desired harmonic confinement and a homogeneous magnetic field is used to tune interatomic interactions via a Feshbach resonance.[2] Then the optical lattice potential is ramped up and further experimental manipulations on the atom clouds are carried out. Finally, absorption or phase-contrast images of the atomic distribution are recorded either after time-of-flight expansion or in-situ. In one experimental cycle images are usually taken along the y- or the z-axis or both (for directions see Fig. 5.2).

The following sections focus on the nonstandard techniques and peculiarities of the experimental setup:

---

[1] It is very convenient that the mixture is almost perfectly balanced after evaporation, even if the initially prepared spin mixture is not exactly 50–50 %. This is caused by the fact that both spin states experience the same dipole trap depth and tend to form an identical Fermi sea. If one spin state has the majority, the excess spins evaporate predominantly.

[2] Details on the creation, stabilization and calibration of magnetic fields for Feshbach resonances in our setup are provided in the Ph.D. thesis of Thorsten Best [3].

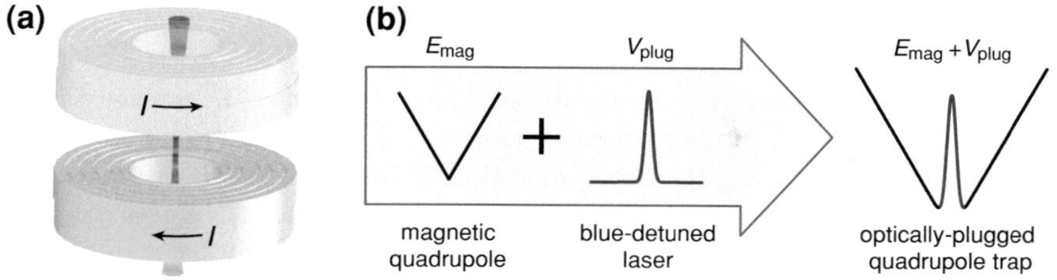

Fig. 5.3 Optically-plugged magnetic quadrupole trap. **a** A coil pair with counterpropagating currents I creates a magnetic quadrupole field that vanishes at the trap center and increases linearly in all directions. A blue-detuned laser is focussed along the symmetry axis. **b** Schematic of the combined magnetic and optical potentials

5.2 Optically-Plugged Magnetic Quadrupole Trap

Neutral atoms in a weak-field seeking Zeeman sublevel can be trapped in inhomogeneous magnetic fields. For such states the spatially dependent Zeeman energy

$$E_{\mathrm{mag}}(\mathbf{r}) = g_F m_F \mu_B |\mathbf{B}(\mathbf{r})| \tag{5.1}$$

constitutes a conservative potential, where g_F is the Landé g-factor for the hyperfine state F, m_F the projection of the angular momentum along the magnetic field axis and μ_B the Bohr magneton. The resulting force is directed towards the magnetic field minimum. Only if the atom moves slowly enough such that the magnetic moment adiabatically follows the direction of the magnetic field, it stays in the weak-field seeking sublevel and remains trapped [5].[3]

Considering a magnetic quadrupole potential (see Fig. 5.3), the adiabaticity condition is impossible to meet at the center, where the magnetic field vanishes and linearly increases in all directions. Here, atoms can undergo Majorana spin flips to untrapped states and get lost from the trap. While this loss mechanism is negligible for laser cooled atoms with a very low density, it becomes more and more severe, when phase space-density increases during evaporation. In our setup, the Majorana hole is plugged by focussing a blue-detuned laser beam along the symmetry axis of the quadrupole field (see Fig. 5.3). The plug laser strongly prevents the atoms from reaching the trap center. The resulting trapping potential has the shape of a cylindrically symmetric torus in the horizontal plane with slight distortions in the vertical direction due the influence of gravity.

For the magnetic part of this hybrid trapping geometry we use a coil pair, which creates a field gradient B' of about 300 G/cm at a current of 35 A. This tightly confining potential offers excellent conditions for rapid thermalization and efficient sympathetic cooling. The optical part is provided by a tapered amplifier seeded by a

[3] Adiabaticity specifically means that the rate of change of the magnetic field direction θ must be smaller than the Larmor frequency ω_{L}: $d\theta/dt < E_{\mathrm{mag}}/\hbar \equiv \omega_{\mathrm{L}}$.

grating stabilized diode laser that operates at 760 nm. The focussed beam has a waist of 20 μm and a power of about 450 mW at the position of the atoms. The detunings of 20 and 7 nm with respect to the D_2-resonances of ^{87}Rb and ^{40}K, respectively, are small enough to ensure a strongly repulsive force and large enough to suppress harmful off-resonant excitation. The strong repulsion actually helps suppressing inelastic photon scattering, because the atoms do not enter the high intensity part of the beam. Nevertheless, care must be taken to prevent resonant photon scattering as diode lasers typically have a broad background of amplified spontaneous emission spanning several tens of nanometers across the whole gain profile [11]. Therefore, we filter out any resonant light by placing two notch filters in the laser beam path.[4]

In view of our demands, the optically-plugged quadrupole trap has several advantages compared to the Ioffe-Pritchard type QUIC trap that was originally installed in the setup [2, 6]. First, the tight confinement allows for much faster RF evaporation (8 s compared to 20 s) and at the same time higher atom numbers are reached at 2 μK (about a factor of 1.5 more for both species). Second, the atom clouds are positioned exactly in the center of the glass cell on the symmetry axis of the quadrupole coil pair. The higher symmetry ensures very good optical access to the atoms and minimizes the risk of creating unwanted additional standing waves inside the glass cell with the dipole trap or lattice beams. Third, the quadrupole coil pair allows to create large homogeneous magnetic fields by switching to copropagating currents, which we use to address Feshbach resonances. The coils are not arranged in perfect Helmholtz configuration such that the resulting magnetic field has a non-vanishing curvature at the center. However, the curvature does not drag the atoms, because they sit on the symmetry axis. Additional technical details on the optically-plugged trap can be found in reference [3].

5.3 Crossed Dipole Trap

The optical dipole trap is formed by two elliptical laser beams that cross each other in the horizontal xy-plane at an angle of about 90° (see Fig. 5.4). The short axis of both beams is aligned along the direction of gravity, which generates a pancake-shaped trap. The trap laser[5] operates at 1,030 nm being far red-detuned both for rubidium and potassium. The polarizations of the beams are chosen orthogonally and a frequency difference of 160 MHz is imprinted using acousto-optical modulators in order to suppress and temporally average unwanted interference. The beam waists are about $w_{\mathrm{hor},x} = 140\,\mu\mathrm{m}$, $w_{\mathrm{vert},x} = 45\,\mu\mathrm{m}$ and $w_{\mathrm{hor},y} = 170\,\mu\mathrm{m}$, $w_{\mathrm{vert},y} = 70\,\mu\mathrm{m}$. The different x- and y-beam parameters are compensated by appropriately adjusting the power levels to yield a symmetric trapping potential in the horizontal plane (see Fig. 5.4a). The power in the beams approximately follows the ratio $P_y/P_x \approx 2$. The

[4] StopLine single-notch filter, NF03-785E-25, by *Semrock*. The notch is tuned to the appropriate wavelength range by changing the angle of incidence.

[5] VersaDisk, by *ELS Elektronik Laser System GmbH*. Ytterbium:YAG disc-laser at 1,030 nm, 18 W, single-mode operation with a linewidth of less than 5 MHz. Production and sales discontinued as of 2010.

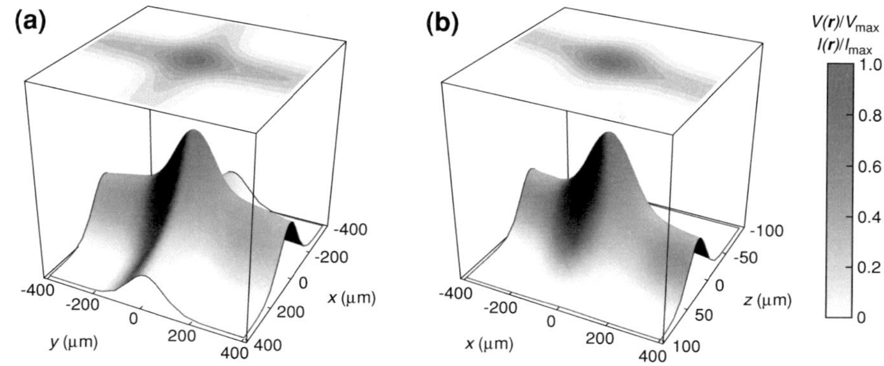

Fig. 5.4 Potential landscape of the crossed dipole trap in the horizontal xy-plane (**a**) and the vertical xz-plane (**b**). The potential is calculated with a numerical model that takes all experimental beam parameters and tilts into account (see main text). Gravity is neglected in the plots for clarity. The difference in the waists of the two beams is clearly visible in panel **a**. A symmetric trapping potential [*top view* of **a**] is achieved by choosing a power ratio of $P_y/P_x \approx 2$. Note that the scaling of the z-axis in panel **b** is stretched by a factor of four

elliptical beam shape results in a tight vertical confinement and ensures good overlap of ^{87}Rb and ^{40}K in the presence of gravity (see Sect. 5.3.1).

In order to avoid the occurrence of standing waves, it is crucial to prohibit direct backreflections of the beams from surfaces of the glass cell or any other optical elements.[6] In combination with geometrical constraints of the vacuum chamber, this requirement is the reason for the surprisingly different waists of the x- and y-beam. Furthermore, the beams are steered through the glass cell under angles of about 3° both in the vertical and horizontal direction. As an experimental check to make sure that the potential is free of standing waves, we provoke Raman-Nath diffraction. To this end, a Bose-Einstein condensate is released from the trap and a single dipole beam is pulsed on with the maximal available power for a few μs during time-of-flight expansion. Our setup is optimized to show no perceptible diffraction peaks.

The optical dipole trap is slightly more red-detuned for ^{40}K than for ^{87}Rb. Therefore, rubidium feels a slightly deeper potential. Neglecting the influence of gravity, the ratio of the potential depths is calculated via Eq. 2.55 (see Appendix A for atomic data) yielding $V_{\mathrm{Rb}}/V_{\mathrm{K}} \approx 1.15$. The corresponding ratio of trap frequencies is given by

$$\frac{\omega_{\mathrm{Rb}}}{\omega_{\mathrm{K}}} = \sqrt{\frac{V_{\mathrm{Rb}}}{V_{\mathrm{K}}}\frac{m_{\mathrm{K}}}{m_{\mathrm{Rb}}}} \approx 0.727. \tag{5.2}$$

[6] When the fraction of backreflected intensity is ρ^2, the modulation depth of the resulting standing wave is 4ρ (see Sect. 2.2.2). Even a tiny fraction of $\rho^2 = 0.01\,\%$ backreflected intensity would lead to a $4\,\%$ modulation!

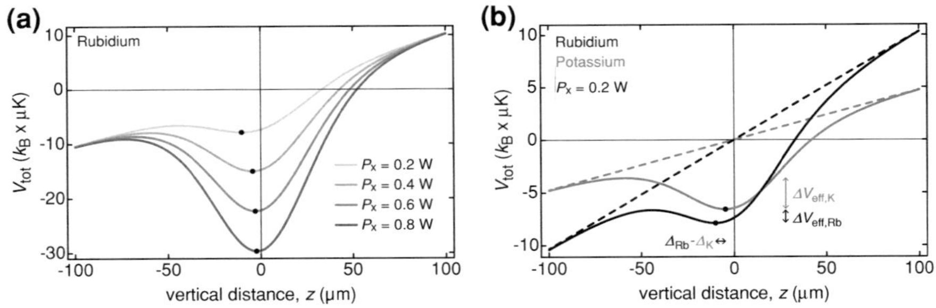

Fig. 5.5 Numerically calculated potential profiles $V_{tot}(0, 0, z)$ of the crossed dipole trap along the vertical axis taking into account gravity and the experimental beam parameters. **a** Potential for rubidium at various laser powers. Gravitational sag strongly pulls the trap minimum (*black dots*) downwards for lower laser powers. **b** Additionally, the effective trap depth for rubidium $\Delta V_{eff,Rb}$ is effectively lower than $\Delta V_{eff,K}$ when the laser power stays below a certain critical level. Therefore, dominantly rubidium is lost when the laser power is reduced during final evaporation. The differential sag is $\Delta_{Rb} - \Delta_K \approx 5\,\mu m$. *Dashed lines* indicate the pure gravitational potential for rubidium and potassium

5.3.1 Gravitational Sag

A good spatial overlap of rubidium and potassium is crucial for efficient thermalization during sympathetic evaporative cooling, as well as for the preparation and investigation of interacting Bose-Fermi mixtures in optical lattices (see Chaps. 8 and 9).

Gravity has a different influence on the potential landscape for ^{87}Rb and ^{40}K owing to their different masses. Assuming a harmonic approximation in the vertical direction of the trap ($x = y = 0$)

$$V_{tot}(0, 0, z) \approx \frac{1}{2} m\omega_z^2 z^2 + mgz, \tag{5.3}$$

the minimum shifts with respect to the unperturbed potential by $\Delta = -g/\omega_z^2$. The mass dependence is contained in the trap frequency ω_z (see Eq. 5.2). Both the absolute gravitational sag Δ and the differential sag between the two species $\Delta_{Rb} - \Delta_K$ can be reduced by choosing a large vertical trap frequency ω_z. This motivates the use of strongly elliptical beams for the dipole trap as specified above. The shift of the trap minimum in the experimental trap setup is illustrated in Fig. 5.5a based on an accurate numerical ab-initio model of the potential.

Gravitational sag is essential to the sympathetic cooling scheme. On the one hand, good spatial overlap is desired to ensure proper thermalization between rubidium and potassium. On the other hand, the rubidium is supposed to be the coolant for potassium: Rubidium should predominantly evaporate, while the potassium atoms should ideally stay in the trap and become colder. At large beam powers, when gravity does not play a role compared to the dipole potential, the trap is slightly deeper for

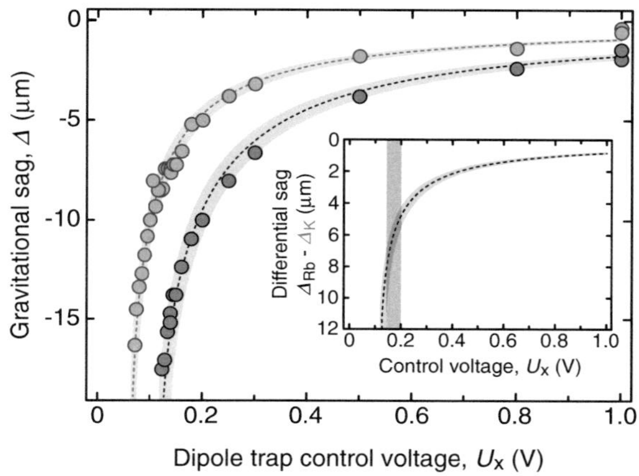

Fig. 5.6 Measurement of the gravitational sag Δ for ^{87}Rb (*dark gray dots*) and ^{40}K (*light gray dots*). The conversion of dipole trap control voltage to beam power is $P_x = U_x \cdot 0.95$ W/V. *Dashed lines* show the results of the numerical trap model, the *gray shaded* areas correspond to a 10 % error in the voltage-to-power conversion. The agreement between measurement and numerical model is excellent. The inset shows the calculated differential sag between ^{87}Rb and ^{40}K corresponding to the distance of the cloud centers in a noninteracting Bose-Fermi mixture. The *gray bar* indicates the operating range of the dipole trap in the experiments with Bose-Fermi mixtures. The typical differential sag is about 6 μm

^{87}Rb (see above) in contrast to the desired evaporation scenario. However, when the beam power is smaller than a certain critical value, gravity effectively reduces the rubidium trap depth below the potassium trap depth (see Fig. 5.5b). At the same time the differential sag remains small enough to ensure proper thermalization over a wide range of powers. For the parameters in Fig. 5.5b, a Bose-Einstein condensate and a degenerate Fermi gas with 3×10^5 atoms have a vertical extent of 9 and 25 μm, respectively (compare Fig. 2.3). The corresponding differential sag of about $\Delta_{\mathrm{Rb}} - \Delta_{\mathrm{K}} \approx 5\,\mu$m is so small that the two clouds still have maximal overlap.

Figure 5.6 shows a measurement of the gravitational sag both for rubidium and potassium together with the results from our numerical trap model. From extrapolation of the theoretical data we obtain a trap bottom of $U_x = 0.116$ V and $U_x = 0.061$ V for rubidium and potassium, respectively, which is in excellent agreement with independent measurements. The trap bottom of the crossed dipole trap is a very sensitive measure for the overlap of the two dipole beams. Therefore, the minimization of the trap bottom is a good strategy for the optimization of the beam overlap during realignment (see Sect. 5.4.2).

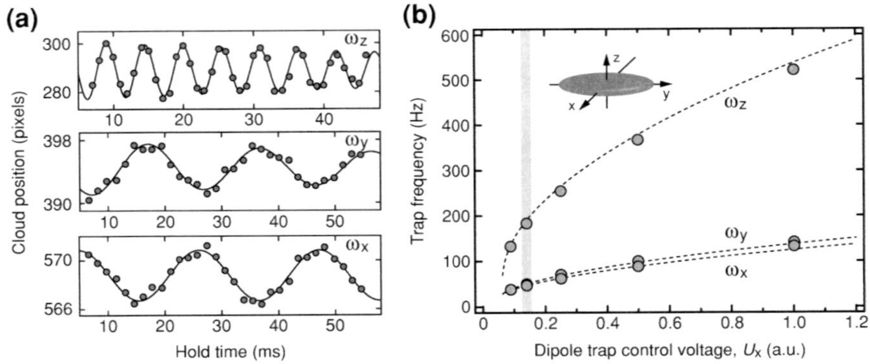

Fig. 5.7 Comprehensive measurement of trap frequencies using spin-polarized ^{40}K. **a** Oscillations along the individual axes at $U_x = 0.14$ V (corresponding to $P_x \approx 135$ mW). The momentum of the oscillating cloud is monitored after time-of-flight expansion. The corresponding oscillation frequencies are obtained from a fit (*solid lines*) yielding $\omega_z = 2\pi \times 185$ Hz (*top*), $\omega_y = 2\pi \times 51$ Hz (*middle*) and $\omega_y = 2\pi \times 47$ Hz (*bottom*). **b** Measurements of the trap frequencies at several power levels yield a full characterization of the trapping potential as a function of the beam power ($P_x \propto U_x$, see above). The *dashed lines* are the result of our numerical ab-initio model. The *gray-shaded* area highlights the data of panel **a**

5.3.2 Characterization

Typically, the extent of the quantum degenerate clouds is much smaller than the waists of the dipole beams. Therefore, the crossed dipole trap can be approximated by a harmonic oscillator in the vicinity of the central potential minimum. Comprehensive measurements of the corresponding harmonic trap frequencies are presented in Fig. 5.7. The frequencies are obtained by inducing dipole oscillations of a spin-polarized ^{40}K cloud (see Fig. 5.7a). After displacement along the axis of gravity and abrupt reduction of the beam power, oscillations of the cloud are initiated along the z-axis. These also couple to the horizontal plane as a result of slight anharmonicities. The atoms are observed after time-of-flight expansion, which reveals the oscillation of the center-of-mass momentum. Time traces with several oscillation cycles are recorded and fitted by a sinusoidal fit model (see Fig. 5.7a).

Figure 5.7b shows the trap frequencies as a function of the beam power. The data is excellently reproduced by the numerical ab-initio model of the trap. For higher beam powers the aspect ratio is approximately $\gamma = \omega_z/\omega_\perp \approx 4$, where $\omega_\perp \approx \omega_x, \omega_y$. For lower beam powers the trap frequencies ω_x and ω_y extrapolate to the origin following a square-root scaling as a function of the beam power, while the vertical trap frequency ω_z becomes quickly smaller when the trap bottom is approached (see previous section). For the data at lowest power we find an aspect ratio of $\gamma \approx 3.5$.

5.4 Three-Dimensional Blue-Detuned Optical Lattice

The blue-detuned optical lattice is one of the most important features of our experimental setup. The section starts with a brief discussion of the pros and cons of different lattice wavelengths for ^{87}Rb and ^{40}K. The reasons for us to choose blue-detuned wavelengths (738 and 755 nm) are discussed. The main part deals with technical requirements, newly developed alignment techniques and calibration methods for the lattice depths and the underlying anticonfinement. The contents of this section are essential to all experiments in this thesis.

5.4.1 What is the Best Lattice Wavelength?

At a given laser wavelength ^{87}Rb and ^{40}K generally feel a different dipole potential owing to the differences in their electronic level structure. Calculated via Eq. 2.55 and shown in Fig. 5.8, this species-dependence offers a variety of interesting choices for the lattice wavelength of the double-species system. Close to the atomic D_1 and D_2 resonances the optical potentials diverge. In between the D-line doublets there is a wavelength at which the lattice effectively vanishes, since the red detuning with respect to D_1 is compensated by the blue detuning with respect to D_2. This destructive interference is located at 768.97 and 790.03 nm for ^{40}K and ^{87}Rb, respectively. Those wavelengths offer the intriguing possibility to create a lattice for one species that is invisible to the other. Nevertheless, the relative proximity of these wavelengths to the atomic resonances, particularly in the first case, bears the problem of strong inelastic scattering of photons and potentially devastating heating [2]. As a rule of thumb, the rate for inelastic scattering at a given lattice depth should not exceed about 10 mHz in order to allow for realistic experimental hold times (about 100 ms).

Most experimental setups use a red-detuned optical lattice. Red detuning bears the advantage that the lattice beams themselves provide confinement and hold the atom clouds without the need of additional potentials (see Sect. 2.2.2). This convenience comes at the cost that the underlying confinement changes as a function of the beam intensity and cannot be independently varied.

In order to gain control of the underlying potential, we use an optical lattice that is blue-detuned for both ^{87}Rb and ^{40}K. The copropagating red-detuned optical dipole trap is used to compensate the anticonfinement of the lattice. This combination allows to realize flexible confining potentials in a broad parameter range, including a homogeneous lattice. Our initial choice for the wavelength has been 755.50 nm, because the lattice depths are equal for rubidium and potassium in units of the respective recoil energy $s_B/s_F = 1$ (see Fig. 5.8b). These conditions are particularly convenient for experiments with Bose-Fermi mixtures, because the Wannier functions of rubidium and potassium have identical shape and the tunneling strengths differ by a constant factor that is given by the mass ratio $J^B/J^F = m_F/m_B$. This wavelength is used in Chap. 8.

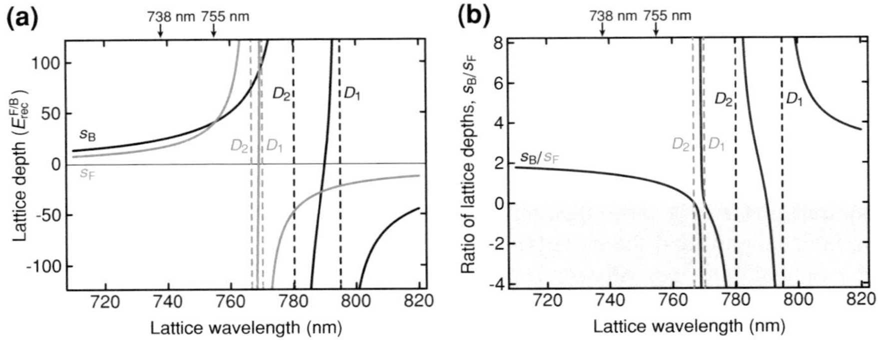

Fig. 5.8 Species-dependence of the optical lattice. **a** Lattice depth for ^{87}Rb (*black*, index B) and ^{40}K (*gray*, index F) as a function of wavelength. The depths are displayed in units of the respective recoil energy. A Gaussian standing wave with a total power of $P = 100\,\mathrm{mW}$ and a beam waist of $w_0 = 170\,\mu\mathrm{m}$ are assumed, corresponding to typical experimental parameters. **b** Ratio s_B/s_F of lattice depths for ^{87}Rb and ^{40}K as a function of the wavelength. Negative values indicate that the lattice has opposite detunings for the two species (*gray shading*). *Dashed lines* indicate the position of the respective D_1 and D_2 lines

In measurements with spin mixtures of fermionic ^{40}K enhanced losses of doubly occupied lattice sites and, even more severely, of weakly bound Feshbach molecules have been observed at 755.50 nm. This problem precluded dependable measurements with Fermi-Fermi mixtures and required a solution. The losses are induced by radiative collisions [12], in which the absorption of a lattice photon excites the molecule to an unbound state in the continuum. Phenomenologically, we found that this process is efficiently suppressed for larger detunings Δ following a $1/\Delta^2$ scaling (see Sect. 6.2.1 and appendix of [13]). Additionally, our investigations showed that the loss rate has local minima at certain intermediate wavelengths. Here, the Condon point of the transition to the continuum state hits a node of the spatial wavefunction of the Feshbach molecule [13]. Eventually, we have opted for a lattice wavelength at about 738 nm being a compromise between the suppression of radiative collisions and available laser power ensuring sufficient lattice depths both for ^{87}Rb and ^{40}K.[7] This wavelength is used in Chaps. 6, 7 and 9.

5.4.2 Adjustment, Tweaking and Calibration

The joint alignment of the red-detuned dipole trap and the blue-detuned lattice is a delicate task: It is crucial to precisely overlap the potential maxima of the anticonfining lattice beams with the potential minimum of the dipole trap. Already slight

---

[7] MBR 110, single-frequency Ti:Sapphire laser, about 3 W output power at the lattice wavelength (738 nm), pumped by Verdi V18, diode-pumped solid-state laser at 532 nm, 18 W output power, by *Coherent, Inc.*

deviations result in an anharmonic combined potential. Large deviations can lead to potential landscapes with several local minima and maxima, for example a Mexican hat potential, precluding a reliable comparison of experimental results to theoretical models.

The technical demand yields a high gain: The combination of the two optical potentials allows to tune the harmonic confinement independent from the lattice depths. This important feature opens the door to new classes of equilibrium and nonequilibrium experiments with ultracold quantum gases, as demonstrated in Chaps. 6, 7 and 9. The three orthogonal lattice axes are formed by round Gaussian laser beams, which have a waist of 170(5) μm at the position of the atoms.

5.4.2.1 Technical Requirements

Several independent laser beams must maintain an accurate alignment with respect to each other. Therefore, all parts of the beam paths should have a high passive stability. A number of measures are taken to achieve this goal: Concerning the optomechanics, all components are mounted on monolithic pedestals, which are adapted to a beam height of 6.8 cm above the surface of the optical table. When beam paths leave the standard height, very rigid second layers or lead-filled boxes are used as mounts for the optics. These measures mainly help to suppress mechanical oscillations.

In order to achieve a good long-term stability of the alignment, the temperature at the experiment table should be stable. Even without active temperature regulation, this goal is approached by removing all heat sources from the vicinity of the experimental setup, apart from those that can intrinsically not be removed (magnetic coils, beam dumps, cameras, etc.). Additionally, the experiment table is surrounded by a protective curtain that minimizes air circulation across the setup. The curtain is rarely opened during operation of the experiments. Remote-controlled motorized mirror mounts (see below) allow to keep it closed even during alignment procedures.

Because the optical lattice is created by a coherent single-mode laser, it is a persistent challenge to avoid false reflections that can lead to the formation of unwanted additional standing waves. For example, the direct backreflection of only 0.01 % of the incoming intensity leads to a standing wave with 4 % modulation depth. However, the glass cell (see Fig. 5.2) is not antireflection coated from the inside and reflections of about 4 % must be expected. Therefore, the formation of unwanted standing waves can only be safely avoided by making it geometrically impossible. Similar to the dipole trap (see Sect. 5.3) the lattice beams are steered through the glass cell under angles of at least 3°. We have further identified several optical elements in the beam paths, such as beam cubes, standard mirrors and dichroic optics, that produce false reflections, for example from their backside. These problems are addressed either by avoiding the use of the respective elements (for example, dichroic optics), minimizing their use whenever possible or blocking the reflections when other solutions are impractical.

All beam paths that need precise alignment (lattice, dipole trap and blue-plug) are equipped with a piezo-actuated mirror mount. These mounts are placed at strategically good positions to ensure a high degree of sensitivity. They are mechanically sturdy and feature reliable actuation.[8]

5.4.2.2 Full Alignment Procedure

The accuracy of the beam alignment is of paramount importance to the experiments of this thesis. In the following, we summarize the individual steps:

- **Optically-plugged quadrupole trap**: The first step is to move the plug beam into the trapping region of laser-cooled rubidium atoms inside the quadrupole trap. To this end, an absorption image of the atom cloud is recorded and its center position is fitted. The plug beam is directly viewed on the CCD camera and stirred to the fitted position. Owing to chromatic shifts this is only a rough alignment, but usually the beam hits the cloud edge. Then, the repulsive force of the blue-detuned light can be identified as a local minimum in the density of the atom cloud. When the beam is stirred towards the cloud center, the suppression of Majorana losses takes effect and leads to larger atom numbers and lower temperatures. The beam position is now accurately scanned with the piezo mirror. Both a maximum in atom number and minimum in temperature reveal the optimal plug position. After optimization with rubidium atoms, the alignment automatically works for the sympathetically cooled potassium.
- **Optical dipole trap**: The procedure starts by aligning one of the dipole trap beams about 50 μm below the center of the quadrupole trap. The beam position is visualized by loading atoms into this single beam trap. This avoids the problem of chromatic shifts that arise when the beam is directly viewed on the camera and compared to the position of the atoms on an absorption image. Then, the second horizontal beam is overlapped with the first one and the trap depth is maximized by minimizing the trap bottom (see Sect. 5.3.1). As soon as the beams are roughly aligned, it is possible to evaporatively cool rubidium atoms towards condensation. The condensate is used to create an atom laser by ramping the dipole trap intensity linearly down across the trap bottom. The higher the trap bottom, the earlier atoms are released and the longer the atom laser. Therefore, the length of the atom laser is a sensitive indicator for the trap bottom. Moving the vertical position of the second beam with the piezo mirror, a minimum in the length of the atom laser is clearly observed and indicates the optimal beam overlap.
- **Optical lattice**: The three axes of the optical lattice are aligned individually. In principle, the procedure is identical for each of them. At first, the mirror for retroreflection is blocked to avoid the formation of a standing wave. The incoming light simply forms a blue-detuned Gaussian laser beam. In order to align the potential maximum with the potential minimum of the dipole trap, a rubidium Bose-Einstein condensate is held in a rather shallow dipole trap. At high power the

[8] Picomotor™ Center Mounts (Model 8807) and Picomotor™ Pint-Sized Center Mounts (Model 8885), by *New Focus*.

blue beam can push the BEC out of its equilibrium position. Using the motorized mirror, the vertical positioning is optimized by maximizing the displacement of the BEC with respect to the equilibrium position. At maximal displacement the BEC is exposed to the highest light intensity corresponding to the vertical center of the blue-detuned Gaussian beam. The horizontal positioning is optimized by finding the position, where the atoms are not displaced from equilibrium to either direction, because the incoming beam hits the cloud head-on. If the intensity of the blue beam is further increased, the BEC splits into two parts. By symmetrizing the atom number in the two halves, the alignment can be further improved. The precise positioning of the incoming lattice beam is crucial, because it also forms the reference for the alignment of the retroreflected beam.

After unblocking the retro-mirror, the BEC is released from a pure dipole trap and in the initial time-of-flight phase exposed to the (possibly poorly modulated) standing wave. Using a short pulse of few tens of μs Raman-Nath diffraction is induced. The power of the lattice beam is adjusted to a level, at which the first-order diffraction peaks are only faintly visible. In this regime, the number of diffracted atoms is a monotonic function of the modulation depth of the standing wave. The position of the retro-mirror is optimized by maximizing the number of diffracted atoms. If second-order diffraction peaks start to appear in the course of optimization, the laser power must be reduced until merely first-order peaks remain. The procedure is iterated until the maximal modulation depth is reached corresponding to optimal overlap of the incoming and the retroreflected beam. Obviously, care must be taken to match the waists of the incoming and the retroreflected beam at the position of the atoms. Furthermore, loss of laser light on the retroreflection beam path should be minimized to ensure a maximally modulated lattice potential.

5.4.2.3 Lattice Calibration

The depth of each lattice axis is calibrated by means of frequency-modulation spectroscopy that probes the energy gap between the first and second lattice band. To this end, a quantum degenerate spin-polarized ^{40}K cloud is prepared in a very shallow dipole trap. One lattice axis is adiabatically ramped up to a power level, at which an energy gap between 70 and 100 kHz between the two lowest bands is expected corresponding to a lattice depth between 20 and 30 $E_{\text{rec}}^{\text{F}}$ (compare Sect. 2.2.3 and Fig. 2.8).

The frequency of the lattice laser ω is sinusoidally modulated by an acousto-optical modulator for about 100 ms with an amplitude $\Delta\omega$ of typically several MHz. This modulation results in a periodic displacement of the lattice, the spatial amplitude of which is determined by the optical path length between the atom cloud and the retro-mirror. The retro-mirror sets the boundary condition to the standing wave and determines the location of the individual lattice wells depending on the frequency of the lattice laser. For our conditions a spatial shaking amplitude Δx_0 of 0.1 to 1.0 % of a lattice wavelength λ can be expected depending on the exact value of $\Delta\omega$ (see inset of Fig. 5.9). The oscillation of a lattice site around its equilibrium position can be expressed by $\Delta x(t) = \Delta x_0 \sin(2\pi f_{\text{FM}} t)$, where f_{FM} is the modulation frequency.

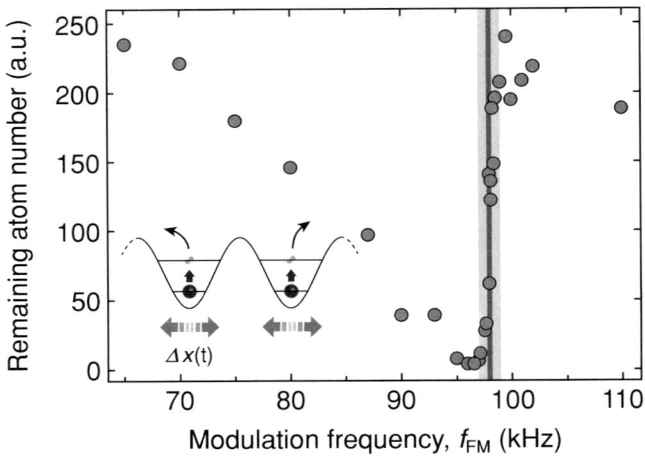

Fig. 5.9 Lattice calibration via frequency-modulation spectroscopy. The data shows a resonance between the first and second band at 98.2 ± 1.0 kHz, which corresponds to a lattice depths of $34.9 \pm 0.6\, E_{\mathrm{rec}}^{\mathrm{F}}$. The *light-gray* shaded area indicates the experimental uncertainty of the resonance position. The *dark-gray* shaded bar represents the width of the resonance as expected from a band structure calculation. The inset illustrates frequency modulation spectroscopy (see main text)

The shaking couples the first and second lattice band. This can be seen in a harmonic oscillator model: The physics at a single lattice site is captured by the Hamiltonian $\hat{H}_0 = \hat{p}^2/(2m_{\mathrm{F}}) + m_{\mathrm{F}}\omega_{\mathrm{lat}}^2\hat{x}^2/2$, where ω_{lat} is the harmonic on-site trap frequency and m_{F} the mass of a potassium atom. Incorporating the time-dependence of the equilibrium position, we obtain

$$\frac{\hat{p}}{2m_{\mathrm{F}}} + \frac{m_{\mathrm{F}}\omega_{\mathrm{lat}}^2}{2}[\hat{x} + \Delta x(t)]^2 \approx \hat{H}_0 + m_{\mathrm{F}}\omega_{\mathrm{lat}}^2\hat{x}\Delta x(t). \tag{5.4}$$

The term proportional to Δx^2 is neglected, because it is small for our conditions. The perturbation $\hat{H}_1 = m_{\mathrm{F}}\omega_{\mathrm{lat}}^2\hat{x}\Delta x_0 \sin(2\pi f_{\mathrm{FM}}t)$ is linear in $\hat{x}$ and therefore couples states of different spatial symmetry. Population is transferred from the first to the second lattice band when the frequency f_{FM} is in resonance with the band gap.[9]

In order to obtain a strong signature in the atom number, the depth of the dipole trap is chosen weak enough that second band population does not stay trapped (compare Fig. 3.7). Resonant excitation is indicated by strong atom loss (see Fig. 5.9). The upper edge of the triangle-shaped feature is given by

$$f_{\mathrm{FM}}^{\max} = \left(E_{\mathrm{q}=0}^{(2)} - E_{\mathrm{q}=0}^{(1)}\right)/h \tag{5.5}$$

[9] This description is analogous to the semi-classical theory of atom-light interactions [14–16].

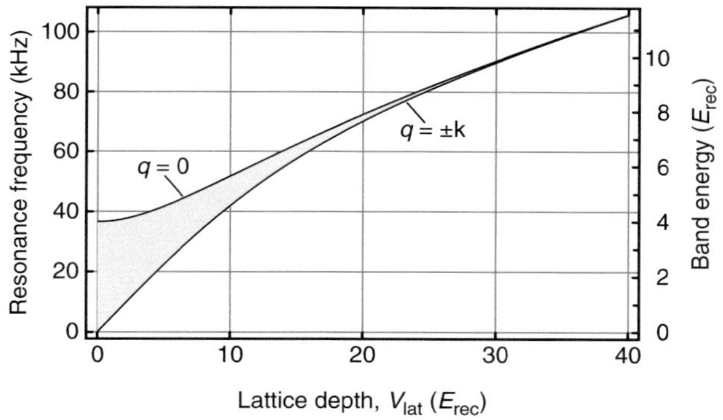

Fig. 5.10 Energy gap between the first and second lattice band $E_q^{(2)} - E_q^{(1)}$ as a function of lattice depth. The scale on the *left* side shows the resonance frequencies for ^{40}K in a 738 nm lattice, while the scale on the *right* side is valid in general. Boundaries of the resonance frequencies are set by the band gaps for $q = 0$ and $q = \pm k$ (see Sect. 2.2.3 and Fig. 2.8)

representing the maximal energy gap between the first and second band (see Sect. 2.2.3). As indicated by the data of Fig. 5.10, the width of the resonance is theoretically expected to be a lot narrower than the observed feature. Owing to the shallow underlying trapping potential, the cloud of fermions has a large extent and samples the inhomogeneous distribution of lattice depths that arises from the Gaussian shape of the laser beams (illustrated in Fig. 2.7). Nevertheless, the sharp upper edge reveals the maximal lattice depth, which is found in the center of the trap, with an accuracy on the level of few per cent. This kind of local resolution distinguishes our method from others that yield an averaged lattice depth sampled by the area of the whole atom cloud, such as in Raman-Nath diffraction [2].

It is advantageous to use spin-polarized fermions for the lattice calibration, because s-wave collisions are suppressed by Pauli's principle and interaction-induced shifts [17] of the calibration can be excluded. The modulation spectra can be safely compared to a single-particle band structure calculation (see Sect. 2.2.3). Based on the measured calibration for ^{40}K we calculate the respective lattice depth for ^{87}Rb using Eq. 2.55 (for spectroscopic data see Appendix A). For a lattice wavelength of 738 nm the conversion of the dimensionless lattice depth is $s_B/s_F = 1.500(15)$ (see also Sect. 5.4.1).

5.4.2.4 Characterization of the Anticonfinement

Accurate knowledge of the anticonfining trap frequencies as a function of the lattice depth is crucial for precise tailoring of the potential landscape. The theoretical background on the anticonfinement in a blue-detuned optical lattice is discussed in Sect. 2.2.2. Substantial effort is made to characterize the corresponding frequencies experimentally.

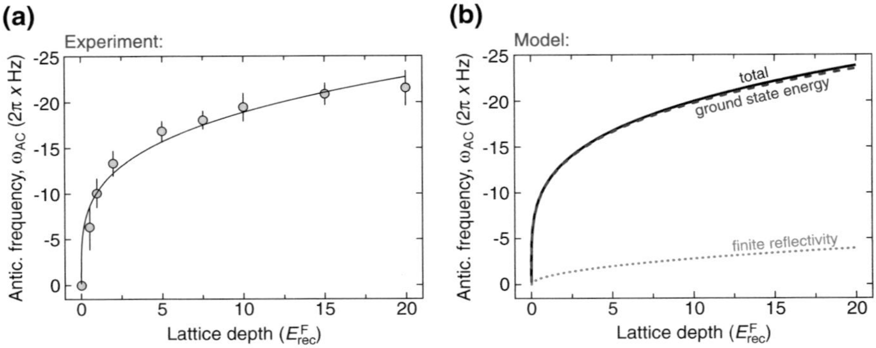

Fig. 5.11 Anticonfinement of the blue-detuned optical lattice in practice. **a** Measured anticonfining trap frequencies for the lattice axis along z-direction. A power law fit ($\omega_{\mathrm{AC}} = as^b$) yields the parameters $a = -10.2(7)$ Hz and $b = 0.27(3)$ (*black line*). The exponent b is in good agreement with the dominant $s^{1/4}$ scaling originating from the varying on-site ground state energy (see Eq. 2.67). **b** Anticonfining frequencies extracted from our numerical ab-initio model. The only free parameter is the reflection coefficient ρ of the retroreflection beam path (see Sect. 2.2.2). Here, $\rho = 0.8$ is chosen to illustrate the effect, while the actual experimental value is rather $\rho \approx 0.9$. The contributions from varying ground state energy (*dashed line*) and finite reflectivity (*dotted line*) are added in quadrature to yield the total anticonfinement (*black line*)

The measurement procedure is similar to the characterization of the bare trapping potential (see Sect. 5.3.2). A spin-polarized ^{40}K cloud is prepared in the dipole trap at a certain depth. One of the three lattice axes is adiabatically ramped up and the atom cloud is displaced (opposite to the direction of gravity) by tightening the dipole trap. When the cloud is released, it oscillates perpendicular to the propagation direction of the lattice. The measured oscillation frequency ω_{osc} contains both the confinement of the dipole trap ω_{dip} and the anticonfinement of the optical lattice ω_{AC}. ω_{dip} is known from the calibration without the lattice (see Sect. 5.3.2). Therefore the anticonfining trap frequency can be extracted via $\omega_{\mathrm{AC}} = -\sqrt{\omega^2 - \omega_{\mathrm{dip}}^2}$. Such measurements are performed for all three lattice axes at several lattice depths yielding a comprehensive characterization.

An exemplary measurement of the anticonfining trap frequency ω_{AC} as a function of the lattice depth is shown in Fig. 5.11a. The fit of a power law model confirms that ω_{AC} approximately scales as $s^{1/4}$. This complies with our theoretical analysis of Sect. 2.2.2, where the local variation of the ground state energy is identified as the main contribution to the anticonfinement. The second contribution originating from the finite reflectivity in the retro-beam path plays a minor role. We have also extracted the anticonfining frequencies from the numerical ab-initio model of the trapping potentials and find a remarkable agreement with the measurements (see Fig. 5.11b). This confirms the accuracy of the model, which is also used to determine the anticonfining frequencies for ^{87}Rb on a theoretical basis.

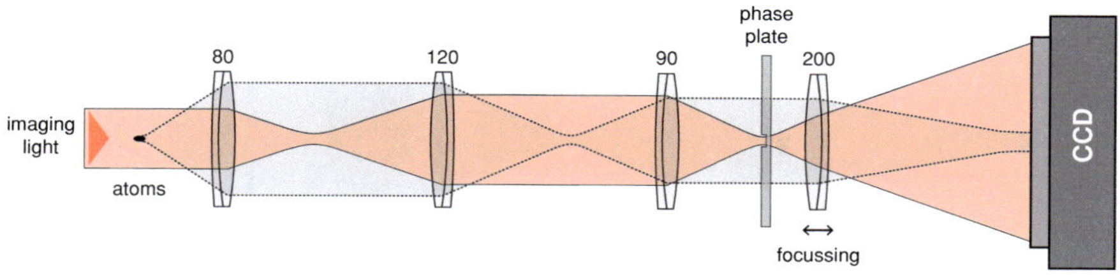

Fig. 5.12 Schematic of the main imaging system for absorption and phase-contrast imaging. The numbers at the lenses denote the respective focal length f. The overall magnification is $m = 3.3$. In phase-contrast imaging the unscattered light passes through the center of the phase plate imprinting a phase shift relative to the coherently scattered light. For absorption imaging the phase plate is laterally moved out of the beam path

5.5 Imaging System

The main imaging system of the experimental setup points along the direction of gravity (z-axis) and allows us to observe atom clouds either in-situ or after time-of-flight expansion. The lens system consists of two telescopes as shown Fig. 5.12: The first one has a magnification of $m_1 = 3/2$ and the second one $m_2 = 20/9$, which gives a total magnification of $m = m_1 m_2 = 3.3$. Due to space limitations the first lens ($f = 80$ mm) has a diameter of $d = 1$ inch, which results in a diffraction-limited resolution of about 6 μm.[10] This value has been experimentally confirmed in absorption images of Bose-Einstein condensates with extremely low atom numbers, where the smallest observable cloud diameters have indeed been 6 μm, demonstrating that the imaging system is diffraction limited. The axial position of the last lens ($f = 200$ mm) is adjustable and allows to adapt the focus to different expansion times. A front-illuminated CCD camera[11] is used to take either absorption or phase-contrast images.

5.5.1 Phase-Contrast Imaging

An introduction to the theory of phase contrast imaging is given in Sect. 4.1.3. The key ingredient of this powerful technique is a phase plate that is placed in a Fourier plane of the beam path to shift the phase of the unscattered light with respect to the scattered light. We use a retarding phase plate (see Fig. 5.13a) that is inserted at the focus of the second telescope. Here the unscattered light traverses the plate through a thinner spot in the center corresponding to a shorter optical path length. To facilitate

[10] The theoretical diffraction-limited resolution is defined as the diameter of the smallest object, whose Airy disc fills the aperture of the first lens, that is $1.22 \cdot 2\lambda f/d$ based on the Fraunhofer approximation [18, 19].

[11] iXon, DV885, by *Andor*, chip size of 1004×1002 pixels with a pixel size of $8\,\mu\text{m} \times 8\,\mu\text{m}$.

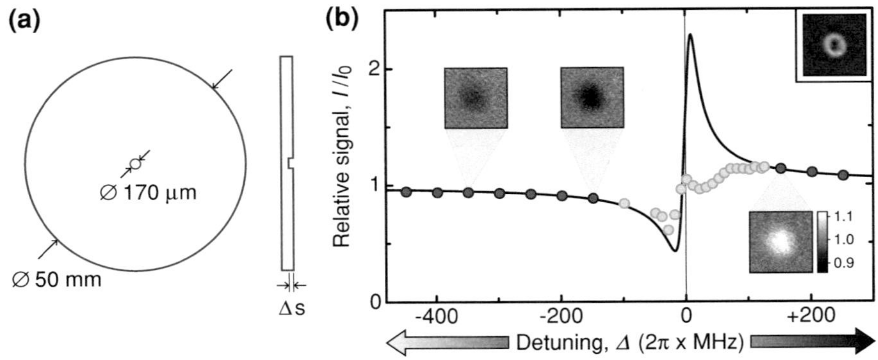

Fig. 5.13 Phase-contrast imaging with a retarding phase plate. **a** The retarding phase plate has a central dimple with a diameter of 170 μm and a depth of $\Delta s = 430$ nm. It induces a phase shift of approximately $\alpha \approx -\pi/2$ for both ^{40}K and ^{87}Rb. **b** Phase-contrast signal versus detuning for ^{40}K atoms. The data shows the relative signal I/I_0 at the center of the cloud. *Dark points* are fitted by the model $I/I_0 = t(n, \Delta)^2 + 2 - 2\cos(\alpha) - 2t(n_{\text{col}}, \Delta)\sqrt{2 - 2\cos(\alpha)}\cos[\phi(n_{\text{col}}, \Delta) + \alpha/2]$ (compare Eq. 4.7) using the column density n_{col} and the phase shift α as fit parameters. *Gray points* are excluded due to proximity to the atomic resonance. The fit yields $n_{\text{col}} = 1.125 \times 10^{13}$ m^{-2} and $\alpha = -1.568(3) \approx -\pi/2$ (*black solid line*). A low atom number is chosen in order to stay in the linear regime $\phi \propto n_{\text{col}}$. Otherwise phase wrapping can occur for small detunings (*upper right* inset, ^{87}Rb Bose-Einstein condensate imaged at $\Delta = +2\pi \times 60$ MHz)

the alignment and ensure reproducibility the phase plate is mounted on a precision xyz-translation stage.

5.5.1.1 Phase Plate and Alignment

Figure 5.13a shows a schematic of the phase plate, for which a fused-silica wafer with broadband antireflection coating forms the basis.[12] The central dimple has been created at the Leibniz-Institute of Surface Modification (Leipzig, Germany) using an ion-etching technique that allows for a surface roughness of better than $\lambda/10$ in the etched region, where λ is the wavelength of the imaging light. For a material with a refractive index n_{ref}, a retardation by $\alpha = -\pi/2$ is achieved when the dimple has a depth of $\Delta s = \lambda/(4n_{\text{ref}} - 4)$. The imaging wavelength is assumed to be $\lambda = 774$ nm, right in between the resonances of ^{40}K and ^{87}Rb. Together with $n_{\text{ref}} = 1.45$ this leads to a depth $\Delta s \approx 430$ nm that creates an approximate phase shift of $\alpha \approx -\pi/2$ for both species.

The choice of the dimple size is a subtle issue. On the one hand, the dimple should not be too small to ensure that all unscattered light safely propagates through the center. On the other hand, the larger the extent of the dimple, the larger the amount of scattered light that is subjected to an unwanted phase shift. This point is a notorious

[12] Fused-silica wafer (Coring HPFS), diameter 2 inch, thickness 1 mm, by *Plan Optik AG*.

problem of the phase-contrast method and the reason for imaging artifacts such as halos or shade-off effects [20–22]. While an exact understanding of the specific artifacts is involved and depends on the details of the imaging system [20], a smaller diameter of the dimple generally helps to suppress them. For the measurements in Chap. 6 we have used a diameter of 300 μm. However, the experimental images show signs of a halo effect, which is accounted for by adding a funnel to the fitting function that is used to extract the cloud radius (see Eq. 6.3). This complication has been reduced by choosing a smaller dimple diameter of 170 μm in later measurements [23].

For alignment, first the axial position of the imaging beam focus is determined using a beam profiling camera and the phase plate is accurately positioned at this axial location. Second, the transverse position is tweaked to thread the imaging beam through the dimple. To this end, we monitor the imaging beam with a beam profiling camera at a distance that is equivalent to the position of the CCD camera. The phase plate is now moved laterally using the xy-directions of the translation stage. When the edge of the dimple crosses the imaging beam, a diffraction pattern appears on the camera. Moving the phase plate further, the pattern will appear again, if the dimple has moved across the focus. This two-fold occurrence of diffraction, while moving the phase plate through the imaging beam, allows to uniquely determine the position of the dimple relative to the beam focus. Once the correct position is found, the phase plate can reproducibly be moved with the translation stage and a rapid switching between absorption and phase-contrast imaging is possible.

5.5.1.2 Phase-Contrast Signal

A characterization of the relative signal of phase-contrast imaging is shown in Fig. 5.13b. Images of spin-polarized ^{40}K clouds are recorded for detunings ranging from the far red to the blue. The insets show that a reduced intensity (black clouds) is detected for red-detuning, while blue-detuning leads to an enhanced intensity (white clouds) as expected for a retarding phase plate. A low number of atoms (about 10^5) is chosen to avoid nonlinearities. The relative signal I/I_0 (see Sect. 4.1.3) at the cloud center is plotted as a function of the detuning and fitted by an ab-initio model corresponding to Eq. 4.7. Data points in the range $-2\pi \times 150\,\text{MHz} < \Delta < +2\pi \times 150\,\text{MHz}$ are excluded, because the clouds suffer from strong incoherent scattering and atom loss. The extracted retardation of $\alpha = -1.568(3)$ is in excellent agreement with the intended value of $\alpha = -\pi/2$.

Instead of shifting the frequency of the imaging beam, the detuning can also be achieved by shifting the atomic levels and leaving the imaging frequency unchanged. This approach is employed in Chap. 6, where Zeeman shifts in high magnetic fields effectively lead to blue detuning in the range $\Delta = 2\pi \times 280$ to 400 MHz (see Appendix B).

References

1. H. Metcalf, P. van der Straten, *Laser Cooling and Trapping* (Springer, Berlin, 1999)
2. T. Rom, *Bosonische und fermionische Quantengase in dreidimensionalen optischen Gittern*. Ph.D. Thesis, Ludwig-Maximilians-Universität München, 2009
3. T. Best, *Interacting Bose-Fermi Mixtures in Optical Lattices*. Ph.D. Thesis, Universität Mainz, 2011
4. W.D. Phillips, Laser Cooling and Trapping of Neutral Atoms, in *Laser Manipulation of Atom and Ions, Processdings of the International School of Physics "Enrico Fermi"Course CXVIII*, ed. by W.D. Phillips, E. Arimondo, F. Strumia (North-Holland, Amsterdam, 1992)
5. W. Ketterle, D. Durfee, D.M. Stamper-Kurn, in *Making, Probing and Understanding Bose-Einstein Condensates*, ed. by S. Stringari, M. Inguscio, C.E. Wieman. Bose-Einstein Condensation in Atomic Gases, (IOS Press, Amsterdam, 1999) pp. 67–176
6. M. Greiner, *Ultracold Quantum Gases in Three-Dimensional Optical Lattice Potentials*. Ph.D. Thesis, LMU München, 2003
7. K.B. Davis, *Evaporative Cooling of Sodium Atoms*. Ph.D. Thesis, Massachusetts Institute of Technology, 1995
8. D.S. Naik, C. Raman, Optically plugged quadrupole trap for Bose-Einstein condensates. Phys. Rev. A **71**, 033617 (2005)
9. M.W. Zwierlein, *High-Temperature Superfluidity in an Ultracold Fermi Gas*. Ph.D. Thesis, Massachusetts Institute of Technology, 2006
10. S. Gupta, Z. Hadzibabic, M.W. Zwierlein, C.A. Stan, K. Dieckmann, C.H. Schunck, E.G.M. van Kempen, B.J. Verhaar, W. Ketterle, Radio-frequency spectroscopy of ultracold fermions. Science **300**, 1723 (2003)
11. Eagleyard Photonics Datasheet, Ridge waveguide laser with AR-coating, EYP-RWE-0790-04000-0750-SOT01-0000
12. V. Vuletić, C. Chin, A.J. Kerman, S. Chu, Suppression of atomic radiative collisions by tuning the ground state scattering length. Phys. Rev. Lett. **83**, 943 (1999)
13. U. Schneider, *Interacting Fermionic Atoms in Optical Lattices—A Quantum Simulator for Condensed Matter Physics*. Ph.D. Thesis, Johannes Gutenberg-Universität Mainz, 2010
14. L. Allen, J.H. Eberly, *Optical Resonance and Two-Level Atoms* (Dover Publications, Inc., New York, 1987)
15. C. Cohen-Tannoudji, J. Dupont-Roc, G. Grynberg, *Atom-Photon Interactions* (Wiley, New York, 1998)
16. W. Ketterle, *Physics 8.421* (*Atomic Physics I*). Lecture Notes, (2006)
17. P. Ernst, S. Götze, J. Krauser, K. Pyka, D.-S. Lühmann, D. Pfannkuche, K. Sengstock, Probing superfluids in optical lattices by momentum-resolved Bragg spectroscopy. Nat. Phys. **6**, 56 (2009)
18. B. Saleh, M. Teich, *Fundamentals of Photonics* (Wiley-Interscience, New York, 1991)
19. E. Hecht, *Optics*, 4th edn. (Addison Wesley, San Francisco, 2002)
20. T. Otaki, Artifact halo reduction in phase contrast microscopy using apodization. Opt. Rev. **7**, 119 (2000)
21. R. Liang, J.K. Erwin, M. Mansuripur, Variation on Zernike's phase-contrast microscope. Appl. Opt. **39**, 2152 (2000)
22. C. Maurer, A. Jesacher, S. Bernet, M. Ritsch-Marte, Phase contrast microscopy with full numerical aperture illumination. Opt. Express **16**, 19821 (2008)
23. U. Schneider, L. Hackermüller, J.P. Ronzheimer, S. Will, S. Braun, T. Best, I. Bloch, E. Demler, S. Mandt, D. Rasch, A. Rosch, Fermionic transport and out-of-equilibrium dynamics in a homogeneous Hubbard model with ultracold atoms. Nat. Phys. **8**, 213 (2012)

Chapter 6
Interacting Fermions in Optical Lattice Potentials

In this chapter I report on the experimental realization of the Fermi-Hubbard model. To this end, a repulsively interacting balanced spin mixture of ultracold ^{40}K atoms is loaded into a three-dimensional optical lattice potential. The emerging quantum phases are probed by measuring the global compressibility of the quantum gas. The compressibility becomes experimentally accessible using in-situ phase-contrast imaging and independent control of external confinement and lattice depth. Comparing the experimental data to calculations based on dynamical mean field theory (DMFT), it is demonstrated that the system evolves from a compressible metallic into an incompressible band insulating state when the external confinement is increased. For strong repulsion evidence for an incompressible Mott insulating phase is found, underlining the great potential of using ultracold fermionic atoms to simulate model systems of condensed-matter theory. Additionally, measurements on attractively interacting spin mixtures are presented that demonstrate the fundamental impact of finite entropy on many-body quantum phases. The chapter concludes with a discussion of strategies to achieve lower entropy in lattice-based quantum gases. Novel routes towards low entropy are intensely investigated, because they will open the field to studies of quantum magnetism. The presentation given here focusses on the experimental peculiarities of the project. A detailed discussion including further theoretical aspects can also be found in the PhD thesis of Ulrich Schneider [1].

6.1 Fermi-Hubbard Model in a 3D Optical Lattice

The general single-band Hubbard model is derived in Chap. 3. Using Eq. 3.28 the Hubbard Hamiltonian [2] for a fermionic spin mixture in a three-dimensional optical lattice with simple cubic symmetry and underlying harmonic potential can be written as

DOI: 10.1007/978-3-642-33633-1_6,

$$\hat{H} = -J \sum_{\sigma} \sum_{\langle \mathbf{i},\mathbf{j} \rangle} \hat{c}^{\dagger}_{\mathbf{i}\sigma} \hat{c}_{\mathbf{j}\sigma} + U \sum_{\mathbf{i}} \hat{n}_{\mathbf{i}\downarrow} \hat{n}_{\mathbf{i}\uparrow} + V_{\mathrm{t}} \sum_{\mathbf{i}} \left(i_x^2 + i_y^2 + \gamma^2 i_z^2 \right) \left(\hat{n}_{\mathbf{i}\downarrow} + \hat{n}_{\mathbf{i}\uparrow} \right). \tag{6.1}$$

Here, the indices $\mathbf{i} = (i_x, i_y, i_z)$ label the lattice sites in three dimensions, $\langle \mathbf{i}, \mathbf{j} \rangle$ denotes nearest-neighbor lattice sites, $\sigma \in \{\downarrow, \uparrow\}$ the two spin states, J the tunneling matrix element, and U the on-site interaction energy. The operators $\hat{c}_{\mathbf{i}\sigma}$ are the annihilation operators of a fermion in spin state σ on the $\mathbf{i}$th lattice site, $\hat{c}^{\dagger}_{\mathbf{i}\sigma}$ are the corresponding creation operators, and $\hat{n}_{\mathbf{i}\sigma}$ measures the number of spin σ fermions on site $\mathbf{i}$.

The presence of the harmonic confining potential is represented by the third term. The parameter $V_{\mathrm{t}} = m\omega_{\perp}^2 a^2/2$ denotes the energy offset between two adjacent lattice sites in the trap center, where m is the mass of a single atom, $a = \lambda/2$ the lattice constant, λ is the wavelength of the lattice laser, and $\omega_{\perp} \equiv \omega_x = \omega_y$ the horizontal trap frequency. The experimental trapping potential is pancake-shaped. Therefore the vertical trap frequency is parametrized using the aspect ratio $\gamma = \omega_z/\omega_{\perp}$. Owing to the single-band restriction and Pauli's principle, each lattice site can be occupied by one atom per spin state at most.

6.1.1 *Metallic, Mott and Band Insulating Phases*

Besides the fundamental influence of finite temperature [3], the phases of the Hubbard model are determined by the interplay of the three energy scales in Hamiltonian 6.1: The kinetic energy, whose scale is set by the bandwidth $12J$ of the lattice (See Sect. 2.2.3), the interaction energy U and the strength of the harmonic confinement. Instead of using V_{t} to quantify the harmonic confinement, it is convenient to define the characteristic trap energy

$$E_{\mathrm{t}} = V_{\mathrm{t}} \left(\frac{\gamma N_{\sigma}}{4\pi/3} \right)^{2/3}, \tag{6.2}$$

which explicitly depends on the atom number N_{σ} per spin state ($N_{\downarrow} = N_{\uparrow}$). In fact, E_{t} is the Fermi energy of a noninteracting fermionic quantum gas in the zero-tunneling limit that is been derived in Eq. 3.39 (see Fig. 6.1a for illustration). The characteristic trap energy is proportional both to the atom number and the trap frequency $E_{\mathrm{t}} \propto \omega_{\perp}^2 N_{\sigma}^{2/3}$ and describes the effective compression of the system. In the experiments it is controlled by the trapping potential (see Sects. 5.3 and 5.4), while the atom number is left as constant as possible.[1]

[1] In typical ultracold atom experiments it is notoriously difficult to vary the atom number from shot to shot in a controlled way. However, our setup offers exceptional control of the harmonic confinement. Therefore we actively vary the harmonic trapping potential, while the atom number is kept constant. The remaining shot-to-shot fluctuations of the atom number are largely suppressed in the experimental data, because all quantities are expressed in rescaled, atom number independent units, such as E_{t} and R_{sc} (see below).

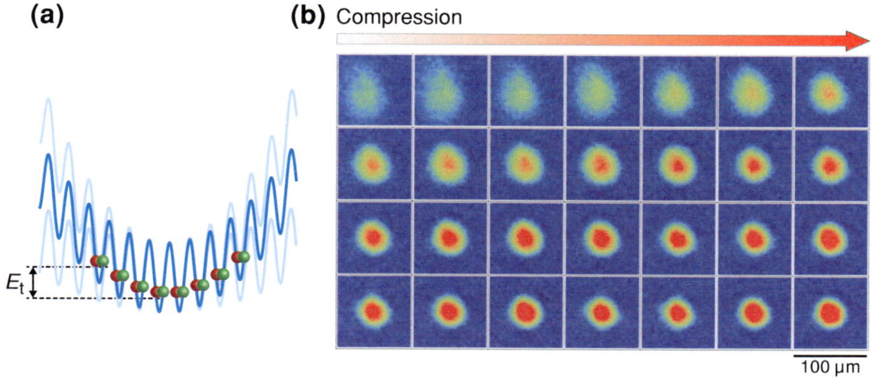

Fig. 6.1 Compression of a fermionic spin mixture. **a** The characteristic trap energy E_t defines the compression of the atom cloud. **b** In-situ phase-contrast images for strong repulsive interactions $U/12J = 1.5$ (see Fig. 6.6, *red* data). The compression of the dipole trap is continuously increased from left to right and top to bottom covering a range of horizontal trap frequencies from $\omega_\perp = 2\pi \times 20$ to $2\pi \times 120$ Hz. The rescaled cloud size $R_{sc} = R/(\gamma N_\sigma)^{1/3}$ is derived by measuring the cloud radius R in phase-contrast images (see main text). The atom number N_σ per spin state σ is extracted from a subsequent time-of-flight image in the same experimental run (see Fig. 6.5)

Figure 6.2 shows the different zero-temperature phases expected in the center of the trap, depending on which term in the Hamiltonian dominates:

- For weak interactions in a shallow trap, $U \ll E_t \ll 12J$, the Fermi energy ϵ_F is smaller than the bandwidth of the three-dimensional lattice, $\epsilon_F < 12J$ (corresponding to region A in Sect. 3.3.3). The atoms are delocalized, minimizing the kinetic energy of the many-body system, and the central filling obeys $n_{\mathbf{0}\sigma} < 1$, where the local filling factor $n_{\mathbf{i}\sigma} = \langle \hat{n}_{\mathbf{i}\sigma} \rangle$ denotes the average occupation per spin state at lattice site **i**. The system is in a compressible metallic state: When the chemical potential is increased, the local filling factor increases continuously. There is no excitation gap.
- Dominating repulsive interactions, $U > 12J$ and $U \gg E_t$, suppress double occupation of lattice sites. This can lead to Fermi liquid ($n_{\mathbf{0}\sigma} < 1/2$) or Mott insulating ($n_{\mathbf{0}\sigma} = 1/2$) states at the trap center. A notable Mott insulating core at experimentally relevant temperatures starts to emerge for interactions $U/12J > 1$ and compressions $E_t/12J \geq 0.3$ (see Fig. 6.3). In the Mott insulating state the addition of a fermion (of opposite spin) to a singly occupied lattice site costs an energy amount U. In order to compress this state, the external potential must overcome the interaction energy, otherwise the system stays constant in size. Therefore, the Mott insulating phase can be viewed as an *interaction* insulator.
- Stronger compression leads to higher filling. For $E_t \gg 12J$ and $E_t \gg U$ this ultimately results in a band insulator with unity filling in the trap center ($n_{\mathbf{0}\sigma} = 1$) at least for vanishing temperature. The band insulating state is

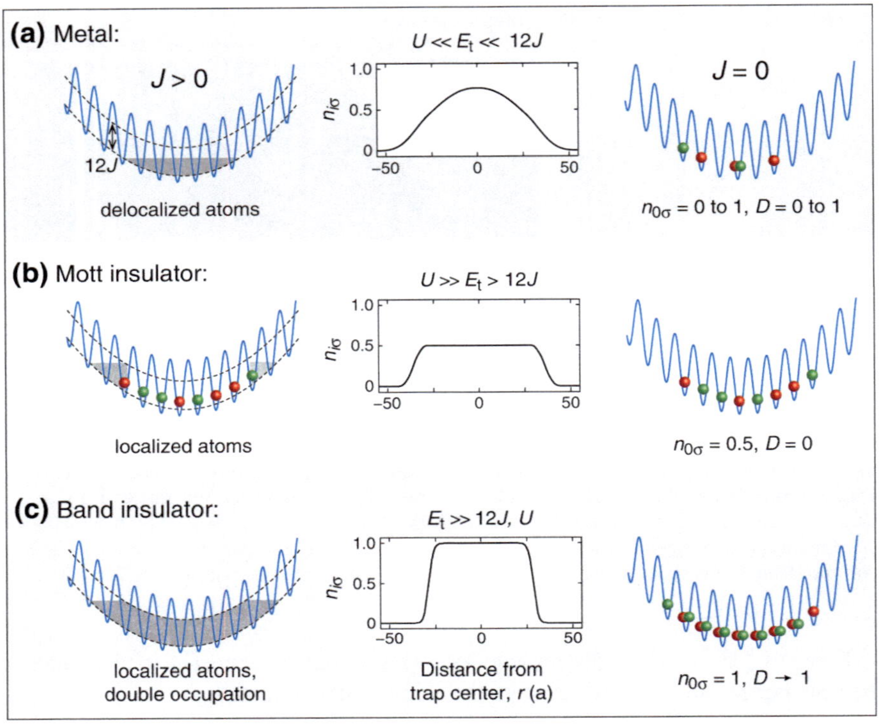

Fig. 6.2 **a–c** Zero-temperature quantum phases of the Hubbard model for a balanced spin mixture in a harmonic confining potential. The *left column* schematically illustrates the equilibrium atom distribution, *dashed lines* indicate the extent of the first band (compare Fig. 3.7). The *center column* displays the corresponding in-trap density profiles. The *right column* shows the distribution of singly and doubly occupied lattice sites after projection into the zero-tunneling limit. Experimentally the projection is achieved by a rapid increase of the lattice depth. $n_{0\sigma}$ denotes the central filling and D the fraction of atoms on doubly occupied lattice sites. Strictly speaking, the Mott insulator (**b**) shows antiferromagnetic ordering at zero temperature (see Sect. 3.3.2). Here, a paramagnetic Mott insulator is displayed, because experimental temperatures are larger than the Néel temperature for magnetic ordering

incompressible, because the external potential must overcome the band gap between the first and second lattice band to achieve a reduction of the system size (see Sect. 3.3.3). Otherwise filling factors larger than unity are prohibited by Pauli's principle. Accordingly, the band insulating state could also be called a *Pauli* insulator.

It is important to note that finite temperature reduces the local fillings and enlarges the cloud size, because the corresponding entropy must be accommodated in the system. Nevertheless, the above characterization of quantum phases remains valid, provided the temperatures are clearly in the quantum degenerate regime ($T \ll T_F$).

In the trapped system the filling varies smoothly from a maximum in the center to zero at the edge of the cloud. This generally leads to a coexistence of several phases. For example, at dominating characteristic trap energy, strongly repulsive interactions and low temperature ($E_t > U > 12J$), a wedding cake-like structure develops (see Fig. 6.3): A band insulating core ($n_{\mathbf{0}\sigma} \approx 1$) is surrounded by a metallic shell ($1/2 < n_{i\sigma} < 1$), a Mott insulating shell ($n_{i\sigma} = 0$) and another metallic shell ($n_{i\sigma} < 1/2$) [5]. It is important to note, that in all cases, independent of interaction, confinement or temperature, the outermost shell is always metallic.

6.2 Experimental Observation of Metallic and Insulating Phases

The measurement idea is to identify the phases of the Hubbard model (Eq. 6.1) by studying the response of the system to changes in the external confinement. To this end noninteracting and repulsively interacting spin mixtures of fermionic ^{40}K are investigated in a three-dimensional optical lattice. Our experimental setup allows to vary the interaction strength, the lattice depth and the external harmonic confinement independent from each other, such that all three terms of the Hamiltonian are individually controlled. The temperatures of the atomic samples are deep in the degenerate regime ($T \ll T_F$) and the corresponding entropies in the lattice are low enough, that the phases of the experimental system are determined by the zero-temperature phases of the Hubbard model outlined above.[2]

On the one hand, the system is probed on a global level by monitoring the in-situ density distribution for increasing harmonic confinement. This allows to directly extract the compressibility of the many-body quantum system and the response of the cloud size uniquely distinguishes compressible metallic from incompressible insulating states. On the other hand, the system is studied on a local level by measuring the fraction of doubly occupied sites for different experimental parameters. Both the local and global observables indicate the entrance into the strongly interacting regime, as increasing repulsion leads to a suppression of double occupancy and larger cloud sizes.

6.2.1 Experimental Techniques: Overcoming the Challenges

In order to realize the above measurement idea, several experimental challenges had to be overcome. In the following the most important ones are summarized:

- **Quantum degenerate fermion clouds**: Initial DMFT calculations had suggested that the fermionic spin mixture should have a temperature $T/T_F \lesssim 0.15$

[2] However, the entropies are not low enough to reach magnetic ordering of the spins, which is discussed later in this chapter (see also Sect. 3.3.2).

prior to lattice loading in order to observe a Mott insulating core after an adiabatic lattice ramp-up. While setups with fermionic ^{6}Li routinely achieve temperatures in this range, it had been a notorious challenge with ^{40}K. The difference is mainly rooted in the low natural abundance of the potassium isotope (resulting in comparably small atomic samples) and less favorable scattering properties.

The temperature goal has eventually been reached by carefully tweaking the cooling sequence in many respects. Starting with the optimization of atom numbers in the double-species magneto-optical trap, over faster sympathetic cooling in the optically-plugged magnetic trap to careful timing of the preparation and subsequent cooling of the fermionic spin mixture in the dipole trap. The individual modifications have usually lead to an increase of the available atom number, which then has been turned into colder temperatures via evaporation.

- **Alignment and calibration of optical potentials**: The combination of the red-detuned dipole trap and the blue-detuned optical lattice required the development of precise alignment techniques for the laser beams creating the potentials. In a red-detuned lattice the atomic cloud is intrinsically dragged to the beam center during loading and the underlying potential can safely be assumed to be harmonic. In our combined setup, the harmonic approximation is only valid, when the individual beams are perfectly centered on each other. This is crucial for a sound comparison to theoretical data. Our techniques for beam alignment and calibration of lattice depth, trap frequencies and anticonfinement are described in Chap. 5 in detail.
- **Radiative collisions in blue-detuned optical lattices**: Generally, atom pairs on the sites of an optical lattice are prone to radiative collisions. In such collisions one of the atoms is excited by a photon of the lattice laser and the atom pair populates a state in the excited molecular potential [6]. Similarly, molecules on the sites of an optical lattice can suffer from photodissociation.

  In a red-detuned lattice, the absorption of a photon can lead to the population of a bound state in the excited molecular potential, provided the photon is resonant with the respective transition. This loss and heating process is unwanted, but it can safely be avoided by shifting the lattice laser frequency by a few GHz to detune the transition.

  In the case of a blue-detuned lattice, radiative collisions are harder to suppress, because absorption of a lattice photon leads to the population of unbound molecular states in the continuum compared to discrete bound states for red-detuning [6]. In detailed studies (see Appendix of Ref. [1]) we have found that the photodissociation rate of molecules decreases for larger detunings, approximately following a $1/\Delta^2$ power law (Δ is the detuning between the D-transitions of ^{40}K and the lattice laser). Furthermore, we have observed a strong reduction of the Frank-Condon overlap between the ground and excited state molecular wavefunctions whenever the Condon point is close to a node of the ground state wavefunction [6]. The Frank-Condon overlap determines the strength of the transitions and the Condon point is the relative distance of the atoms, at which resonant excitation is possible for a given laser frequency. By choosing a lattice wavelength at about 738 nm, it has been possible to suppress the loss of doubly occupied sites and molecules below a critical level. This was a cornerstone to unleash the full potential of the combined use of a red-detuned dipole trap and a blue-detuned lattice.

- **Adiabatic lattice loading**: Adiabatic loading generally requires a slow ramp-up of the lattice depth to make sure that the sample stays in its many-body ground state at any point of the ramp. However, the time frame for the experimental sequence is limited, because heating rates continuously increase the entropy in the system [7]. The heating has technical origins, ranging from spontaneous light scattering over radiative collisions to collisions with the background gas in the vacuum chamber.

  The time scale for adiabaticity can be significantly shortened by choosing an intelligent ramping sequence that aims at the least complicated thermodynamic path through phase space. One step to reach this goal is mode-matched loading, which minimizes the mass transport on the way from the initial state in the harmonic trap to the final strongly interacting state in the lattice. Mass transport in a lattice and the decay of doubly occupied sites have been shown to be extremely slow, already at weak interactions ($U/12J > 0.2$) [1, 8, 9]. This suggests to first increase the harmonic confinement and then raise the lattice depth in a compression measurement.

  However, there is another important aspect that must be taken into account in the context of adiabatic preparation: In a three-dimensional optical lattice the first and the second band have a partial overlap until a depth of 2.2 E_{rec} is reached. A gap only opens up for deeper lattices (see Sect. 2.2.3). When the atom cloud is compressed prior to lattice loading, there is a risk that states in the second band get populated, which only slowly relax back into the first band (see Fig. 2.10). If that happened, the interpretation of experimental results in terms of the single-band Hubbard model would be hindered. Therefore, a preramp of the lattice to a depth of 1 E_{rec} is performed, then the harmonic trap is compressed and, finally, the lattice depth is risen to the desired value.[3]

- **In-situ imaging**: The measurement of the in-situ cloud sizes necessitates a reliable imaging technique to detect optically dense atomic samples inside the optical lattice. Initially, detuned absorption imaging as well as saturated imaging with high light intensities have been tested. The former was prone to lensing, as expected for imaging systems with limited numerical aperture [10], and the latter showed strong nonlinearity between the atomic and the optical density [11]. Finally, dependable measurements of in-situ cloud sizes have become

---

[3] Actually, it seems desirable to perform a preramp to a lattice depth above 2.2 E_{rec} in order to ensure a real band gap. However, this would require a very slow compression beyond the time scale that is affordable due to technical heating. In this respect the preramp to 1 E_{rec} is a already compromise.

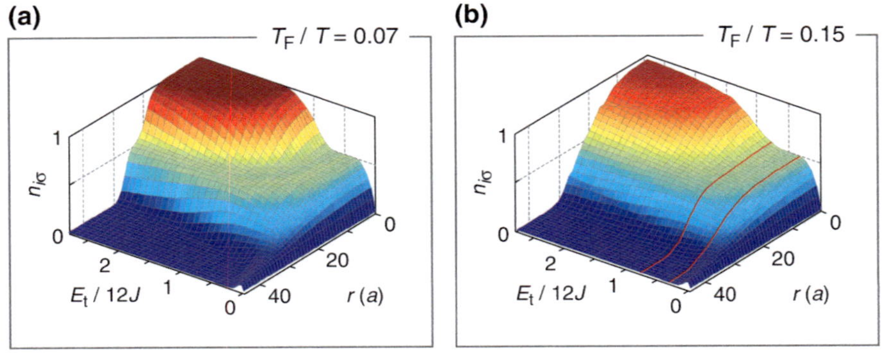

Fig. 6.3 Theoretical in-trap density distributions at strongly repulsive interactions ($U/12J = 1.5$). The radial distance from the trap center is denoted by r given in units of the lattice constant $a = \lambda/2$. While for lower temperature the different phases can be clearly distinguished in the radial profile (**a**), higher temperature washes out the boundaries (**b**). The *red solid lines* enclose the region, where a Mott insulating core has formed. In this regime the global compressibility is reduced as shown in Fig. 6.7d. Please note that the dimensionless temperatures T/T_F given here, refer to the Fermi gas in the harmonic trap prior to lattice loading. The DMFT calculations have been performed by Theo Costi, Achim Rosch and collaborators [4]

possible by an elegant implementation of high-field phase-contrast imaging (see Sects. 4.1.3, 5.5 and Appendix B).

6.2.2 *Experimental Parameters*

The experiment is performed with a balanced, quantum degenerate mixture of fermionic ^{40}K atoms in the two hyperfine states $|F, m_F\rangle = |9/2, -9/2\rangle \equiv |\downarrow\rangle$ and $|9/2, -7/2 \equiv |\uparrow\rangle$. The mixture is prepared in a pancake-shaped optical dipole trap ($\lambda_{\text{dip}} = 1030$ nm) formed by two elliptical beams in the horizontal plane. Measurements of the trap frequencies in the vertical direction and the horizontal plane yield an aspect ratio of $\gamma = \omega_z/\omega_\perp \approx 4$, in compliance with a numerical trap model that is based on the geometrical beam sizes and the power balance in the beams (see Sect. 5.3).

After sympathetic cooling of ^{87}Rb and ^{40}K down to a temperature of about $T = 2\,\mu$K in the magnetic trap, both species are loaded into the dipole trap. Rubidium is transferred to the absolute ground state $|F, m_F\rangle = |1, +1\rangle$ and the potassium spin mixture is created. By lowering the intensity of the trapping laser in an exponential ramp, sympathetic cooling is continued. After the trap bottom for ^{87}Rb is crossed, the rubidium atoms are gone. Then the interacting ^{40}K spin mixture further cools evaporatively, thermalizing via s-wave collisions. With this procedure samples with 1.5×10^5 to 2.5×10^5 potassium atoms at $T/T_F = 0.15(3)$ are created. The tempera-

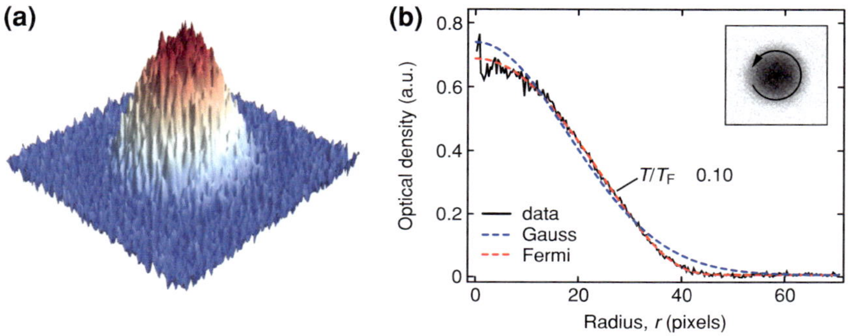

Fig. 6.4 Ultracold fermionic spin mixture. **a** Column density of a noninteracting ^{40}K spin mixture after 10 ms time-of-flight expansion. The vertical axis corresponds to the optical density (a.u.). **b** The azimuthally averaged raw data (*solid line*, see inset) is fitted by a Gaussian (*blue dashed line*) and a Fermi-Dirac model (*red dashed line*) omitting the noisy data of the central pixels. The Fermi fit yields a temperature of $T/T_F \approx 0.10$, which is close to the reliability limit of this thermometry method for the given cloud sizes [12]

tures are measured by fitting the momentum distribution recorded after time-of-flight with a Fermi-Dirac model (see Fig. 6.4 and Appendix C for further details).

The Feshbach resonance located at a magnetic field $B = 202.1$ G (see Sect. 2.3) is used to tune the scattering length a_{FF}, providing direct control of the on-site interaction energy U. The preparation of the spin mixture and final evaporation either happen above the resonance (220 G) or below the resonance (165 G). While the former gives access to noninteracting (209.9 G) and repulsively interacting samples with $a_{FF} \leq 150\ a_0$, the latter allows to reach scattering lengths of up to $a_{FF} = 300\ a_0$ (191.3 G) by approaching the resonance from below. A further approach of the resonance is not possible owing to enhanced losses and heating that probably originate from a p-wave resonance in striking distance (see Sect. 2.3). The desired magnetic field is chosen directly after evaporation (see Fig. 6.5).

In a first step, the optical lattice ($\lambda_{lat} = 738$ nm) is raised to a depth of $V_{lat} = 1\ E_{rec}$. This preramp is performed to separate the first and second lattice band (see above). Subsequently, the external harmonic confinement is tuned to the desired value by changing the dipole trap depth within 100 ms. Horizontal trap frequencies in the range from $\omega_\perp \approx 2\pi \times 20$ Hz to $2\pi \times 120$ Hz are accessed. The preparation is completed by increasing the lattice depth to $V_{lat} = 8\ E_{rec}$ in a linear ramp within 50 ms.

After a hold time of 12 ms we record an in-situ image of the cloud along the vertical axis of the trap using phase-contrast imaging. Phase-contrast imaging with a retarding phase plate requires blue-detuned laser light in order to ensure linearity between the imaging signal and the atomic density (see Sect. 4.1.3 and Fig. 4.1b). This detuning is elegantly generated by taking the image at the same magnetic field that is initially used to adjust the interaction. The imaging laser, which is resonant at zero magnetic field, is effectively blue-detuned with respect to the Zeeman-shifted transition by $\Delta = 2\pi \times 280\,\text{MHz}$ to $2\pi \times 400\,\text{MHz}$ depending on the exact value of the magnetic field (see Appendix B). The phase-contrast image is used to extract the cloud size $R = \sqrt{\langle r_\perp^2 \rangle} = \sqrt{\langle x^2 + y^2 \rangle}$ with a two-dimensional fitting function[4]

$$F(x, y) = A\,\mathrm{Li}_2\left(-100\,e^{-\frac{(x-x_0)^2}{2\sigma_x^2}-\frac{(y-y_0)^2}{2\sigma_y^2}}\right) + B\sqrt{\frac{(x-x_0)^2}{\sigma_x^2} + \frac{(y-y_0)^2}{\sigma_y^2}} + C, \tag{6.3}$$

with the free parameters x_0, y_0 (cloud center); σ_x, σ_y (widths); A, B and C. The second term accounts for a funnel that has a technical reason rooted in phase-contrast imaging. The Fermi-type distribution $F(x, y)$ fits the in-situ clouds better than a Gaussian distribution (see supplementary material of [4]). The resulting cloud radius is given by $R = \sqrt{1.264^2(\sigma_x^2 + \sigma_y^2) - \eta^2}$, where $\eta \approx 3\,\mu\text{m}$ accounts for the finite resolution of our imaging system (see Sect. 5.5).

Because very few photons are scattered in phase-contrast imaging, almost all atoms remain in the sample and the momenta are only marginally modified. This allows to measure the quasi-momentum distribution in the same experimental run using a band-mapping technique [13–16]. To this end, the lattice is adiabatically ramped down in 200 μs and a standard resonant absorption image is taken after 10 ms time-of-flight.

6.2.3 Probing the Global Compressibility

In order to benchmark the experimental data, we compare it to accurate DMFT calculations [5, 17–19] by Theo Costi at the Institute for Advanced Simulation in Jülich and the group of Achim Rosch at the University of Cologne. As a first step in the calculations, the homogeneous Hubbard model is solved for a wide range of temperatures and chemical potentials using a numerical renormalization group approach [20–22]. In a second step, the trapped system is locally approximated by uniform miniature systems through a local density approximation (LDA). This procedure yields reliable results even close to the boundary between metal and insulator, which has been validated in Refs. [5, 23]. Density profiles for various compressions and temperatures are calculated (see Fig. 6.3) and, finally, theoretical line-of-sight integrated density profiles are derived in analogy to the experimental data.

The comparison of theoretical and experimental results is facilitated by expressing the cloud size R in rescaled units $R_{\text{sc}} = R/(\gamma N_\sigma)^{1/3}$ and the harmonic confinement in terms of the dimensionless compression $E_t/12J$. In these units, the cloud size only depends on the interaction strength $U/12J$ and the average entropy per particle S/N, but not on the absolute atom number N. The theoretical calculations are based on the entropy per particle that is experimentally determined by the temperature T/T_F

---

[4] For a Gaussian distribution this definition of the cloud radius yields the standard deviation.

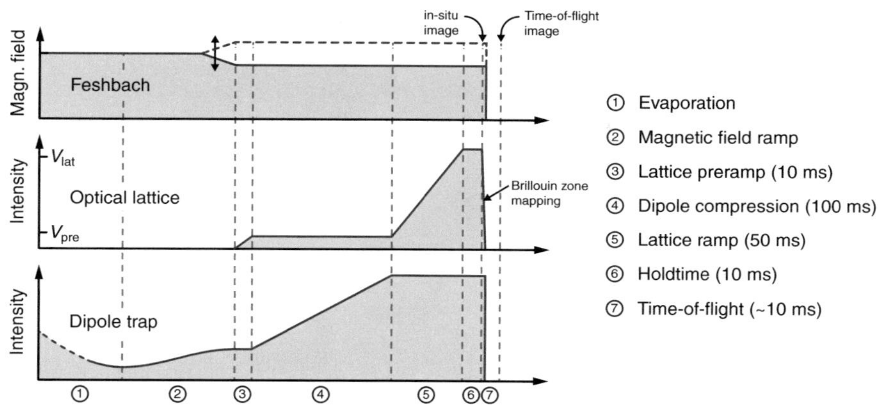

Fig. 6.5 Overview of the experimental sequence as described in the main text. Remarkable features are the preramp of the optical lattice ③, the nondestructive in-situ phase-contrast image after ⑥ and the subsequent time-of-flight absorption image after ⑦ mapping out the Brillouin zone

of a noninteracting Fermi gas in the harmonic trap (see Eq. 2.25) and the additional assumption of adiabatic lattice loading.

The rescaled units necessitate the precise knowledge of several experimental parameters. The interaction energy U and the tunneling matrix element J are derived from the lattice depth, the trap frequencies ω_α ($\alpha = x, y, z$) are precisely measured using the methods presented in Sects. 5.3 and 5.4. These parameters and the temperature T/T_F are known with a relative uncertainty around or below 10 %. However, it is notoriously difficult to determine the absolute atom number $N = N_\uparrow + N_\downarrow$ with a similar precision. Several calibration methods are compared to each other: First, a theoretical calibration of the potassium number is obtained by analyzing the transition strength for the given polarization of the imaging light (see Sect. 4.1.2). Second, we measure the in-situ cloud size of a noninteracting harmonically trapped spin mixture. Based on the Thomas-Fermi radius (see Eq. 2.30) the atom number can be extracted. However, owing to the weak dependence of the radius on the atom number $R \propto N^{1/6}$ the method suffers from a rather large uncertainty. Third, we make use of the favorable size scaling $R \propto N^{1/3}$ of a band insulator of noninteracting fermions in the lattice. The band insulating regime is reached for compressions around $E_t/12J = 2$ and the expected cloud sizes can be calculated by relatively simple means (see Sect. 3.3.3). Comparison of the three methods yields consistent results, suggesting a high accuracy of the atom number calibration.

Figure 6.6 shows the quantitative comparison between the measured and the numerically calculated cloud sizes R_{sc} as a function of the compression $E_t/12J$. In addition, the global compressibility of the system

$$\kappa_{R_{sc}} = -\frac{1}{R_{sc}^3}\frac{\partial R_{sc}}{\partial(E_t/12J)} \tag{6.4}$$

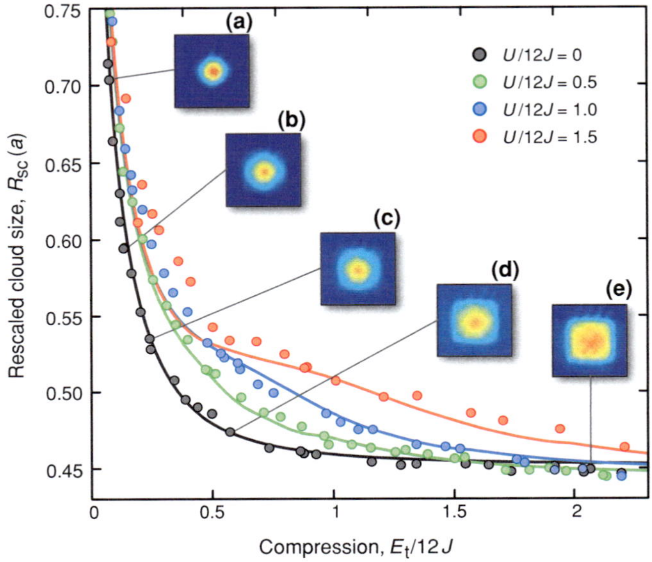

Fig. 6.6 Response of the interacting spin mixture to compression of the underlying harmonic confinement. The rescaled cloud sizes R_{sc} in a $V_{lat} = 8\ E_{rec}$ lattice are shown as a function of the external trapping potential for interactions $U/12J$ ranging between 0 and 1.5. *Circles* denote individual experimental shots, while *lines* show the theoretical expectations obtained by DMFT for an initial temperature $T/T_F = 0.15$ prior to lattice loading. The insets **a–e** show the quasi-momentum distributions for the noninteracting case obtained via Brillouin zone mapping (averaged over several experimental shots)

is extracted from the experimental data by means of linear fitting of four consecutive data points to determine the derivative.

6.2.3.1 Noninteracting System

For the noninteracting system we find that the cloud sizes decrease continuously (see Fig. 6.6, black data points) until the characteristic trap energy roughly equals the lattice bandwidth ($E_t/12J \approx 1$) (see Sect. 3.3.3). For stronger compression, the cloud size essentially remains constant and the compressibility (see Fig. 6.7a) approaches zero. At this point almost all atoms are in the band insulating regime, while the surrounding metallic shell has a negligible effect on the cloud size.

This is supported by the measured distributions of the quasi-momenta, which are shown as insets in Fig. 6.6. The distributions change gradually from a partially filled first Brillouin zone, as expected for a metal, to an almost evenly filled first Brillouin zone for increasing compression, as expected for a band insulator [16]. Nevertheless, experimental band (or Brillouin zone) mapping can only reveal the relative occupations of the extended Bloch states. Information on the absolute occupation

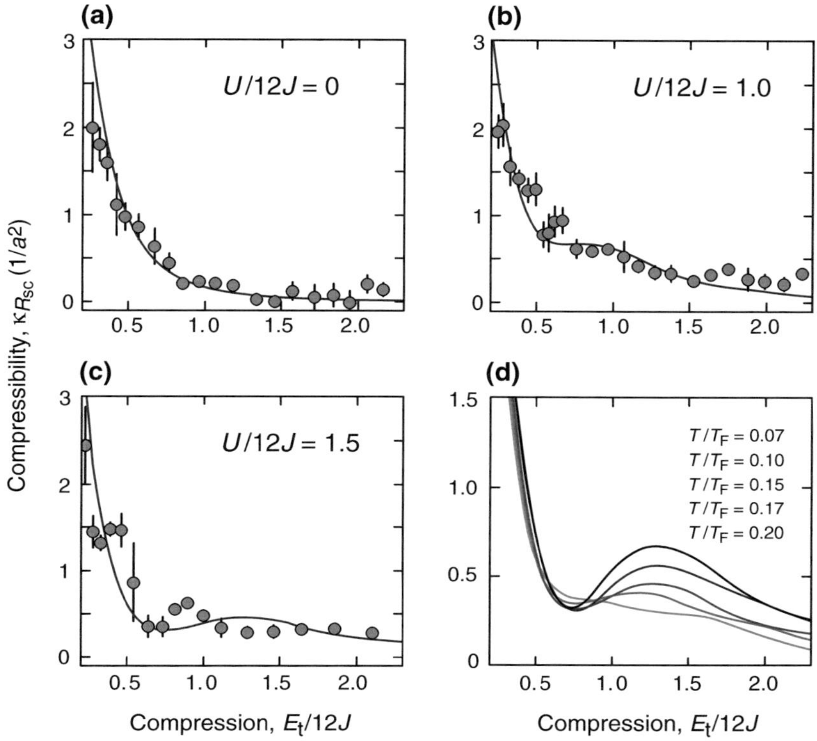

Fig. 6.7 **a–c** Global compressibility $\kappa_{R_{sc}}$ as a function of the compression for several interactions. Each data point corresponds to the linear slope of four consecutive points in the cloud size data (see Fig. 6.6). Error bars represent the uncertainty of the fit. *Solid lines* show the theoretically expected results for an initial temperature $T/T_F = 0.15$. Panel **d** illustrates the influence of the initial temperature on the calculated compressibility at strong repulsion $U/12J = 1.5$

of the Bloch states cannot be extracted. For example, localized atoms or a heated sample would both show a homogeneously filled Brillouin zone. In contrast, our measurement of compressibility directly demonstrates the incompressibility of the fermionic band insulator in excellent agreement with the theoretical expectation for a noninteracting Fermi gas (see Figs. 6.6 and 6.7a, black data).

6.2.3.2 Repulsively Interacting System

For moderately repulsive interactions ($U/12J = 0.5$ and 1.0, green and blue data in Fig. 6.6), the cloud size is clearly bigger than in the noninteracting case, but eventually reaches the size of the band insulator. For stronger repulsion ($U/12J = 1.5$, red data in Fig. 6.6) only a marginal decrease of the cloud size is observed in the compression range $0.5 < E_t/12J < 0.7$, while for stronger confinements the compressibility increases again. This behavior is consistent with the formation of a Mott insulating core that is surrounded by a metallic shell, as shown in the theoretical in-trap profiles of Fig. 6.3. The increase of compressibility for strong confinements is due to the emergence of an additional metallic core ($1/2 < n_{\mathrm{i}\sigma} < 1$) in the center of the trap.

We point out that a local minimum in the global compressibility is a genuine characteristic of a Mott insulating core. At large U and low temperature T, the theoretical analysis shows that the minimum of global compressibility decreases in the Mott region as $1/U^2$. The remaining compressibility originates from the surrounding metallic shell, which becomes smaller for larger interaction. The data shows an indication of this decrease in the compressibility minimum (compare Fig. 6.7b, c). For strong repulsion the location of the minimum (see Fig. 6.7c) at $E_{\mathrm{t}}/12J \approx 0.6$ approximately coincides with theory, while the subsequent increase happens slightly earlier. When the system is further compressed, all cloud sizes eventually approach the band insulating state and the corresponding compressibilities tend to zero.

The general agreement between the measured cloud sizes and the theoretical calculations is remarkably good for the range of interactions investigated here. Nevertheless, we observe that the cloud sizes for repulsive interactions and intermediate compression (around $E_{\mathrm{t}}/12J \approx 0.5$) are larger than predicted by theory and the discrepancies become more pronounced for stronger interactions. While the exact origin of the deviation is not yet known, a likely reason is nonadiabatic lattice loading. The data in Fig. 6.8 shows that the atomic cloud shrinks considerably during the loading ramp. However, the transport of atoms is substantially slowed down already at moderate interactions [9], which can lead to larger cloud sizes. Furthermore, at low compressions the atomic cloud has a radial extent of up to 100 μm (on the order of the radius of the lattice beams) and suffers from gravitational sag. The inhomogeneous lattice depths across the cloud as well as trap anharmonicities may further account for the observed deviations. Finally, there may be effects that reach beyond the single-band Hubbard model or the DMFT formalism.

6.2.3.3 Lattice Loading, Adiabaticity and Heating

In order to make sure that the lattice loading ramp is as adiabatic as possible, the in-situ cloud sizes are measured as a function of lattice ramp time at a compression $E_{\mathrm{t}}/12J \approx 0.4$ (see Fig. 6.8). This is the most interesting compression regime, because here the Mott insulating core starts to form and the discrepancy between experiment and theory is most pronounced. The measurement shows that for very fast ramping the cloud does not shrink to its minimal size, indicating nonadiabatic loading. For the noninteracting case, an increase of the cloud size is observed for very slow ramping. Interestingly, this is not the case for repulsive interactions (see Fig. 6.8, blue data) albeit heating should play a similar role. This might suggest that a second, longer adiabaticity time scale is present for stronger interactions, obscuring a clear minimum in the cloud size. Nevertheless, based on these data we have decided to use a lattice ramp time (from 1 E_{rec} to 8 E_{rec}) of 50 ms in the experiment, supposedly being as adiabatic as possible.

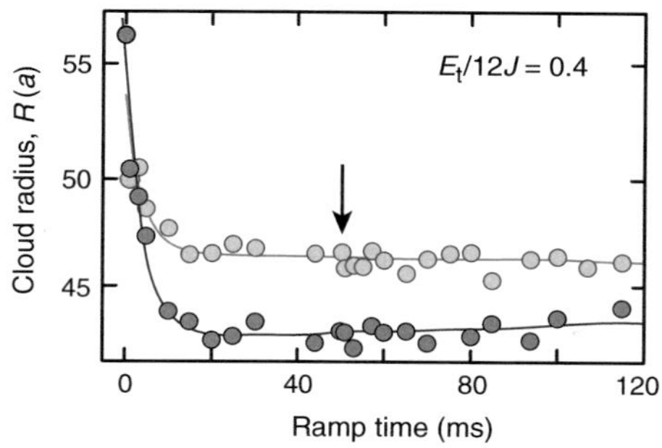

Fig. 6.8 Experimental determination of the optimal ramp time for lattice loading. The in-situ cloud size is measured at intermediate compression ($E_t/12J = 0.4$) as a function of the ramp time, both for a noninteracting (*black*) and an interacting ($U/12J = 1.0$, *gray*) spin mixture. The rapid initial decrease of the cloud size can be attributed to improved spatial redistribution of atoms during the ramp and corresponds to better adiabaticity. The slowly increasing radii for longer ramp times mainly originate from heating effects in the optical lattice. The *arrow* indicates the ramp time of 50 ms used in the experiment, which is a compromise between adiabaticity and heating

We have further investigated the role of heating in the experimental sequence. To this end, the temperature before lattice loading is compared to the temperature that is measured after loading and unloading the lattice in a reversed sequence. The duration of this round trip is exactly twice as long as the actual measurement sequence. We observe a rise in temperature T between $0.010(5)\ T_F$ for a noninteracting cloud and $0.05(2)\ T_F$ for medium repulsion of $U/12J = 1$ at compressions around $E_t/12J \approx 0.5$. Before lattice loading, typical temperatures lie in the range $T/T_F = 0.15(3)$. Given that the theoretical calculations, which assume initial temperatures of $T/T_F = 0.15$ and adiabatic preparation, agree well with the experimental data, the actual initial temperatures are expected to lie at the low end of the measured temperature range.

The theoretical calculations in Fig. 6.7d show, that the minimum in the local compressibility, which signals the Mott insulating state, starts to appear at initial temperatures in the range of $0.15 < T/T_F < 0.2$. At these temperatures the entropy per particle amounts to $1.5k_B < S/N < 2k_B$, which is much higher than the entropy per particle of a Mott insulator in a homogeneous system. In fact, it even exceeds the highest possible value in the half-filled homogeneous Hubbard model ($k_B \ln(4) \approx 1.4k_B$).[5] How is it possible to observe interesting quantum phases at these entropies?

[5] At half-filling, the maximal entropy is accommodated in the case of vanishing interactions. In a system of N lattice sites and $N/2$ atoms per spin state, each configuration j has the same probability $p_j = \left[\frac{N!}{(N/2)!(N/2)!}\right]^{-2}$. With $S = -k_B \sum_j p_j \ln(p_j)$ and Stirling's formula $\ln(N!) \approx N \ln(N)$ the entropy per particle $S/N = k_B \ln(4)$ is readily obtained. For strong repulsion each lattice site is occupied by exactly one atom. Here, all configurations have equal probability $p_j^\infty = \left[\frac{N!}{(N/2)!(N/2)!}\right]^{-1}$ leading to $S/N = k_B \ln(2)$. In the case of long-range antiferromagnetic order, there are only two configurations for the global quantum state. In the limit of large N this results in vanishing entropy per particle $S/N \gtrsim 0$.

The key to this question is the presence of the underlying trapping potential. It induces a strong variation of the local filling, and the local filling crucially determines the entropy capacity. The entropy capacity typically increases for lower filling factors. Therefore, an enormous fraction of the entropy is accumulated in the metallic shell at the edges of the cloud. Here, diluted atoms carry a large amount of configurational entropy as shown in Fig. 6.12. This ensures that the entropy per particle can locally approach the value $k_B \ln(2) \approx 0.69 k_B$ that is required to observe a Mott insulating phase free of thermal excitations.

6.2.4 Measuring the Double Occupancy

In addition to the global compressibility, the fraction of atoms on doubly occupied lattice sites D is measured [24]. To this end, atoms on doubly occupied lattice sites are converted into molecules using an adiabatic magnetic field ramp (0.2 ms/G) across the Feshbach resonance [25] (see Sect. 2.3.3). This technique can only be applied for field values above the Feshbach resonance, in our case corresponding to the interactions $U/12J = 0$, 0.5 and 1. The initial preparation is identical to the compression measurements described above, followed by a jump of the lattice depth to 20 E_{rec} (within 200 μs) to prevent tunneling of atoms during the magnetic field ramp. After the ramp (during 15 ms hold time) the lattice depth is reduced to zero within 200 μs and an absorption image is recorded after time-of-flight. The image shows the number of atoms on singly occupied sites, while the converted molecules remain invisible, because their resonance is shifted with respect to the bare atomic transition. The difference of the atom number in runs with and without the magnetic field ramp, normalized to the total atom number, yields the double occupancy D. Correction factors are applied to account for the losses of doubly occupied sites during the hold time of 15 ms in the deep lattice (see Sect. 6.2.1).

The global double occupancy D provides insight into the local on-site properties of the many-body state. It is mathematically defined by

$$D = \frac{2}{N} \sum_{\mathbf{i}} \langle \hat{n}_{\mathbf{i}\uparrow} \hat{n}_{\mathbf{i}\downarrow} \rangle, \tag{6.5}$$

where the sum runs over all lattice sites $\mathbf{i}$ and the expectation value $\langle \ldots \rangle$ is calculated within the many-body quantum state. In a homogeneous lattice all sites are equivalent and the site index can be suppressed. The double occupancy takes the form $D_{\text{hom}} = 2\langle \hat{n}_\uparrow \hat{n}_\downarrow \rangle / \langle \hat{n}_\uparrow + \hat{n}_\downarrow \rangle = \langle \hat{n}_\uparrow \hat{n}_\downarrow \rangle / n$ with the additional assumption of a balanced mixture $n = \langle \hat{n}_\uparrow \rangle = \langle \hat{n}_\downarrow \rangle$. For a homogeneous system, three limiting cases can be distinguished:

- At strongly attractive interactions ($|U| \gg 12J$ and $|U| \gg k_B T$) all atoms form pairs to minimize the interaction energy. Therefore, $D_{\text{hom}}^{-\infty} = 1$.

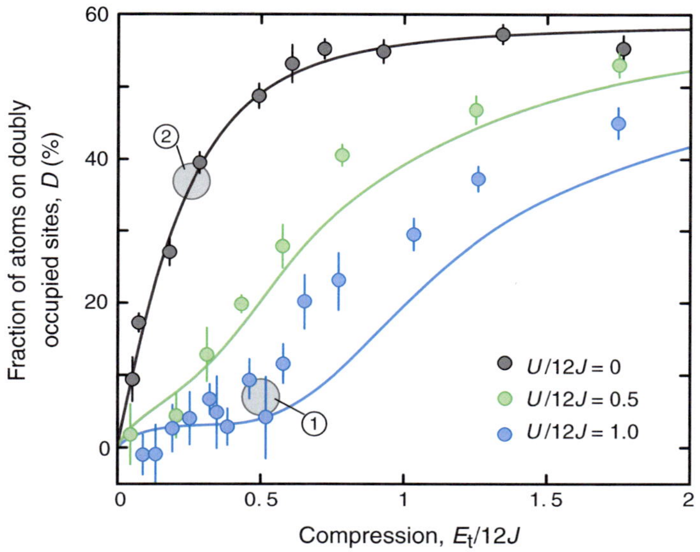

Fig. 6.9 Double occupancy D as a function of compression for the interaction strengths $U/12J = 0$, 0.5 and 1. The *shaded circles* ① and ② indicate the fraction of atoms on doubly occupied sites for a constant cloud size $R_{sc} \approx 0.5$ (see Fig. 6.6), which illustrates the strong suppression of double occupancy in the interacting case. Error bars denote the standard deviation of at least four measurements

- At vanishing interactions the spin states do not mutually influence their local occupation numbers. Therefore, $\langle \hat{n}_\uparrow \cdot \hat{n}_\downarrow \rangle = n^2$ and $D^0_{\text{hom}} = n$.
- At strongly repulsive interactions ($U \gg 12J$ and $U \gg k_B T$) two regimes must be distinguished: For $n \leq 0.5$, double occupancy is perfectly suppressed $D^\infty_{\text{hom}} = 0$ in order to avoid the energy cost of U. For $n > 0.5$, each atom beyond $n = 0.5$ creates a doubly occupied site on a perfectly half-filled background. Therefore, $D^\infty_{\text{hom}} = 2(n - 0.5)/n$ starting at a local filling of $n = 0.5$.

The third case constitutes the most extreme example for suppressed double occupancy due to repulsive interactions. The experimental curves are expected to be smoothened by the finite value of the repulsion and the inhomogeneous system. In combination with the in-situ size, the double occupancy D can be compared for different interaction strengths at constant cloud size R_{sc}. Then, the suppression of double occupancy serves as an important indicator for strong correlations.

Figure 6.9 shows that D approaches zero in the limit of weak confinement, because the cloud is large and diluted, regardless of the interactions. For intermediate compression, the fraction of doubly occupied sites crucially depends on the interaction. Choosing $R_{sc} \approx 0.5$ as a constant cloud radius, a double occupancy of 40 % is observed for the noninteracting system and of about 5 % for intermediate repulsive interaction $U/12J = 1.0$ (shaded circles in Fig. 6.9). Here, repulsive interactions energetically favor the reduction of double occupancy despite the cost of additional potential and kinetic energy.[6] For very strong compression, the system develops a band insulating core and double occupancy becomes comparable irrespective of the interaction. We find that D is ultimately limited to slightly below 60 %. The finite entropy per particle intrinsically reduces the filling in the band insulating core and, additionally, gives rise to extended surrounding metallic shells (see Fig. 6.3).

Although the noninteracting and the repulsively interacting curve $U/12J = 0.5$ match the DMFT results for an initial temperature of $T/T_F = 0.15$, notable deviations are observed for stronger repulsion $U/12J = 1.0$. The measured pair fraction is about 10 % higher than predicted by theory. Similar to the deviations in the cloud size discussed above, nonadiabatic lattice loading likely explains this behavior. In the early stage of the lattice ramp a small interaction parameter $U/12J$ favors a larger double occupancy D. When the lattice depth is further increased, transitional double occupancy must decay into singly occupied sites, which is hindered by very slow time scales [8, 9]. Nevertheless, additional experimental studies are needed to investigate the mechanisms behind the deviations in further detail.

The measurement of double occupancies yields valuable, complementary information to the cloud size measurements presented above. It is an important check for consistency. However, the suppression of doubly occupied sites compared to the noninteracting case is not a unique feature of the Mott insulating state. On the contrary, the suppression of double occupancy generally occurs when the interaction is strongly repulsive and the thermal energy in the lattice $k_B T_{\mathrm{lat}}$ is much smaller than U. In particular, even for a purely metallic phase with $n_{\mathbf{i}\sigma} < 0.5$ the double occupancy vanishes in the limit of strong repulsive interactions as discussed above.

6.3 Attractively Interacting Spin Mixtures in an Optical Lattice

So far, we have focussed on the repulsive Hubbard model. Owing to the versatility of Feshbach resonances, it is experimentally straightforward to access the domain of attractive interactions by choosing magnetic field values, at which the scattering length is negative. Applying the same preparation and detection techniques as in the repulsive case, we have studied the first realization of the attractive Hubbard model with ultracold atoms [26]. The most important results are briefly summarized in this section.

The attractive Hubbard model plays an important role in solid state physics, because lattice deformations (phonons), collective charge oscillations (plasmons) or spin fluctuations (magnons) can effectively lead to attractive interactions between electrons overcoming the intrinsic electrostatic repulsion [27]. In Sect. 3.3.2 the resulting phases are reviewed for the case of half-filling. In the context of ultra-

[6] Localized wavefunctions have a stronger curvature, which corresponds to higher momentum contributions and, consequently, higher kinetic energy.

cold atoms, several theoretical works have addressed the regime of preformed pairs [28–31].

Here, we are not concerned with the zero-temperature phases, but observe a counterintuitive thermodynamic effect that arises from the interplay of strong correlations and finite entropy: When loading an attractively interacting spin mixture at finite temperature into a three-dimensional optical lattice, the in-situ cloud size increases as a function of attraction instead of contracting. This phenomenon is similar to the well-known Pomeranchuk effect [32], which occurs in the liquid-to-solid transition of ^{3}He: Solid helium has a larger entropy capacity than the liquid phase owing to randomly oriented spins. Therefore, ^{3}He freezes into a solid during adiabatic squeezing, while absorbing heat and cooling the surrounding liquid.

6.3.1 Experimental Results

A fermionic mixture with $1.6(2) \times 10^5$ atoms per spin state and a temperature of $T/T_F = 0.15(3)$ is prepared as described above. The scattering length is tuned in the range between $a_{FF} = -400\ a_0$ to $+150\ a_0$ and the external confinement is adjusted with the dipole trap within 100 ms. Subsequently, the optical lattice is slowly ramped to depths ranging from 0 to 9 E_{rec} with a rate of 7 ms per E_{rec}. The in-situ cloud size is measured using phase-contrast imaging (along the vertical z-axis) and the double occupancy is extracted via molecule conversion analogous to the measurements on the repulsive side.

Figure 6.10a shows a contraction of the gas for weak attractive interactions, as one might expect intuitively. However, when the attraction exceeds a critical value, typically at a scattering length between $|a_{FF}| \approx 20\ a_0$ to $40\ a_0$, the cloud shows an anomalous expansion. The effect is the more pronounced, the deeper the lattice and the stronger the localization of the atoms. In addition, the double occupancy D steeply increases when the interactions become more attractive (see Fig. 6.10b). Despite the expansion of the cloud and atomic densities much lower than two atoms per site, D reaches up to 80 % for strong attraction and deep lattices, which indicates that the system is in a preformed-pair regime [31] (see Sect. 3.3.2). Without the optical lattice, the anomalous expansion disappears and the cloud size remains constant. The double occupancy only shows a moderate increase to about 40 %, which is measured by quickly projecting the harmonically trapped cloud into a deep lattice.

Apparently, localization of atoms and strong correlations are necessary to evoke the counterintuitive expansion of the atomic cloud at increasing attraction. Without the lattice, these correlations are not strong enough to lead to a significant change in the density of states. However, what is the physical mechanism behind the anomalous expansion?

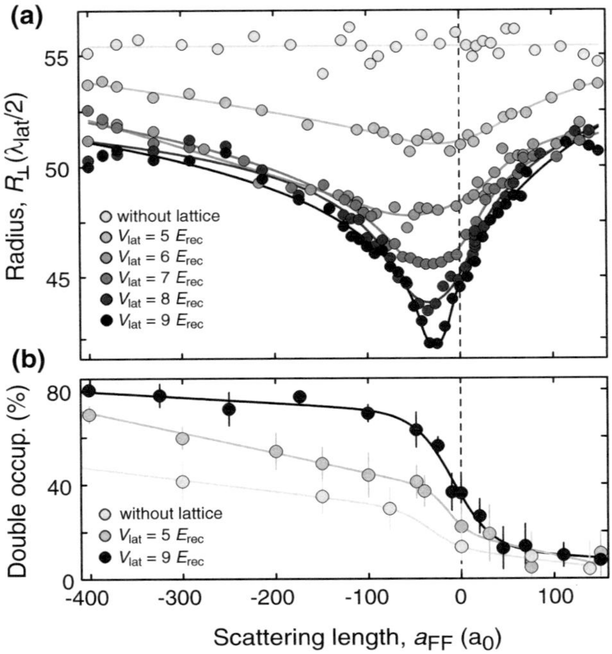

Fig. 6.10 Measured cloud size and double occupancy. **a** The cloud radius in the horizontal plane $R_\perp$ and **b** the fraction of atoms on doubly occupied lattice sites as a function of scattering length are shown for several lattice depths. The data points in **a** correspond to a moving average over three consecutive points to smoothen the noise. Each point in **b** shows the average of at least five measurements at identical parameters; the error bars denote the standard deviation. The data is recorded at a fixed horizontal (vertical) trap frequency $\omega_\perp = 2\pi \times 25\,\text{Hz}$ ($\omega_z \approx 2\pi \times 100\,\text{Hz}$)

6.3.2 A Basic Explanation

We assume that the total entropy in the system is identical for all scattering lengths, since the initial preparation is identical for all samples and all manipulations are considered adiabatic. Hence, for all interactions a certain amount of entropy must be accommodated in the lattice system (see Fig. 6.11).

Basic insight is obtained in the zero-tunneling limit, in which the atoms are fully localized and each site can be considered individually. Within a grand-canonical ensemble, the on-site physics is characterized by the probabilities for zero, single and double occupancy. In principle, the zero-tunneling limit allows to exactly calculate the thermodynamic properties for arbitrary interactions. Such calculations have been carried out by Belen Paredes and Maria Moreno-Cardoner and are presented in Ref. [26].

Here, only the entropy capacities per lattice site at zero and infinitely attractive interactions are considered. Generally, the entropy is given by $S = -k_B \sum_j p_j \ln(p_j)$, where p_j denotes the probability of finding the system in the state j and the sum runs

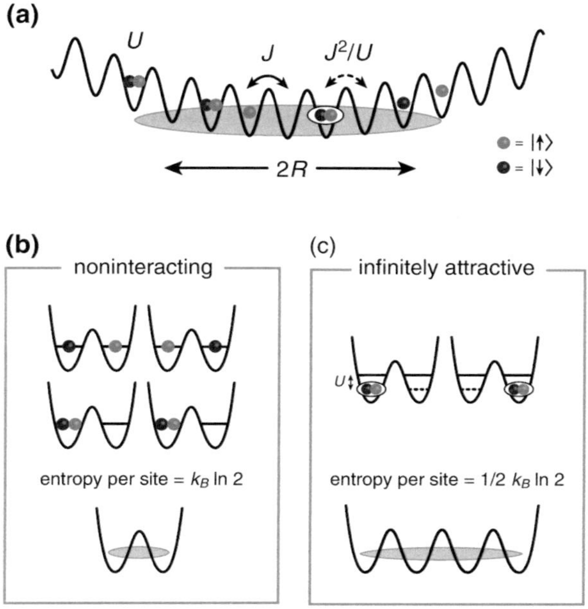

Fig. 6.11 Attractively interacting fermionic spin mixture and finite entropy. **a** The cloud size R is determined by the interplay between entropy and several energy scales: The interaction energy U, the tunneling matrix element J and the characteristic trap energy E_t. Additionally, the tunnel coupling J^2/U of bound atom pairs becomes relevant at strongly attractive interactions. **b**, **c** Configurations of two fermions (spin up and down) in a zero-tunneling double well potential. For vanishing interactions there are four possibilities to arrange the atoms. In the limit of infinitely attractive interactions the atoms are tightly bound, merely allowing for two configurations. Therefore, the entropy per lattice site is reduced by a factor of two. The atom cloud is forced to expand in order to accommodate the total entropy present in the finite-temperature system

over all allowed states. For a spin mixture of noninteracting fermions the entropy at site **i** is given by

$$s_{\mathbf{i}}^{0} = -2k_B \left[n_{\mathbf{i}} \ln(n_{\mathbf{i}}) + (1 - n_{\mathbf{i}}) \ln(1 - n_{\mathbf{i}})\right], \tag{6.6}$$

because the local filling per spin state $n_{\mathbf{i}} = \langle n_{\mathbf{i}\uparrow}\rangle = \langle n_{\mathbf{i}\downarrow}\rangle$ corresponds to the probability of finding a spin up or spin down atom at this site. The factor of two accounts for the contribution of both spin states. In the limit of infinite attraction, on-site pairs form in order to gain the large negative interaction energy. These pairs behave as hard-core bosons and lead to a local entropy of

$$s_{\mathbf{i}}^{-\infty} = -k_B \left[n_{\mathbf{i}} \ln(n_{\mathbf{i}}) + (1 - n_{\mathbf{i}}) \ln(1 - n_{\mathbf{i}})\right]. \tag{6.7}$$

Comparing the expressions of the two limiting cases, we observe that correlations can significantly reduce the local entropy density. The system with strongly attractive interactions needs to occupy twice as many lattice sites to store the same amount of entropy as the noninteracting system. This result is also illustrated in Fig. 6.11 using a simple double well system.

Within the zero-tunneling model the minimal cloud size is expected to occur at vanishing interactions $a_{\mathrm{FF}} = 0$. In the experimental data, however, the minimum is shifted towards slightly attractive interactions (see Fig. 6.10a). This shift originates from a competition between energy minimization through delocalization and reduced entropy capacity. Assuming zero temperature and vanishing interactions, the cloud will be larger at finite tunneling than in the zero-tunneling limit, because the delocalization of atoms minimizes the total energy. If in this $T = 0$ scenario the attraction between the atoms is increased, the cloud size shrinks further until all states in the trap center are occupied by a spin up and a spin down atom.[7] If finite temperature is introduced again, the shrinking of the cloud at attractive interactions is counteracted by the need to accommodate the finite entropy. The entropy effect will dominate only beyond a critical attraction and therefore the minimum in the cloud size is slightly shifted. Based on these considerations one can expect a larger shift of the size minimum towards attractive interactions, the more tunneling and delocalization dominate over the harmonic confinement. This behavior has also been experimentally confirmed [26].

This brief overview on our studies of the attractive Hubbard model underlines the importance of the interplay between finite entropy, interaction and confinement in real experimental systems. In any finite-sized quantum system at finite temperature unusual thermodynamics can emerge. In fact, related arguments have helped to understand the loss of bosonic coherence in a weakly interacting Bose-Fermi mixture discussed in Chap. 8. Generally, the finite entropy effects presented here could help to detect transitions between quantum many-body phases with different entropy capacity. This might be particularly promising for transitions between topological phases [33], whose different topology can lead to very distinctive ways of storing entropy [34].

6.4 Conclusion and Outlook

In this chapter various regimes of the Fermi-Hubbard model have been studied with fermionic ^{40}K spin mixtures in a fully tunable optical lattice potential. For the case of repulsive interactions, we have extracted the global compressibility of the quantum system by measuring the in-situ cloud sizes using phase-contrast imaging. This has allowed us to observe the evolution from a compressible Fermi liquid over an incompressible Mott to a band insulating phase, when the confinement and the interactions are increased. By measuring the suppression of double occupancy, it has been possible to identify regimes of strong correlations. Our studies of attractively interacting

[7] In the limit of strong compression, this corresponds to the maximally packed state of an ideal band insulator.

spin mixtures have revealed an anomalous expansion of the cloud size for increasing attraction. Here, the formation of atom pairs reduces the local entropy capacity, which entails the expansion of the system to accommodate the total entropy.

Together with the works in Tilman Esslinger's group [8, 16, 24, 35, 36], our studies constitute the first steps in the experimental investigation of Fermi-Hubbard models with optical lattices [4, 9, 26]. The observation of the fermionic Mott insulator has been an important achievement. The next milestone will be the creation of low temperature phases of the Fermi-Hubbard model [37–39], namely the BCS-BEC crossover for attractive and antiferromagnetic order for repulsive interactions (see Sect. 3.3.2). The latter will open the path to investigations of quantum magnetism with ultracold atoms [40]. This could ultimately lead to the observation of a d-wave superconducting phase [41, 42] that is believed to emerge within the two-dimensional Hubbard model.

The prospects of fermionic quantum gases in optical lattices are truly amazing, but a number of challenges are awaiting solutions. In order to enter the antiferromagnetically ordered phase, it will be necessary to achieve entropies lower than $S/N \lesssim k_B \ln(2)$. Only below this level excitations on the energy scale J^2/U are suppressed, which result from the superexchange coupling that mediates an effective spin-spin interaction between neighboring lattice sites [19, 43–45]. Such entropies are predicted to require initial temperatures of less than $T/T_{\mathrm{F}} \approx 0.06$ [44] in a homogeneous system. This is about a factor of two lower than the current state of the art. In an optimistic scenario, the entropy redistribution in the inhomogeneous lattice system might actually help to locally generate regions with sufficiently low entropies, although the global entropy per particle may not fall below the required level (see Fig. 6.12). In this case the temperature criterion would be relaxed by a bit, but the antiferromagnetically ordered domains may be small and their detection rather challenging.

Besides the demand for lower initial entropies, it will be important to minimize unwanted excitations of the quantum state during the experimental sequence [7]. On the one hand, technical heating could be minimized. On the other hand, lattice loading should be performed as adiabatically as possible. This has turned out to be surprisingly delicate due to slow time scales for mass redistribution already at moderate interactions [8, 9]. Apart from slower ramping, which has the potentially devastating disadvantage of more technical heating, better adiabaticity may be achieved by developing ramp sequences that minimize mass redistribution (in the presence of interactions) during preparation.

6.4.1 Creation of Low Entropy States

The DMFT analysis in the context of our experimental work has quantitatively revealed remarkable variations in the spatial entropy distribution of the strongly interacting fermions. In large part the entropy is carried by the metallic phases at the edge of the cloud, while the Mott and, even more pronounced, the band insulating

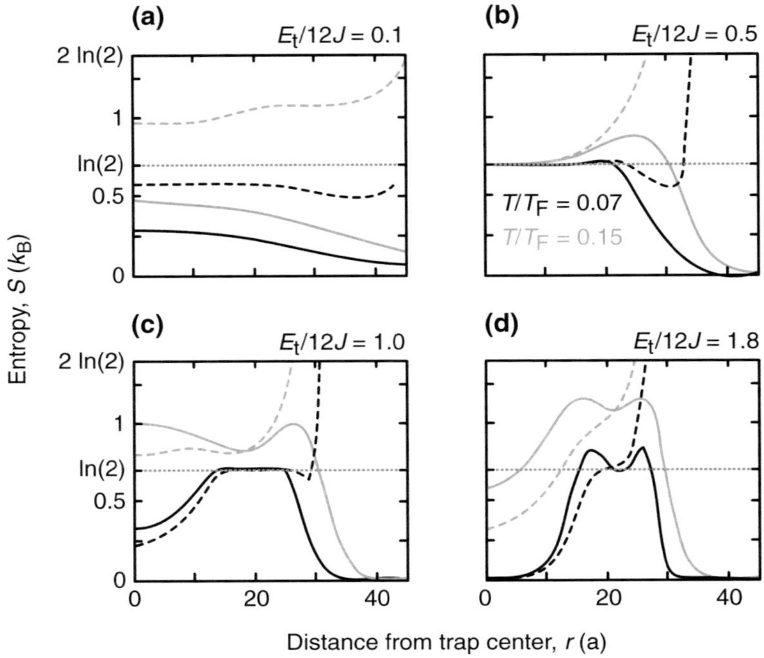

Fig. 6.12 **a–d** Calculated spatial entropy distribution of a balanced spin mixture for various compressions and fixed global atom number N ($N_\uparrow = N_\downarrow = N/2$). The interaction parameter is strongly repulsive $U/12J = 1.5$. The *solid* (*dashed*) *lines* show the entropy per lattice site (per particle) for initial temperatures of $T/T_F = 0.07$ (*black*) and $T/T_F = 0.15$ (*gray*). Identical entropy per particle and lattice site of ln(2) indicates a Mott insulating phase with unity filling, but without magnetic ordering. A Mott insulating core is clearly visible in **b** for both temperatures. The calculations have been performed by Theo Costi, Achim Rosch and collaborators published in the supplementary online material of [4]

regions store relatively little entropy (see Fig. 6.12). This entropic shell structure induced by the underlying trapping potential suggests novel approaches to lower the entropy in an optical lattice.[8]

One possibility might be the creation of a band insulator in the center of the trap and subsequent active removal of the outer high entropy regions [46, 47]. Then, a suitable potential must adiabatically be build around the remaining band insulating core, because otherwise a highly excited many-body state would remain in the trap center. Alternatively, a combined potential of a shallow harmonic trap and a tightly confining dimple could help to create a central region with ultra-low entropy, while the dilute gas in the shallow outer part will carry almost all the entropy [48]. It has

[8] To my knowledge, the first qualitative arguments pointing to these possibilities have been made by Nikolay Prokof'ev during a DARPA OLE team meeting in the summer of 2008 in Cambridge, MA.

been proposed to adiabatically remove the dilute outer shell via clever reshaping of the potentials [49]. The dynamical scheme of quantum distillation [50] suggests to utilize the small tunneling rate of doubly occupied sites: While the low-entropy band insulating core will stay stable [8, 9], the outer metallic shell with high entropy will quickly move away from the trap center upon relaxing the external confinement. Apart from the above schemes for spatial filtering, it might also be promising to perform local atom number filtering. Because much of the entropy is carried by number fluctuations, the selective removal of doubly occupied sites in a region that is expected to be half-filled at zero temperature can effectively cool the sample, as experimentally shown in a recent experiment [51].

Filtering schemes of the above type predict a reduction of the entropy per particle by about a factor of ten. Nevertheless, their implementation is challenging in typical experimental setups that rely on the use of Gaussian beams for the creation of optical potentials and employ imaging systems with medium resolution. One anticipates that it will be much easier to create the required potential landscapes in novel setups with single-site resolution [52, 53] by holographic projection of optical potentials [54]. Such setups will also simplify the observation of low entropy regions, because antiferromagnetic ordering can even be identified if the spatial extent of the ordered region is limited [54, 55].

Further interesting proposals for the creation of low entropy states suggest to bring the lattice quantum gas into thermal contact with a second atomic species (ideally bosons) for which the lattice is transparent (see Sect. 5.4.1). The additional species is supposed to serve as a coolant, which is evaporated from the system after absorbing entropy from the fermionic spin mixture [56–58]. Being similar to the principle of usual sympathetic cooling, the basic idea seems appealing. However, the details of thermalization and the risk of enhanced inelastic three-body collisions between the species are yet to be explored. An alternative route towards studying the low temperature phases of the repulsive Hubbard model makes use of the fact that the lowest temperatures in fermionic quantum gases have been reached for attractive interactions. The study of the attractive side of the phase diagram (see Sect. 3.3.2) might help to also understand the effects on the repulsive side using a mapping based on particle-hole transformation [59].

6.4.2 Detection of Magnetic Ordering

When approaching the ultra-low entropy phases of the repulsive Hubbard model, powerful methods for the detection of antiferromagnetic spin order are needed. Because magnetic ordering happens in real space (see Fig. 3.6), usual time-of-flight imaging, which reveals the momentum distribution, will not be useful. Additionally, it has recently been pointed out that any spin imbalance will lead to the formation of a canted antiferromagnet, in which the staggered order forms in the plane perpendicular to the quantization axis [60, 61]. Slight spin imbalances can hardly be avoided in experiments. Therefore, it will be necessary to reorient the spins along the quantization axis using a $\pi/2$ pulse, which is likely to reduce the "visibility" of the order [62]. In any case, powerful approaches to the detection of antiferromagnetic phases are much in demand.

A meaningful extension to time-of-flight imaging is the analysis of density-density correlations in the quantum noise of absorption images [63] (see Sect. 4.2.3). This technique has been validated for a Mott insulator of bosons [64], a band insulator of noninteracting fermions [65] and patterned loading of bosons in an optical superlattice [66]. For the case of staggered order the emergence of correlation peaks is expected at distances corresponding to integer multiples of $\mathbf{k}_\perp = (\pi/a, \pi/a)$ [as opposed to $(2\pi/a, 2\pi/a)$, see Sect. 4.2.3], where a is the lattice constant [67]. Remarkably, quantum Monte Carlo simulations have shown that shallow precursors of these peaks already appear above the Néel temperature $T_\mathrm{N} \approx 4J^2/U$ heralding the onset of singlet formation of neighboring spins [67]. Nevertheless, noise correlations are a tough experimental observable, because many images recorded under identical conditions (on the order of a hundred) must be averaged to obtain unambiguous signals.

Observables that yield information on the in-situ order in a single experimental shot are highly desirable. One possibility to achieve this goal might be light scattering from the lattice quantum gas, which acts as a refractive medium [55, 62]. This technique would be analogous to neutron or X-ray scattering in solid state physics. Alternatively, external light fields could also be used to probe the excitation spectrum of the quantum gas by means of Bragg [68] or Raman scattering [69]. A variation of this approach is the application of lattice modulation [24, 70, 71] (compare Sect. 5.4.2) and subsequent measurement of the double occupancy [8, 36, 72]. This scheme relies on the assumption that the creation of doubly occupied sites via lattice modulation is enhanced, when the many-body quantum state approaches antiferromagnetic order. One of the advantages of the method is the sensitivity to nearest-neighbor correlations that are expected to emerge already above T_N (see above). However, it is very indirect and requires an accurate quantitative evaluation of double occupancy, which must be compared to a theoretical model.

A model-independent smoking gun for antiferromagnetic order could be obtained via in-situ imaging with single-site and single-atom resolution. Important technical challenges have been mastered in setups that use bosonic ^{87}Rb [52, 53, 55]. There are several more to be mastered in upcoming setups for fermionic species. A future spin resolved in-situ detection will unambiguously identify staggered order simply by looking at it [55, 73]. In particular, surrounding metallic shells in the trapped geometry will not spoil the signal as the magnetic order can be identified entirely on a local level. Above the Néel temperature the dilute metallic wings of the sample may even be used to perform thermometry, for example, by measuring the equation of state [74–77].

References

1. U. Schneider, *Interacting Fermionic Atoms in Optical Lattices—A Quantum Simulator for Condensed Matter Physics*. Ph.D. Thesis, Johannes Gutenberg-Universität Mainz, (2010)
2. J. Hubbard, Electron correlations in narrow energy bands. Proc. R. Soc. Lond. Ser. A **276**, 238 (1963)
3. S. Trotzky, L. Pollet, F. Gerbier, U. Schnorrberger, I. Bloch, N. Prokof'ev, B. Svistunov, M. Troyer, Suppression of the critical temperature for superfluidity near the Mott transition. Nat. Phys. **6**, 998 (2010)
4. U. Schneider, L. Hackermüller, S. Will, T. Best, I. Bloch, T.A. Costi, R.W. Helmes, D. Rasch, A. Rosch, Metallic and insulating phases of repulsively interacting fermions in a 3D optical lattice. Science **322**, 1520 (2008)
5. R.W. Helmes, T.A. Costi, A. Rosch, Mott transition of fermionic atoms in a three-dimensional optical trap. Phys. Rev. Lett. **100**, 056403 (2008)
6. V. Vuletić, C. Chin, A.J. Kerman, S. Chu, Suppression of atomic radiative collisions by tuning the ground state scattering length. Phys. Rev. Lett. **83**, 943 (1999)
7. H. Pichler, A.J. Daley, P. Zoller, Nonequilibrium dynamics of bosonic atoms in optical lattices: decoherence of many-body states due to spontaneous emission. Phys. Rev. A **82**, 063605 (2010)
8. N. Strohmaier, D. Greif, R. Jördens, L. Tarruell, H. Moritz, T. Esslinger, R. Sensarma, D. Pekker, E. Altman, E. Demler, Observation of elastic doublon decay in the fermi-Hubbard model. Phys. Rev. Lett. **104**, 080401 (2010)
9. U. Schneider, L. Hackermüller, J.P. Ronzheimer, S. Will, S. Braun, T. Best, I. Bloch, E. Demler, S. Mandt, D. Rasch, A. Rosch, Fermionic transport and out-of-equilibrium dynamics in a homogeneous Hubbard model with ultracold atoms. Nat. Phys. **8**, 213 (2012)
10. W. Ketterle, D. Durfee, D.M. Stamper-Kurn, Making, probing and understanding Bose-Einstein condensates, in *Bose-Einstein Condensation in Atomic Gases*, ed. by S. Stringari, M. Inguscio, C.E. Wieman (IOS Press, Amsterdam, 1999), pp. 67–176
11. G. Reinaudi, T. Lahaye, Z. Wang, D. Guéry-Odelin, Strong saturation absorption imaging of dense clouds of ultracold atoms. Opt. Lett. **32**, 3143 (2007)
12. C. Regal, *Experimental Realization of BCS-BEC Crossover Physics with a Fermi Gas of Atoms*. Ph.D. Thesis, University of Colorado, (2005)
13. A. Kastberg, W.D. Phillips, S.L. Rolston, R.J.C. Spreeuw, P.S. Jessen, Adiabatic cooling of cesium to 700 nK in an optical lattice. Phys. Rev. Lett. **74**, 1542 (1995)
14. M. Greiner, I. Bloch, O. Mandel, T.W. Hänsch, T. Esslinger, Exploring phase coherence in a 2D lattice of Bose-Einstein condensates. Phys. Rev. Lett. **87**, 160405 (2001)
15. M. Greiner, *Ultracold Quantum Gases in Three-Dimensional Optical Lattice Potentials*. Ph.D. Thesis, LMU München, (2003)
16. M. Köhl, H. Moritz, T. Stöferle, K. Günter, T. Esslinger, Fermionic atoms in a three dimensional optical lattice: observing Fermi surfaces, dynamics, and interactions. Phys. Rev. Lett. **94**, 080403 (2005)
17. A. Georges, G. Kotliar, W. Krauth, M.J. Rozenberg, Dynamical mean-field theory of strongly correlated fermion systems and the limit of infinite dimensions. Rev. Mod. Phys. **68**, 13 (1996)
18. G. Kotliar, D. Vollhardt, Strongly correlated materials: insights from dynamical mean-field theory. Phys. Today **57**, 53 (2004)
19. L. De Leo, C. Kollath, A. Georges, M. Ferrero, O. Parcollet, Trapping and cooling fermionic atoms into Mott and Néel states. Phys. Rev. Lett. **101**, 210403 (2008)
20. R. Bulla, T.A. Costi, D. Vollhardt, Finite-temperature numerical renormalization group study of the Mott transition. Phys. Rev. B **64**, 045103 (2001)
21. R. Bulla, T.A. Costi, T. Pruschke, Numerical renormalization group method for quantum impurity systems. Rev. Mod. Phys. **80**, 395 (2008)
22. R.W. Helmes, *Dynamical Mean Field Theory of Inhomogeneous Correlated Systems*. Ph.D. Thesis, University of Cologne, (2008)
23. R.W. Helmes, T.A. Costi, A. Rosch, Kondo proximity effect: how does a metal penetrate into a Mott insulator? Phys. Rev. Lett. **101**, 066802 (2008)

24. R. Jördens, N. Strohmaier, K. Günter, H. Moritz, T. Esslinger, A Mott insulator of fermionic atoms in an optical lattice. Nature **455**, 204 (2008)
25. C.A. Regal, C. Ticknor, J.L. Bohn, D.S. Jin, Creation of ultracold molecules from a Fermi gas of atoms. Nature **424**, 47 (2003)
26. L. Hackermüller, U. Schneider, M. Moreno-Cardoner, T. Kitagawa, T. Best, S. Will, E. Demler, E. Altman, I. Bloch, B. Paredes, Anomalous expansion of attractively interacting fermionic atoms in an optical lattice. Science **327**, 1621 (2010)
27. A. Auerbach, *Interacting Electrons and Quantum Magnetism* (Springer, New York, 1994)
28. M. Randeria, N. Trivedi, A. Moreo, R.T. Scalettar, Pairing and spin gap in the normal state of short coherence length superconductors. Phys. Rev. Lett. **69**, 2001 (1992)
29. A. Toschi, P. Barone, M. Capone, C. Castellani, Pairing and superconductivity from weak to strong coupling in the attractive Hubbard model. New J. Phys. **7**, 7 (2005)
30. C.-C. Chien, Q. Chen, K. Levin, Fermions with attractive interactions on optical lattices and implications for correlated systems. Phys. Rev. A **78**, 043612 (2008)
31. T. Paiva, R. Scalettar, M. Randeria, N. Trivedi, Fermions in 2D optical lattices: temperature and entropy scales for observing antiferromagnetism and superfluidity. Phys. Rev. Lett. **104**, 066406 (2010)
32. R.C. Richardson, The Pomeranchuk effect. Rev. Mod. Phys. **69**, 683 (1997)
33. C. Nayak, S.H. Simon, A. Stern, M. Freedman, S. Das Sarma, Non-Abelian anyons and topological quantum computation. Rev. Mod. Phys. **80**, 1083 (2008)
34. N.R. Cooper, A. Stern, Observable bulk signatures of non-Abelian quantum Hall states. Phys. Rev. Lett. **102**, 176807 (2009)
35. N. Strohmaier, Y. Takasu, K. Günter, R. Jördens, M. Köhl, H. Moritz, T. Esslinger, Interaction-controlled transport of an ultracold Fermi gas. Phys. Rev. Lett. **99**, 220601 (2007)
36. R. Jördens, L. Tarruell, D. Greif, T. Uehlinger, N. Strohmaier, H. Moritz, T. Esslinger, L. De Leo, C. Kollath, A. Georges, V. Scarola, L. Pollet, E. Burovski, E. Kozik, M. Troyer, Quantitative determination of temperature in the approach to magnetic order of ultracold fermions in an optical lattice. Phys. Rev. Lett. **104**, 180401 (2010)
37. R. Micnas, J. Ranninger, S. Robaszkiewicz, Superconductivity in narrow-band systems with local nonretarded attractive interactions. Rev. Mod. Phys. **62**, 113 (1990)
38. P.A. Lee, N. Nagaosa, X.-G. Wen, Doping a Mott insulator: physics of high-temperature superconductivity. Rev. Mod. Phys. **78**, 17 (2006)
39. W. Hofstetter, J.I. Cirac, P. Zoller, E. Demler, M.D. Lukin, High-temperature superfluidity of fermionic atoms in optical lattices. Phys. Rev. Lett. **89**, 220407 (2002)
40. M. Lewenstein, A. Sanpera, V. Ahufinger, B. Damski, A. Sen, U. Sen, Ultracold atomic gases in optical lattices: mimicking condensed matter physics and beyond. Adv. Phys. **56**, 243 (2007)
41. S. Trebst, U. Schollwöck, M. Troyer, P. Zoller, d-Wave resonating valence bond states of fermionic atoms in optical lattices. Phys. Rev. Lett. **96**, 250402 (2006)
42. A.M. Rey, R. Sensarma, S. Fölling, M. Greiner, E. Demler, M.D. Lukin, Controlled preparation and detection of d-wave superfluidity in two-dimensional optical superlattices. Europhys. Lett. **87**, 60001 (2009)
43. F. Werner, O. Parcollet, A. Georges, S.R. Hassan, Interaction-induced adiabatic cooling and antiferromagnetism of cold fermions in optical lattices. Phys. Rev. Lett. **95**, 056401 (2005)
44. A. Koetsier, R.A. Duine, I. Bloch, H.T.C. Stoof, Achieving the Néel state in an optical lattice. Phys. Rev. A **77**, 023623 (2008)
45. M. Snoek, I. Titvinidze, C. Tőke, K. Byczuk, W. Hofstetter, Antiferromagnetic order of strongly interacting fermions in a trap: real-space dynamical mean-field analysis. New J. Phys. **10**, 093008 (2008)
46. T. Esslinger, Fermi-Hubbard physics with atoms in an optical lattice. Annu. Rev. Condens. Matter Phys. **1**, 129 (2010)
47. D.C. McKay, B. DeMarco, Cooling in strongly correlated optical lattices: prospects and challenges. Rep. Prog. Phys. **74**, 054401 (2011)
48. D.M. Stamper-Kurn, H.-J. Miesner, A.P. Chikkatur, S. Inouye, J. Stenger, W. Ketterle, Reversible formation of a Bose-Einstein condensate. Phys. Rev. Lett. **81**, 2194 (1998)

49. J.-S. Bernier, C. Kollath, A. Georges, L. De Leo, F. Gerbier, C. Salomon, M. Köhl, Cooling fermionic atoms in optical lattices by shaping the confinement. Phys. Rev. A **79**, 061601 (2009)
50. F. Heidrich-Meisner, S.R. Manmana, M. Rigol, A. Muramatsu, A.E. Feiguin, E. Dagotto, Quantum distillation: dynamical generation of low-entropy states of strongly correlated fermions in an optical lattice. Phys. Rev. A **80**, 041603 (2009)
51. W.S. Bakr, P.M. Preiss, M.E. Tai, R. Ma, J. Simon, M. Greiner, Orbital excitation blockade and algorithmic cooling in quantum gases. Nature **480**, 500 (2011)
52. W. Bakr, J. Gillen, A. Peng, S. Fölling, M. Greiner, A quantum gas microscope for detecting single atoms in a Hubbard-regime optical lattice. Nature **462**, 74 (2009)
53. J. Sherson, C. Weitenberg, M. Endres, M. Cheneau, I. Bloch, S. Kuhr, Single-atom-resolved fluorescence imaging of an atomic Mott insulator. Nature **467**, 68 (2010)
54. W.S. Bakr, A. Peng, M.E. Tai, R. Ma, J. Simon, J.I. Gillen, S. Fölling, L. Pollet, M. Greiner, Probing the superfluid-to-Mott insulator transition at the single-atom level. Science **329**, 547 (2010)
55. C. Weitenberg, M. Endres, J. Sherson, M. Cheneau, P. Schauß, T. Fukuhara, I. Bloch, S. Kuhr, Single-spin addressing in an atomic Mott insulator. Nature **471**, 319 (2011)
56. T.-L. Ho, Q. Zhou, Squeezing out the entropy of fermions in optical lattices. Proc. Natl. Acad. Sci. U S A **106**, 6916 (2009)
57. J. Catani, G. Barontini, G. Lamporesi, F. Rabatti, G. Thalhammer, F. Minardi, S. Stringari, M. Inguscio, Entropy exchange in a mixture of ultracold atoms. Phys. Rev. Lett. **103**, 140401 (2009)
58. D. McKay, B. DeMarco, Thermometry with spin-dependent lattices. New J. Phys. **12**, 055013 (2010)
59. A.F. Ho, M.A. Cazalilla, T. Giamarchi, Quantum simulation of the Hubbard model: the attractive route. Phys. Rev. A **79**, 033620 (2009)
60. T. Gottwald, P.G.J. van Dongen, Antiferromagnetic order of repulsively interacting fermions on optical lattices. Phys. Rev. A **80**, 033603 (2009)
61. B. Wunsch, L. Fritz, N.T. Zinner, E. Manousakis, E. Demler, Magnetic structure of an imbalanced Fermi gas in an optical lattice. Phys. Rev. A **81**, 013616 (2010)
62. T.A. Corcovilos, S.K. Baur, J.M. Hitchcock, E.J. Mueller, R.G. Hulet, Detecting antiferromagnetism of atoms in an optical lattice via optical Bragg scattering. Phys. Rev. A **81**, 013415 (2010)
63. E. Altman, E. Demler, M.D. Lukin, Probing many-body states of ultracold atoms via noise correlations. Phys. Rev. A **70**, 013603 (2004)
64. S. Fölling, F. Gerbier, A. Widera, O. Mandel, T. Gericke, I. Bloch, Spatial quantum noise interferometry in expanding ultracold atom clouds. Nature **434**, 481 (2005)
65. T. Rom, T. Best, D. van Oosten, U. Schneider, S. Fölling, B. Paredes, I. Bloch, Free fermion antibunching in a degenerate atomic Fermi gas released from an optical lattice. Nature **444**, 733 (2006)
66. S. Fölling, *Probing Strongly Correlated States of Ultracold Atoms in Optical Lattices*. Ph.D. Thesis, Johannes Gutenberg-Universität Mainz, (2008)
67. G.M. Bruun, O.F. Syljuåsen, K.G.L. Pedersen, B.M. Andersen, E. Demler, A.S. Sørensen, Antiferromagnetic noise correlations in optical lattices. Phys. Rev. A **80**, 033622 (2009)
68. P. Ernst, S. Götze, J. Krauser, K. Pyka, D.-S. Lühmann, D. Pfannkuche, K. Sengstock, Probing superfluids in optical lattices by momentum-resolved Bragg spectroscopy. Nat. Phys. **6**, 56 (2009)
69. J.T. Stewart, J.P. Gaebler, D.S. Jin, Using photoemission spectroscopy to probe a strongly interacting Fermi gas. Nature **454**, 744 (2008)
70. T. Stöferle, H. Moritz, C. Schori, M. Köhl, T. Esslinger, Transition from a strongly interacting 1D superfluid to a Mott insulator. Phys. Rev. Lett. **92**, 130403 (2004)
71. C. Kollath, A. Iucci, I.P. McCulloch, T. Giamarchi, Modulation spectroscopy with ultracold fermions in an optical lattice. Phys. Rev. A **74**, 041604 (2006)
72. D. Greif, L. Tarruell, T. Uehlinger, R. Jördens, T. Esslinger, Probing nearest-neighbor correlations of ultracold fermions in an optical lattice. Phys. Rev. Lett. **106**, 145302 (2011)

73. J. Simon, W. Bakr, R. Ma, E. Tai, P. Preiss, M. Greiner, Quantum simulation of an antiferromagnetic spin chain in an optical lattice. Nature **472**, 307 (2011)
74. T.-L. Ho, Q. Zhou, Obtaining the phase diagram and thermodynamic quantities of bulk systems from the densities of trapped gases. Nat. Phys. **6**, 131 (2010)
75. P.N. Ma, L. Pollet, M. Troyer, Measuring the equation of state of trapped ultracold bosonic systems in an optical lattice with in situ density imaging. Phys. Rev. A **82**, 033627 (2010)
76. S. Nascimbéne, N. Navon, K.J. Jiang, F. Chevy, C. Salomon, Exploring the thermodynamics of a universal Fermi gas. Nature **463**, 1057 (2010)
77. N. Navon, S. Nascimbene, F. Chevy, C. Salomon, The equation of state of a low-temperature Fermi gas with tunable interactions. Science **328**, 729 (2010)

Chapter 7
Quantum Revival Spectroscopy and Multi-Body Interactions

Interactions lie at the heart of correlated many-body quantum phases [1–3]. Typically, the interactions between microscopic particles, such as electrons, atoms, or molecules, are described as two-body interactions. Only in rare cases, such as strongly interacting nuclear matter, higher-body interactions are considered by introducing a new energy scale, whenever three or more particles collide. Recent theoretical studies have shown that such multi-body interactions may give rise to novel, exotic many-body quantum phases in systems of ultracold quantum gases [4–6]. So far, higher-body interactions in ultracold atomic systems have only been observed as *inelastic* loss resonances in three- and four-body recombinations of atom-atom and atom-dimer collisions [7–11]. *Elastic* multi-body interactions, however, have been elusive, because a method to detect them has not been available.

In this chapter, I report on a novel method to precisely measure the interactions among ultracold atoms residing on the sites of an optical lattice: quantum revival spectroscopy. The technique resolves the number of atoms involved into on-site collisions and detects the corresponding interaction strengths on an absolute energy scale. The energy difference between the interacting and the noninteracting system is directly accessed as a result of an intrinsic interference effect.

Equipped with these capabilities, the presence of effective multi-body interactions is revealed in systems of bosonic atoms that are tightly confined at single lattice sites. Such multi-body interactions emerge from two-body collisions through virtual transitions of atoms from the lowest to higher vibrational states [12]. This process underlines the relevance of multi-band physics for interacting atoms in optical lattices. Additionally, the precision measurement of interaction energies provides input on the question, how accurately optical lattice systems are described by single-band Hubbard models or, vice-versa, how well single-band Hubbard models are realized. This assessment is crucial for the comparison of optical lattice quantum simulators with many-body quantum theory [13].

In the first part of this chapter, the origin of quantum phase revival dynamics is explained on general grounds using a simplified single-orbital single site model. For the case of ideal coherent states it is shown, that collapse and revival dynamics

S. Will, *From Atom Optics to Quantum Simulation*, Springer Theses,
DOI: 10.1007/978-3-642-33633-1_7,

arise, when the eigenenergies of the underlying Fock states depend nonlinearly on the particle number [14–20]. In the second part, I discuss how these dynamics can be practically realized and observed with ultracold bosonic atoms in an optical lattice [21–23]. By establishing a direct connection between experimental and theoretical observables, foundations are laid for quantum revival spectroscopy that allows to accurately measure the interaction energies of individual Fock states. It is additionally shown that the atom number statistics at a lattice site can be observed, similar to foundational experiments in cavity quantum electrodynamics that yield the statistics of a cavity photon field [24–26]. Subsequently, two theoretical models for the on-site Fock state energies are discussed and it is demonstrated, how multi-orbital physics gives rise to effective multi-body interactions. The third part reports on the observation of long-lived collapse and revival dynamics of bosonic ^{87}Rb on the sites of an optical lattice. Fourier transform of the resulting time traces yields the energies of individual Fock states with high spectral resolution. Comparing the measured Fock state energies to calculations that rely on the exact diagonalization of a multi-orbital single site system, we find excellent agreement. Based on these data the strengths of the multi-body interactions are accurately extracted. Quantum revival spectroscopy is also used to track the change in the on-site atom number statistics from Poissonian to number-squeezed when approaching the Mott insulator transition. The chapter concludes with two excursions that propose the creation and detection of a Schrödinger cat and exotic condensate states by means of collapse and revival dynamics.

7.1 Quantum Evolution of Coherent States

Coherent states represent the most stable field solution in physics [25, 27]: Formed by a distinct superposition of number states (Fock states) they are insensitive to the removal of particles. Coherent states are characterized by a single amplitude and phase and are used extensively in the quantum description of classical coherent fields, ranging from laser light to coherent matter waves in superconductors, superfluids and atomic Bose-Einstein condensates (BECs). Their remarkable properties also play a crucial role in the path integral formalism of many-body perturbation theory [1, 28].

When coherent states are used to describe matter wave fields, such as Bose-Einstein condensates, the interactions between the underlying particles can give rise to a nonequilibrium phase evolution of the Fock states constituting the coherent state. When this phase evolution depends nonlinearly on the particle number, the coherent state undergoes a sequence of collapses and revivals [21–23]. In such a sequence, the quantum state first evolves into a highly correlated and entangled state in which at the time of the collapse the field amplitude vanishes. However, when the evolution progresses, the entanglement is unravelled again and the original coherent field is ideally recreated. This section explains the inner workings of the collapse and revival phenomenon focussing on the case of interacting atoms [14, 15, 20]. However, remarkable examples of such collapse and revival dynamics have also been observed for a coherent light field interacting with a single atom in cavity quantum electrodynamics [24] or for a classical oscillation of a single ion stored in a trap [29].

7.1.1 Coherent States

We consider a confining potential, such as a box or a harmonic oscillator, whose ground state is occupied by n bosonic atoms (see Chap. 2). The higher lying excited states of the potential are omitted, which means the system is treated in a single-mode approximation. The occupation of the ground state mode is described by Fock states $|n\rangle$, where n can be any positive integer number including zero, constituting the basis of the Fock space. The operators $\hat{a}$ and $\hat{a}^\dagger$ denote the annihilation and creation operators, respectively, which reduce and increase the occupation of the ground state mode by one atom, They are defined by

$$\hat{a}|n\rangle = \sqrt{n}|n-1\rangle, \tag{7.1}$$

$$\hat{a}^\dagger|n\rangle = \sqrt{n+1}|n+1\rangle. \tag{7.2}$$

Fock states are the eigenstates of the atom number operator $\hat{n} = \hat{a}^\dagger\hat{a}$ and fulfill $\hat{n}|n\rangle = n|n\rangle$ as follows directly from the above relations.

Coherent states, on the other hand, are defined by the property of being eigenstates of the annihilation operator $\hat{a}$,

$$\hat{a}|\alpha\rangle = \alpha|\alpha\rangle. \tag{7.3}$$

Thus the action of the annihilation operator, which corresponds to the removal of an atom, leaves a coherent state unchanged besides multiplication with the eigenvalue α. Generally, this eigenvalue is a complex number $\alpha = |\alpha|e^{i\varphi}$ and the coherent state $|\alpha\rangle$ is uniquely defined by the amplitude $|\alpha|$ and the phase φ. Using the defining Eq. 7.3, the structure of coherent states in the Fock basis can be straightforwardly derived. The general form of a coherent state is given by a coherent superposition of Fock states,

$$|\alpha\rangle = e^{-|\alpha|^2/2}\sum_{n=0}^{\infty}\frac{\alpha^n}{\sqrt{n!}}|n\rangle, \tag{7.4}$$

where the statistical weight of the Fock states $|n\rangle$ follows a Poisson distribution $P(n) = e^{-|\alpha|^2}|\alpha|^{2n}/n!$ and the exact atom number remains uncertain. The mean atom number is given by $\langle\hat{n}\rangle = |\alpha|^2 \equiv \bar{n}$ and the variance $\langle\hat{n}^2\rangle - \langle\hat{n}\rangle^2 = \langle\hat{n}\rangle = \bar{n}$ characterizes the fluctuations. Accordingly, we may rewrite the eigenvalue as $\alpha = \sqrt{\bar{n}}\cdot e^{i\varphi}$.

Coherent states have originally been introduced in the quantum description of electro-magnetic waves. The operator of the electric field of a single light mode is given by $\hat{E}(\chi) \propto \hat{a}e^{-i\chi} + \hat{a}^\dagger e^{i\chi}$ [30]. When the corresponding mode is occupied by a coherent state, the expectation value of the electric field $\langle\alpha|\hat{E}(\chi)|\alpha\rangle$ is a cosine wave $|\alpha|\cos(\chi-\varphi)$ with a well-defined macroscopic phase corresponding to a classical coherent wave. The intensity of the light field is proportional to $|\alpha|^2=\bar{n}$, which is the mean number of photons occupying the mode.

A Bose-Einstein condensate is the matter wave analogue of a classical coherent wave. When a Bose-Einstein condensate of *noninteracting* atoms is abruptly split by deforming the confining potential into a double well, the ground state orbital of each well is occupied by a superposition of Fock states, which ideally corresponds to a coherent state. Both parts may be viewed as classical waves and their coherence can be probed in an interference experiment; when the phase of the interference pattern is stationary in consecutive realizations of the experiment, the matter wave fields have a fixed relative phase [31–33].[1] When a Bose-Einstein condensate of *interacting* atoms is split, the quantum states of the two parts do not necessarily correspond to ideal coherent states, but are composed of a superposition of Fock states with a narrower "squeezed" atom number distribution [33].

The general atom number superposition state

$$|\psi\rangle=\sum_{n=0}^{\infty}c_n|n\rangle \tag{7.5}$$

can describe pure Fock states as well as coherent and number-squeezed states. The coefficients c_n denote the probability amplitude of the corresponding Fock state $|n\rangle$ and fulfill the normalization condition $\sum_{n=0}^{\infty}|c_n|^2=1$.

7.1.2 Interactions and Time Evolution of Fock States

Interactions between ultracold atoms are typically described by binary s-wave collisions that are parametrized by a single parameter, the s-wave scattering length a_s (see Sect. 2.3). Higher partial waves can be neglected due to low collision energies and inelastic three-body losses can be minimized by choosing low atom densities and short interaction times. The ^{87}Rb atoms in the $|F,m_F\rangle=|1,+1\rangle$ hyperfine state used in the experiments of this chapter interact repulsively with a positive s-wave scattering length of about $a_s\approx 102\,a_0$ [34, 35].

When two identical repulsively interacting atoms occupy the ground state of the confining potential, the interatomic collisions raise the total energy of the system by an amount U. This interaction energy per atom pair is generally given by

$$U=\frac{4\pi\hbar^2a_s}{m}\int|\phi(\mathrm{r})|^4\,d^3r, \tag{7.6}$$

---

[1] In case the phase relation between the two halves is not fixed, the phase will fluctuate between consecutive realizations of the experiment [20].

where m is the mass of the colliding atoms and $\phi(\mathbf{r})$ the spatial orbital of the atoms. The integral $\int |\phi(\mathrm{r})|^4 \, d^3r$ quantifies the overlap of the atomic densities, $n(\mathbf{r}) = |\phi(\mathrm{r})|^2$.

Assuming weak interactions, the spatial orbital $\phi(\mathbf{r})$ is often identified with the single-particle ground state orbital $\phi_0(\mathbf{r})$ of the confining potential. This is justified when U is much smaller than the energy gap between the ground state and the higher lying excited states, such that multi-orbital effects can be neglected; specifically for a harmonic oscillator potential with frequency ω the condition $U \ll \hbar\omega$ must be fulfilled. In the case of a single site of an optical lattice, the ground state orbital $\phi_0(\mathbf{r})$ is often identified with the maximally localized wavefunction in the first lattice band, the Wannier function $w(\mathbf{r})$.

Based on this single-orbital approximation it is straightforward to derive the total interaction energy when a Fock state $|n\rangle$ occupies the ground state. The interaction of $n(n-1)/2$ atom pairs gives rise to the single-orbital Fock state energy (see Fig. 7.1a)

$$E_n^{\mathrm{SO}} = \frac{U}{2} n(n-1). \tag{7.7}$$

This is the eigenenergy of the Hamiltonian

$$\hat{H}^{\mathrm{SO}} = \frac{U}{2} \hat{n}(\hat{n}-1), \tag{7.8}$$

which is identical to the zero tunneling limit ($J \to 0$) of the Bose-Hubbard Hamiltonian at a single lattice site [2, 36, 37] (see Chap. 3). The condition for the single-orbital approximation to hold for larger atom numbers n is even stricter, as E_n^{SO} must be much smaller than the energy gap between ground and excited states to ensure that U is independent of the atom number. For a harmonic potential this means $E_n^{\mathrm{SO}} \ll \hbar\omega$. We note that the total energy of a Fock state $|n\rangle$ comprises not only E_n^{SO} but also the zero point energy of the given confining potential must be added; e.g. for the case of a three-dimensional isotropic harmonic potential the zero point energy $n \cdot 3\hbar\omega/2$ must be taken into account even when the atoms do not interact. However, this offset can typically be neglected as it does not have an observable impact on the dynamical evolution of the Fock states, which will become clear below.

Within the single-orbital approximation the time evolution of a Fock state $|n\rangle$ is determined by the Hamiltonian $\hat{H}^{\mathrm{SO}}$ with $\hat{H}^{\mathrm{SO}}|n\rangle = E_n^{\mathrm{SO}}|n\rangle$ and we get

$$|n\rangle(t) = e^{-iE_n^{\mathrm{SO}}t/\hbar}|n\rangle = e^{-iUn(n-1)t/2\hbar}|n\rangle. \tag{7.9}$$

Hence, the collisional phase shift acquired over time is quadratic in the atom number. For a single Fock state this merely leads to an unobservable global phase, but for a superposition of Fock states the nonlinearity in the atom number gives rise to an intriguing dynamical evolution.

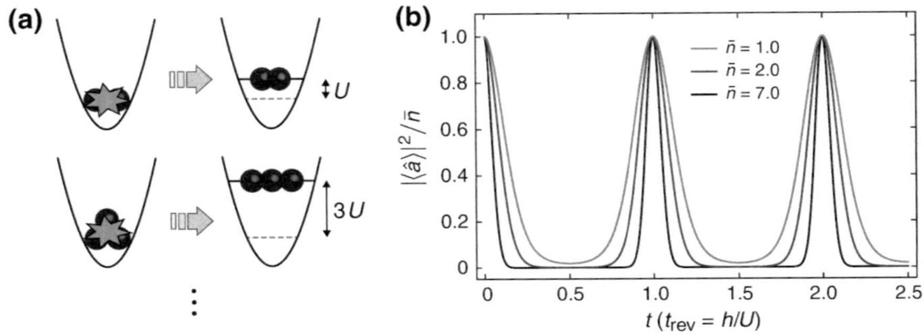

Fig. 7.1 Interactions in a single-orbital system and quantum phase revival dynamics. **a** Interacting atoms in the ground state of a confining potential undergo coherent collisions. If only the single ground state orbital is taken into account, the interaction energy per atom pair U is independent of the filling n as given by Eq. 7.7. **b** Under such conditions a coherent matter wave field undergoes perfectly periodic collapses and revivals. The revival time $t_{\text{rev}} = h/U$ is inversely proportional to the interaction energy U, while the collapse time $t_{\text{coll}} = \hbar/(\sqrt{\bar{n}}U)$ is additionally inversely proportional to $\sqrt{\bar{n}}$, which is the standard deviation of the Poisson distribution

7.1.3 Time Evolution of Coherent States

Coherent states are formed by a superposition of Fock states and as such they are no eigenstates of the Hamiltonian governing the system (Eq. 7.8). In fact, each Fock state individually evolves according to Eq. 7.9 and the complete evolution is given by

$$|\psi_\alpha(t)\rangle = e^{-|\alpha|^2/2} \sum_{n=0}^{\infty} \frac{\alpha^n}{\sqrt{n!}} \cdot e^{-iUn(n-1)t/2\hbar} |n\rangle. \tag{7.10}$$

The quadratic dependence of the collisional phase on the atom number causes the Fock states to dephase and $|\psi_\alpha(t)\rangle$ evolves away from a coherent state loosing its initially well-defined macroscopic phase. This would not be the case if the eigenenergies had a linear atom number dependence. Here the phase factor of each Fock state could be included into the definition of α, giving rise to a time dependent macroscopic phase, while the state itself would remain coherent.

The dynamical evolution of $|\psi_\alpha(t)\rangle$ cannot be observed directly; it is only accessible through the time dependence of a physical observable. In analogy to the quantum description of the electric field (see above), the property that quantifies the presence of a macroscopic phase is given by the expectation value of the field operator, $\langle\hat{a}\rangle(t) = \langle\psi_\alpha(t)|\hat{a}|\psi_\alpha(t)\rangle$. This expectation value, which is called the field amplitude in the following, generally has a complex value and reads [15, 19]:

$$\langle\hat{a}\rangle(t) = \alpha \cdot e^{-\bar{n}} \sum_{n=0}^{\infty} \frac{\bar{n}^n}{n!} \cdot e^{-inUt/\hbar}. \tag{7.11}$$

We show below, that the matter wave dynamics of a coherent state can be experimentally detected as the squared modulus of $\langle\psi_\alpha(t)|\hat{a}|\psi_\alpha(t)\rangle$, complying with the requirement that physical observables assume real values. The dynamical evolution is observed as

$$|\langle\hat{a}\rangle(t)|^2 = \bar{n}\cdot e^{-2\bar{n}}\cdot\exp\left(\bar{n}\left(e^{-iUt/\hbar}+e^{iUt/\hbar}\right)\right)$$
$$= \bar{n}\cdot\exp\left(2\bar{n}\left(\cos\left(\frac{Ut}{\hbar}\right)-1\right)\right). \qquad (7.12)$$

This result describes an oscillatory behavior (see Fig. 7.1b) that is governed by the fundamental frequency U/h and its higher harmonics $n\cdot U/h$.

Collapse

For short times ($t \ll h/U$) Eq. 7.12 can be approximated by

$$|\langle\hat{a}\rangle(t)|^2 \simeq \bar{n}\cdot\exp\left(-2\bar{n}U^2t^2/\hbar^2\right), \qquad (7.13)$$

expressing the collapse of the matter wave field. The width of this Gaussian decay defines the characteristic collapse time $t_{\text{coll}} = \hbar/(\sqrt{\bar{n}}U)$, where $\sqrt{\bar{n}}$ is the standard deviation of the Poisson distribution. This result is intuitively clear as the mutual dephasing of Fock states is expected to happen the faster the broader the atom number distribution is, as illustrated in Fig. 7.1b.

Revival

Due to the discreteness of the atom number distribution, the coherent field is fully restored at the revival time $t_{\text{rev}} = h/U$ when each Fock state has acquired a phase shift which is an integer multiple of 2π. According to Eq. 7.9 at t_{rev} the Fock states with zero and one atom have not evolved at all, the state with two atoms has gained a phase of 2π, the state with three atoms $3\cdot 2\pi$ and so on. The phases of all Fock states are identical modulo 2π and the initial coherent state is recreated, $|\psi_\alpha(t_{\text{rev}})\rangle = |\psi_\alpha(0)\rangle \equiv |\alpha\rangle$.[2] The rephasing repeats itself after integer multiples of t_{rev} and gives rise to strictly periodic collapse and revival dynamics of the matter wave field amplitude.

---

[2] This statement is actually not correct when the zero point energy of each Fock state is taken into, e.g. $n\cdot 3\hbar\omega/2$ for a three-dimensional harmonic confinement. However, the corresponding phase shifts drop out when the time evolution $|\langle\hat{a}\rangle(t)|^2$ is calculated due to the linearity in n.

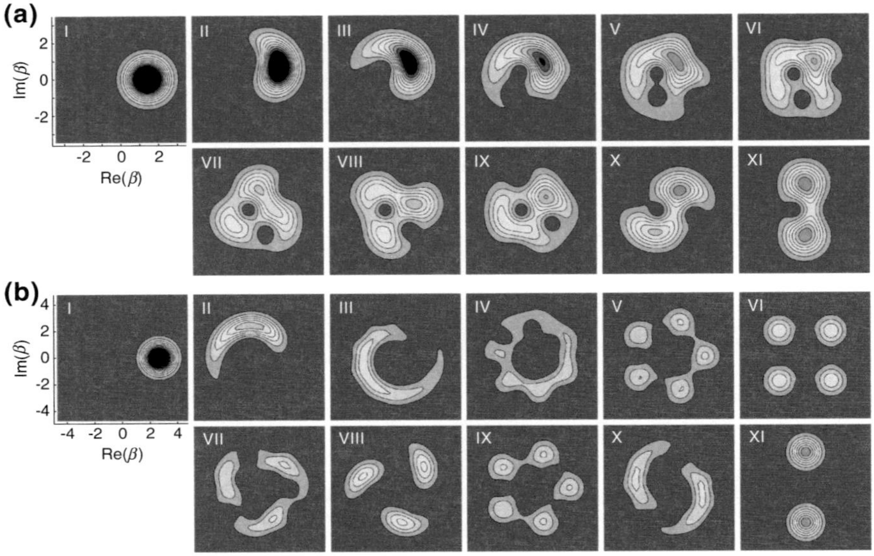

Fig. 7.2 Matter wave dynamics of a coherent state in a single-orbital system visualized by the Q-function. The coherent states $|\beta\rangle$ are characterized by points in the complex plane. The contour plots show their overlap with $|\psi_\alpha(t)\rangle$ at discrete times t ranging from 0 (*I*) to $t_{\text{rev}}/2$ (*XI*) scanned in steps of $t_{\text{rev}}/20$. $|\psi_\alpha(t)\rangle$ evolves away from a coherent state until it forms a Schrödinger cat-type superposition of two quasi-coherent states with a phase difference of π at the time of maximal collapse ($t_{\text{rev}}/2$). Panel **a** shows the evolution for $\bar{n} = 2.0$ and **b** for $\bar{n} = 7.0$, where the latter reveals even more complex superpositions of quasi-coherent states at intermediate times. Contour lines are spaced by 0.02 and the maximal value of the Q-function is $1/\pi \approx 0.32$ by definition

7.1.4 Visualization of the Time Evolution

In quantum optics the Q-function is commonly used as a visual representation of quantum fields [25, 38]. Here we employ

$$Q(\beta, t) = \frac{|\langle\beta|\psi_\alpha(t)\rangle|^2}{\pi} \tag{7.14}$$

to visualize the collapse and revival dynamics of the matter wave field $|\psi_\alpha(t)\rangle$. It quantifies the overlap between the evolving quantum state $|\psi_\alpha(t)\rangle$ and a coherent state $|\beta\rangle$ with a mean atom number $\bar{n}_\beta = |\beta|^2$ and phase $\varphi_\beta = \arg(\beta)$. The Q-function has the maximum value $1/\pi$ for perfect overlap.

Figure 7.2 shows the Q-function of the state $|\psi_\alpha(t)\rangle$ at discrete times of the evolution. In the beginning, $|\psi_\alpha(0)\rangle$ corresponds to a coherent state with $\alpha = \sqrt{\bar{n}}$, where we have chosen $\varphi = 0$ without loss of generality. The Q-function is maximal for $\beta = \alpha$, but also has finite values in a circular region around α. The finite overlap with neighboring coherent states signifies that coherent states are quasi-orthogonal forming an overcomplete basis of the Fock space.

During the quantum evolution, $|\psi_\alpha(t)\rangle$ moves away from a coherent state until the macroscopic phase becomes maximally uncertain at $t = t_{\text{rev}}/2$. Here, the state forms a coherent superposition of two quasi-coherent states $|\alpha_1\rangle$ and $|\alpha_2\rangle$ with complex phases $\alpha_{1,2} = \pm i\alpha$.[3] Because the eigenvalues of the superimposed states have the same absolute value, but an opposite sign, the expectation value $\langle\hat{a}\rangle$ vanishes. The superposition of two macroscopic quantum fields at this point of maximal collapse can be viewed as a Schrödinger cat state that is discussed in more detail in Sect. 7.4. In Fig. 7.2b even more complex superpositions of quasi-coherent states can be identified; those appear at higher mean fillings and can also be found for noninteger values of $\bar{n}$. For example, four quasi-coherent states with relative phases of $\pi/2$ are superimposed at $t = t_{\text{rev}}/4$ [see Fig. 7.2b (VI)].

7.2 Quantum Revival Spectroscopy

When a superfluid of bosonic atoms is loaded into a shallow optical lattice, the quantum state at a single lattice site is effectively indistinguishable from a coherent state. The individual sites can be isolated from each other by abruptly increasing the lattice depth and a huge array of coherent states is formed. The collapse and revival dynamics are realized on many thousand lattice sites in parallel.

In this section, we discuss how the dynamics can be detected on a macroscopic scale after time-of-flight expansion and identify, which conclusions about the on-site dynamics can be drawn from this global view. These considerations form the foundations for using quantum phase revivals as a spectroscopic technique. In addition, we highlight that the on-site physics in an optical lattice can generally not be understood in the single-orbital framework of the previous section, because the band structure gives rise to multi-orbital effects. Interactions can promote atoms to higher-lying orbitals and effectively change the shape of the on-site wavefunction as a function of the atom number. We present an instructive physical picture of the process that is provided by effective field theory: Virtual transitions of atoms to excited orbitals generate effective multi-body interactions as higher-order corrections to the single-orbital two-body interaction per atom pair U. It is shown that quantum phase revivals are an ideal experimental tool to detect such multi-body interactions.

---

[3] It is worth mentioning that the superimposed states are not perfectly coherent states, which is also signaled by the fact that the Q-function does only reach half its maximal value. The exact nature of the state is discussed in Sect. 7.4.

7.2.1 Coherent and Number-Squeezed States in Optical Lattices

A coherent state of massive particles cannot be realized in an isolated potential, because it is not compatible with particle number conservation. However, by splitting a Bose-Einstein condensate into two or more parts it is possible to create an array of states that are essentially indistinguishable from coherent states. The splitting procedure should be nonadiabatic, such that each atom remains in a delocalized state with a finite probability to be found in any part. The many-body state of the global system is highly entangled and the coupling between the parts should be fully suppressed to avoid relaxation. If one only takes a look at one of the parts, the local quantum state appears to be constructed of a superposition of Fock states that resembles a coherent state with a well-defined amplitude and phase.

In the experiment we split a Bose-Einstein condensate into many thousand parts using a three-dimensional optical lattice. Initially the condensate is prepared in a harmonic trap and all (or most) atoms occupy the macroscopic ground state wavefunction in the harmonic potential. The atoms are delocalized and can be found in a huge spatial volume. When a shallow optical lattice is switched on the atoms remain delocalized; for vanishing on-site interaction $U/J \approx 0$ all atoms occupy the Bloch state with the lowest energy $\phi_{q=0}(\mathbf{r})$, which extends macroscopically over the whole lattice (see Sect. 2.2.3 and Fig. 3.7). Picking out a single lattice site, there is a finite probability for each atom to be found there. In the Wannier basis the many-body wavefunction for a system of N atoms on M lattice sites can expressed as [39, 40]

$$\begin{aligned} |\Psi_{\text{SF}}\rangle_{U/J\approx 0} &\propto \left(\frac{1}{\sqrt{M}} \sum_{i=1}^{M} \hat{a}_i^\dagger \right)^N |0\rangle \\ &= \frac{1}{\sqrt{M^N}} \sum_{n_1,\dots,n_M} \sqrt{\frac{N!}{n_1!\cdots n_M!}}\, |n_1,\dots,n_M\rangle, \end{aligned} \tag{7.15}$$

where the occupation numbers n_i in the second line have to comply with the additional constraint $\sum_{i=1}^{M} n_i = N$. This multinomial state (Eq. 7.15) directly signifies the entanglement and the atom number fluctuations between the lattice sites. There is even a very small, albeit finite probability to find all N atoms at a single lattice site.

Relaxing the constraint of a fixed global atom number, that is assuming a grand canonical ensemble, the many-body wavefunction 7.15 can be factorized into a product of on-site wavefunctions $|\psi_i\rangle$,

$$|\Psi_{\text{SF}}\rangle_{U/J\approx 0} \approx \prod_{i=1}^{M} |\psi_i\rangle. \tag{7.16}$$

Here, the intricate entanglement between lattice sites is absent, which simplifies many calculations without affecting the qualitative and quantitative results provided N and M are large [40, 41]. For the case of a homogeneous lattice with a fixed local filling

$|\alpha|^2 = \bar{n} = N/M$, the on-site wavefunctions $|\psi_i\rangle$ are essentially indistinguishable from a coherent state

$$|\psi_i\rangle = |\alpha\rangle = e^{-|\alpha|^2/2} \sum_{n=0}^{\infty} \frac{\alpha^n}{\sqrt{n!}} |n\rangle. \tag{7.17}$$

When the Bose-Einstein condensate is split with finite interactions $U/J > 0$, the on-site wavefunction will show sub Poissonian number statistics. The repulsion between the atoms limits the delocalization, which effectively squeezes the statistical distribution of Fock states [42, 43]. For very large interactions $U/J \to \infty$, the system is in the Mott insulator phase, where each lattice site is occupied by a fixed number of atoms, and the many-body wavefunction at commensurate integer filling $\bar{n}$ is given by a product of local Fock states

$$|\Psi_{\mathrm{MI}}\rangle_{U/J\to\infty} \approx \prod_{i=1}^{M} |\bar{n}\rangle. \tag{7.18}$$

Generally, the global many-body state $|\Psi\rangle$ in an optical lattice, which has been adiabatically raised up to a certain ratio U/J, can be conveniently captured by a product of on-site wavefunctions that are formed by a superposition of Fock states [36, 39]

$$|\Psi\rangle \approx \prod_{i=1}^{M} \sum_{n=0}^{\infty} c_n^{(i)} |n\rangle_i = \prod_{i=1}^{M} |\psi_i\rangle. \tag{7.19}$$

The coefficients $c_n^{(i)}$ denote the probability amplitudes for a Fock state $|n\rangle$ at lattice site i and fulfill the normalization $\sum_{n=0}^{\infty} |c_n^{(i)}| = 1$. This wavefunction corresponds to the Gutzwiller ansatz that, for example, can be used to calculate the coefficients $c_n^{(i)}$ for a given mean filling $\bar{n}$ within the Bose-Hubbard model [44–46].

After creating an array of coherent or number-squeezed states of the general type 7.19, we rapidly increase the lattice depth in a nonadiabatic process to switch off the tunnel coupling between the lattice sites ($J \to 0$). Because the lattice sites are isolated from each other, the local Fock states $|n\rangle_i$ are the new eigenstates of the system and their eigenenergies $E_n^{(i)}$ determine the dynamical evolution. Accordingly, the time dependent many-body state can be written as

$$|\Psi(t)\rangle = \prod_{i=1}^{M} |\psi_i(t)\rangle \quad \text{with} \quad |\psi_i(t)\rangle = \sum_{n=0}^{\infty} c_n^{(i)} e^{iE_n^{(i)}t/\hbar} |n\rangle_i. \tag{7.20}$$

In the following section, we discuss the impact of this evolution on the matter wave interference pattern observed after time-of-flight expansion.

7.2.2 Evolving Momentum Distribution in Homogeneous Lattices

Quantum phase revival dynamics do not change the atom distribution on the individual lattice sites as the evolving phases do not change the probabilities $|c_n^{(i)}|$. However, they leave strong signatures in the momentum distribution, which can be observed after sufficiently long time-of-flight expansion of the atomic sample in the far-field.[4] The momentum distribution is generally given by

$$n(\mathbf{k}) = |w(\mathbf{k})|^2 \sum_{i,j} e^{-i\mathbf{k}\cdot(\mathbf{r}_i-\mathbf{r}_j)} \langle \hat{a}_i^\dagger \hat{a}_j \rangle,$$
$$= |w(\mathbf{k})|^2 \left(\sum_{i=j} \langle \hat{n}_i \rangle + \sum_{i\neq j} e^{-i\mathbf{k}\cdot(\mathbf{r}_i-\mathbf{r}_j)} \langle \hat{a}_i^\dagger \hat{a}_j \rangle \right) \tag{7.21}$$

where $\mathbf{r}_i$ denotes the coordinate of lattice site i and $\hat{a}_i$ ($\hat{a}_i^\dagger$) the corresponding annihilation (creation) operator for an atom. The envelope $|w(\mathbf{k})|^2$ is determined by the Fourier transform of the on-site Wannier function $w(\mathbf{r})$, which together with the factor $\sum_{i=j}\langle \hat{n}_i \rangle$ gives rise to the so-called Wannier background. The term $e^{-i\mathbf{k}\cdot(\mathbf{r}_i-\mathbf{r}_j)}$ is responsible for the emergence of the interference pattern and the factor $\langle \hat{a}_i^\dagger \hat{a}_j \rangle$ measures the correlation between the lattice sites i and j. Quantifying the coherence in the system, this factor determines how well the interference pattern is modulated. The integral of $|w(\mathbf{k})|^2 \sum_{i\neq j} e^{-i\mathbf{k}\cdot(\mathbf{r}_i-\mathbf{r}_j)} \langle \hat{a}_i^\dagger \hat{a}_j \rangle$ over complete momentum space vanishes identically.[5] Nevertheless, this term encodes the dynamical evolution of the individual lattice sites. Using the general many-body wavefunction (Eq. 7.19), the time dependence of the correlations ($i \neq j$) can be derived as

$$\langle \Psi(t)|\hat{a}_i^\dagger \hat{a}_j|\Psi(t)\rangle = \langle \psi_i(t)|\hat{a}_i^\dagger|\psi_i(t)\rangle \cdot \langle \psi_j(t)|\hat{a}_j|\psi_j(t)\rangle. \tag{7.22}$$

This induces a temporal modulation of the interference pattern. In order to derive an analytical expression for the time dependent momentum distribution $n(\mathbf{k}, t)$, we make the following simplifying assumptions about the general on-site state $|\psi_i(t)\rangle$ shown in Eq. 7.19:

- The energies of individual Fock states have identical values at each lattice site $E_n^{(i)} \equiv E_n$. This assumption is justified when the lattice is homogeneous over the extent of the atom cloud. We note that the lattice is homogeneous in experimental

[4] Typical experimental expansion times (10 to 20 ms) are not long enough to exactly transform the in-situ density into the momentum distribution. This has been investigated in detail by Gerbier et al. [47]. However, in the analysis of the experimental images we use observables (visibility and fraction of coherent atoms) that are fairly insensitive to the details of the time-of-flight pattern (see Sect. 4.2). Therefore it is justified to assume infinite time-of-flight for practical purposes.

[5] Note that $\int d\mathbf{k}\, |w(\mathbf{k})|^2 e^{-i\mathbf{k}\cdot(\mathbf{r}_i-\mathbf{r}_j)} = \delta_{ij}$, because the Wannier function is symmetric and its squared modulus assumed to be normalized to unity [48].

realizations, when two conditions are fulfilled: First, the lattice depth must be uniform over the extent of the atom cloud and, second, the underlying potential should be neither confining or anticonfining, but a flat. Both conditions are largely met in the experiments of this chapter.

- The on-site atom number statistics are assumed to be identical at each lattice site such that $c_n^{(i)} \equiv c_n$, which also implies a uniform mean filling $\langle \hat{n}_i \rangle = \bar{n}$ for all indices i. We note that this requirement is hard to meet with present day technology. During the initial loading of a shallow lattice, the cloud is typically confined in a harmonic trap which gives rise to mean on-site fillings ranging between zero and a maximal value $\langle \hat{n}_i \rangle = \bar{n}_{\max}$ in the center of the trap. To achieve uniform number statistics across the lattice, however, a box-like underlying potential would be necessary.

Under these assumptions, the index of the on-site wavefunctions can be dropped and we obtain

$$\langle \Psi(t)|\hat{a}_i^\dagger \hat{a}_j|\Psi(t)\rangle \xrightarrow{|\psi_i(t)\rangle \equiv |\psi(t)\rangle} |\langle \psi(t)|\hat{a}|\psi(t)\rangle|^2 = |\langle \hat{a}\rangle(t)|^2, \tag{7.23}$$

where $|\psi(t)\rangle = \sum_{n=0}^{\infty} c_n e^{iE_n t/\hbar}|n\rangle$. Therefore we can write down a simplified expression for the time dependent momentum distribution in a lattice with M sites

$$n(\mathbf{k}, t) = |w(\mathbf{k})|^2 \left(M\bar{n} + |\langle \hat{a}\rangle(t)|^2 \sum_{i \neq j} e^{-i\mathbf{k}\cdot(\mathbf{r}_i - \mathbf{r}_j)} \right). \tag{7.24}$$

The amplitude of the interference term is given by the squared modulus of the field amplitude, which refers to a single lattice site similar to the dynamics discussed in Sect. 7.1. For the evaluation of the experimental images we use the visibility of the interference pattern (see Sect. 4.2). The visibility has a monotonic relation to $|\langle \hat{a}\rangle(t)|^2$ and therefore constitutes an appropriate experimental observable for the quantum dynamics.

The evolution of the matter wave field amplitude can be expressed in general terms by

$$|\langle \hat{a}\rangle(t)| = \sum_{n=0}^{\infty} \sqrt{n+1}\, c_n^* c_{n+1}\, e^{-i(E_{n+1}-E_n)t/\hbar} \tag{7.25}$$

such that the squared modulus reads

$$|\langle \hat{a}\rangle(t)|^2 = \sum_{m,n=0}^{\infty} \sqrt{m+1}\sqrt{n+1}\, c_m c_{m+1}^* c_n^* c_{n+1}\, e^{i(E_{m+1}-E_m-E_{n+1}+E_n)t/\hbar}. \tag{7.26}$$

For the specific case of a coherent state with the probability amplitudes $c_n = e^{-|\alpha|^2/2}\alpha^n/\sqrt{n!}$ and the eigenenergies $E_n^{\mathrm{SO}} = Un(n-1)/2$ of a single-orbital model, a straightforward calculation recovers the dynamics of Eq. 7.12. Therefore periodic collapses and revivals of the visibility at the frequencies U/h and its higher harmonics would be expected in the experiment (see Fig. 7.3c, grey solid line).

On the one hand, Eq. 7.26 can be used to calculate the quantum dynamics for given Fock state energies E_n and coefficients c_n. On the other hand, it can also be used in the reverse way forming the basis of quantum revival spectroscopy. From this equation we know that the dynamical evolution $|\langle \hat{a} \rangle (t)|^2$ is generally composed of oscillations at frequencies

$$f_{mn} = (E_{m+1} - E_m - E_{n+1} + E_n)/h. \tag{7.27}$$

Each frequency has a spectral weight of

$$C_{mn} = \sqrt{m+1}\sqrt{n+1}\, c_m c^*_{m+1} c^*_n c_{n+1}, \tag{7.28}$$

where the indices m and n can take any integer value including zero. When the quantum evolution $|\langle \hat{a} \rangle (t)|^2$ is accurately measured in an experiment, the oscillation frequencies f_{mn} and their spectral weight C_{mn} can be extracted, for example, using a Fourier transform. It is convenient to determine the energetic reference point by setting the Fock state energies of the noninteracting system to zero, which corresponds to omitting the on-site zero point energy. Using this convention, the Fock state energies for zero atoms and one atom vanish, $E_0 = E_1 = 0$, and the general frequencies of order U/h are given by

$$\begin{aligned} f_{10} &= E_2/h && \text{with} \quad C_{10} = \sqrt{2}\, c_0\, |c_1|^2\, c_2^* \\ f_{21} &= (E_3 - 2E_2)/h && \text{with} \quad C_{21} = \sqrt{6}\, c_1\, |c_2|^2\, c_3^* \\ f_{32} &= (E_4 - 2E_3 + E_2)/h && \text{with} \quad C_{32} = \sqrt{12}\, c_2\, |c_3|^2\, c_4^* \end{aligned} \tag{7.29}$$

and so forth. Because the Fock state energy E_2 is obtained directly, it is possible to iteratively extract all energies E_n with $n > 2$ from the experimental data. Furthermore, the frequencies of order $2U/h$ and their spectral weights read

$$\begin{aligned} f_{20} &= (E_3 - E_2)/h && \text{with} \quad C_{20} = \sqrt{3}\, c_0^*\, c_1\, c_2\, c_3^* \\ f_{31} &= (E_4 - E_3 - E_2)/h && \text{with} \quad C_{31} = \sqrt{8}\, c_1^*\, c_2\, c_3\, c_4^* \\ f_{42} &= (E_5 - E_4 - E_3 + E_2)/h && \text{with} \quad C_{42} = \sqrt{15}\, c_2^*\, c_3\, c_4\, c_5^* \end{aligned} \tag{7.30}$$

and so forth. Analogously, higher frequency components of order $\ell\, U/h$ are given by $f_{n+\ell,n}$ and the corresponding weights by $C_{n+\ell,n}$, where ℓ is a positive integer.

Based on the above derivation, we conclude that quantum revival spectroscopy offers a unique tool to measure the energy of individual Fock states on an absolute scale. The inference of different Fock states *intrinsically* ensures that the energy difference between the interacting and the noninteracting state is measured. In addition, information on the probability amplitudes c_n can be obtained via the spectral

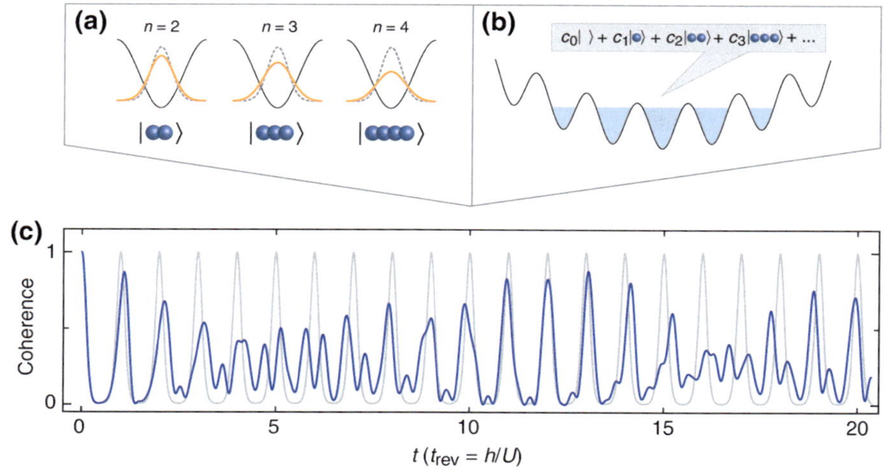

Fig. 7.3 Signature of multi-body interactions in quantum phase revivals. **a** For repulsive interactions, virtual transitions to higher lattice orbitals broaden the ground-state wavefunction at a lattice site depending on the atom number (*orange solid lines*). This gives rise to characteristic Fock state energies, which can be described by effective multi-body interactions. **b** A BEC loaded into a weak optical lattice forms a superfluid in which each atom is delocalized over several lattice sites. The quantum states at each site can be expressed as a superposition of Fock states, $|n\rangle$, with amplitudes c_n. **c** Quantum phase revivals of a coherent state of interacting atoms in the multi-orbital system of a deep lattice well (*blue solid line*). The beat signal indicates coherent multi-body interactions. The dynamics are markedly different from the monochromatic evolution expected in a single-orbital model with a single two-body interaction energy, U (*gray solid line*)

weights C_{mn}. Equations 7.29 and 7.30 explicitly show that the evolution of $|\langle\hat{a}\rangle(t)|^2$ is only nontrivial when the many-body quantum state shows number fluctuations. Only if the probability amplitudes c_n of at least three Fock states do not vanish, an oscillatory behavior can arise. On the other hand, the mixing of several Fock states can "amplify" the signal of a Fock state with a small probability amplitude, because the statistical weight may be increased by large coefficients c_n of the other contributing Fock states. Amazingly, quantum revival spectroscopy is also sensitive to the probability amplitude c_0 of the Fock state with zero atoms, which is the occupation of the vacuum state.

7.2.3 Quantum Phase Revivals with Fixed Global Atom Number

We have derived the quantum phase revival dynamics assuming a grand canonical ensemble and a factorized global many-body state composed of independent on-site states. While these assumptions considerably simplify calculations, they do not resemble experimental reality. Here the total atom number is rather fixed and the many-body state is a highly entangled multinomial superposition of global many-body Fock states (see Eq. 7.15). It is an important question whether the dynamics change when these realistic conditions are taken into account.

The question has only recently been answered by Schachenmayer et al. [41], who have analytically calculated the quantum dynamics in a lattice system with N atoms, a mean on-site filling of $\bar{n}$ and single-orbital eigenenergies $E_n^{\mathrm{SO}} = Un(n-1)/2$. They obtain

$$|\langle\hat{a}\rangle(t)|^2 = \bar{n}\cdot\left(1+\frac{2\bar{n}}{N}\left(\cos\left(\frac{Ut}{\hbar}\right)-1\right)\right)^{(N-1)}, \tag{7.31}$$

which differs from Eq. 7.12 for small atom numbers N. However, in the limit of large atom numbers $N \gg \bar{n}$, this expression reduces to Eq. 7.12 as $\lim_{N\to\infty}(1+x/N)^N = e^x$. Already for a system with $N = 5$ and $\bar{n} = 1$, the result of the exact treatment is almost indistinguishable from the approximate calculation employed above. Experimental lattice systems consist of typically $N = 10^5$ atoms with a local filling on the order of $\bar{n} \approx 1$ and therefore the simplified coherent state treatment is safely applicable.

7.2.4 *Multi-Orbital Physics, Fock State Energies and Multi-Body Interactions*

Using quantum revival spectroscopy we will see in the experimental data (Sect. 7.3) that the single-orbital approximation employed above breaks down at rather moderate interactions and lattice depths. In a multi-orbital system, atom-atom interactions can promote particles to higher-lying orbitals; the admixture of these orbitals modifies the shape of the spatial on-site wavefunctions and gives rise to renormalized Fock state energies [4, 49–57].

In this section, we present two approaches to treat the on-site multi-orbital system theoretically. The first one is based on exact diagonalization of a finite multi-orbital system in analogy to the so-called configuration interaction method used in quantum chemistry [46, 58, 59]. The second one uses the techniques of effective field theory to derive a perturbation expansion for the Fock state energies [12, 60, 61].

7.2.4.1 Calculation of Fock State Energies Using Exact Diagonalization

It is intuitively understandable that interactions between atoms can lead to a modification of the ground state wavefunction at a lattice site. For the case of repulsive interactions a broadening of the on-site ground state wavefunction is expected (Fig. 7.3a). Because the interaction energy per atom pair is proportional to the overlap of the atomic densities (Eq. 7.6), this will reduce the energy of the Fock state E_2. The

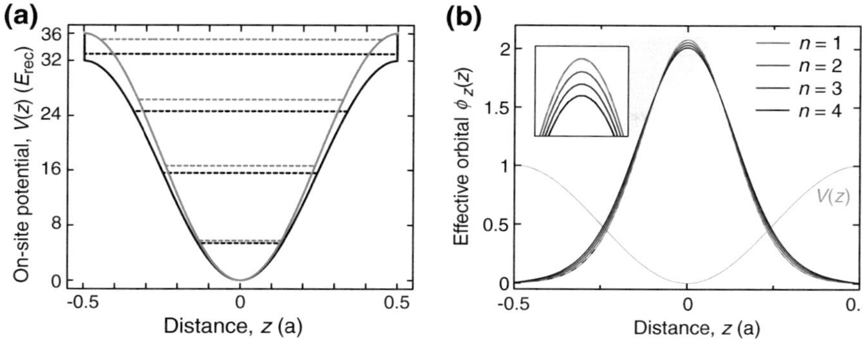

Fig. 7.4 **a** On-site potential at lattice depth $V_{lat} = 32\ E_{rec}$ and $V_{lat} = 36\ E_{rec}$ with hard boundary conditions. The *dashed lines* show the lowest lying single-particle energies with three and four classically bound states, respectively. **b** The effective one-dimensional ground state wavefunction in a $V_{lat} = 40\ E_{rec}$ lattice is modified as a function of the filling $n = 1$ to 4 for interacting ^{87}Rb atoms. The *black solid line* shows the single-particle on-site wavefunction $\phi_1(z)$ calculated on an individual lattice site with hard boundary conditions (see (**a**)). $\phi_1(z)$ is essentially identical to the Wannier function $w(z)$ of the lowest lattice band (*dashed line*). The figure is displayed with kind permission of Dirk-Sören Lühmann [46]

reduction can be expected to be more and more pronounced the higher the number of atoms at the lattice site.

To calculate the Fock state energies E_n in deep three-dimensional lattices V_{lat} using exact diagonalization, it is convenient to focus on a single lattice site with hard boundary conditions (Fig. 7.4a). The considerations are further restricted to the $s = 3$ or 4 lowest lying orbitals in each of the three dimensions, because those orbitals—being classically bound—are least prone to tunneling.[6] This gives rise to a finite Hilbert space with $m = 3^3$ or 4^3 single-particle states. The corresponding three-dimensional orbitals are constructed from the one-dimensional eigenfunctions in x, y, and z direction via

$$\phi_{\alpha\beta\gamma}(\mathbf{r}) = \phi_\alpha(x)\,\phi_\beta(y)\,\phi_\gamma(z), \tag{7.32}$$

with $\alpha, \beta, \gamma = 1, \ldots, s$. The occupation numbers $n_{\alpha\beta\gamma}$ of the orbitals $\phi_{\alpha\beta\gamma}(\mathbf{r})$ define the on-site many-body states $|N\rangle = |n_{111}, n_{211}, \ldots, n_{sss}\rangle$, which we label by a number N for simplicity. Assuming that n atoms occupy the lattice site, the number of mutually orthogonal many-body states $|N\rangle$ amounts to

$$N_{max} = \frac{(n+m-1)!}{n!(m-1)!} \tag{7.33}$$

[6] We will discuss in Sect. 7.3 that the energy cutoff that is effectively introduced by the restriction to those orbitals is justified by the experimental scenario.

and defines the dimension of the Hilbert space, where $m = s^3$ is the number of single-particle orbitals. We note that the uncorrelated noninteracting ground state with n bosonic atoms is given when all atoms occupy the lowest orbital $\phi_{111}(\mathbf{r})$, which is $|1\rangle = |n, 0, \ldots, 0\rangle$ in the notation employed here. This corresponds to the Fock state $|n\rangle$ within the single-orbital Bose-Hubbard model.

Repulsive interactions between the n atoms induce the admixture of energetically higher lying many-body states $|N\rangle$. Those admixtures can be exactly calculated through the diagonalization of the Schrödinger equation of the n atom system in its matrix representation

$$\sum_{N'} \langle N|\hat{H}|N'\rangle \, c_{N'} = \tilde{E}_n \, c_N. \tag{7.34}$$

Defining the field operator $\hat{\phi}(\mathbf{r}) = \sum_{\alpha\beta\gamma} \phi_{\alpha\beta\gamma}(\mathbf{r}) \, \hat{a}_{\alpha\beta\gamma}$, the Hamiltonian takes the form

$$\hat{H} = \int d^3r \, \hat{\phi}^\dagger(\mathbf{r}) \left(\frac{\hat{\mathbf{p}}^2}{2m} + V(\mathbf{r}) + \frac{g}{2} \hat{\phi}^\dagger(\mathbf{r}) \hat{\phi}(\mathbf{r}) \right) \hat{\phi}(\mathbf{r}), \tag{7.35}$$

where $V(\mathbf{r})$ is the on-site trapping potential and $g = 4\pi\hbar^2 a_s/m$ the atom-atom contact interaction characterized by the s-wave scattering length a_s.

The lowest eigenvalue $\tilde{E}_n^{\min}$ of $\hat{H}$ corresponds to the ground state energy of the interacting n-atom system, which we call the multi-orbital Fock state energy $\tilde{E}_n^{\min} = E_n^{\mathrm{MO}}$. The energy offset is ideally chosen such that the ground state energy of the noninteracting n atom system vanishes. It is important to remark that the multi-orbital Fock state energy E_n^{MO} is the total energy of the Fock state $|n\rangle$ in the interacting system and corresponds to the energy that is measured in a quantum revival experiment. It does not only include interaction energy, but also the kinetic and potential energy carried by the admixed higher lying many-body states $|N\rangle$ (see Fig. 7.5). As shown in the following section, E_n^{MO} is significantly lower than the single-orbital Fock state energy $E_n^{\mathrm{SO}} = \frac{gn(n-1)}{2} \int d^3r \, |\phi_{111}(\mathbf{r})|^4$. However, E_n^{MO} is larger than the pure interaction energy $E_{\mathrm{int},n}^{\mathrm{MO}} = \frac{g}{2} \int d^3r \, \langle \hat{\phi}^\dagger(\mathbf{r}) \hat{\phi}^\dagger(\mathbf{r}) \hat{\phi}(\mathbf{r}) \hat{\phi}(\mathbf{r}) \rangle$ that accounts for the interaction energy, but neglects the additional potential and kinetic energy of the excited many-body states [59, 62].

7.2.4.2 Emergence of Multi-Body Interactions Using Effective Field Theory

A treatment of the multi-orbital system of interacting atoms within the framework of effective quantum field theory illustrates the physical processes pictorially: Atom-atom interactions induce virtual transitions to higher-lying excited orbitals at a lattice site. Those transitions appear as higher-order terms in the perturbation expansion and generate effective multi-body interactions. The resulting multi-orbital Fock state energies in an individual lattice well are given by the expansion

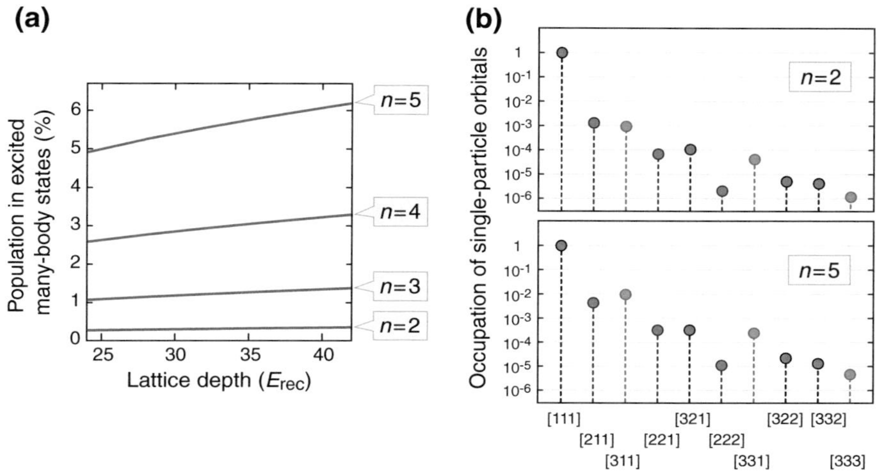

Fig. 7.5 Occupation of many-body states on a lattice site. **a** Population that is *not* found in the many-body ground state of the noninteracting n-atom system $|1\rangle = |n, 0, \ldots, 0\rangle$ as a function of lattice depth and atom number assuming $a_s = 102\, a_0$ and 4^3 single-particle orbitals. **b** Occupation of single-particle orbitals for Fock states with $n = 2$ and $n = 5$ atoms (3^3 orbitals, $V_{\rm lat} = 40\, E_{\rm rec}$, $a_s = 102\, a_0$). The notation $[\alpha\beta\gamma]$ comprises all permutations of α, β, and γ. All orbitals corresponding to a set $[\alpha\beta\gamma]$ are energetically degenerate and their occupations have been added up. The energy of the orbitals increases from *left* to *right*. Occupation of the second band, while energetically favorable compared to the third band, is suppressed due to conservation of parity

$$E_n^{\rm MO} = \frac{\tilde{U}_2}{2}n(n-1) + \frac{\tilde{U}_3}{6}n(n-1)(n-2) + \frac{\tilde{U}_4}{24}n(n-1)(n-2)(n-3) + \ldots \quad (7.36)$$

which represents the eigenenergies of the corresponding effective Hamiltonian [12]. The characteristic strength of the effective m-body interaction is denoted by $\tilde{U}_m$.

We briefly sketch the emergence of multi-body interactions following the derivation given in Johnson et al. [12]. The eigenenergy of a Fock state with n atoms is given by a perturbation expansion in higher-order terms

$$E_n^{\rm MO} \simeq E_n^{(0)} + E_n^{(1)} + E_n^{(2)} + \ldots \quad (7.37)$$

The zeroth-order term corresponds to the energy of the noninteracting system that can be set to zero $E_n^{(0)} = 0$ without loss of generality. The first-order term represents the pure two-body interaction $E_n^{(1)} = \tilde{U}_2 n(n-1)/2$ that happens within the lowest single-particle orbital. The coupling to excited single-particle orbitals is introduced by the second-order term

$$E_n^{(2)} = -\frac{\tilde{U}_2^2}{4}\sum_{\mu\geq\nu}^{\Lambda} \frac{s_{\mu\nu}K_{\mu\nu 00}^2 |\langle n;\, \mu\nu|\hat{a}_\mu^\dagger \hat{a}_\nu^\dagger \hat{a}_0 \hat{a}_0|n;\, 00\rangle|^2}{E_{\mu\nu}} + \frac{A}{2}n(n-1) \quad (7.38)$$

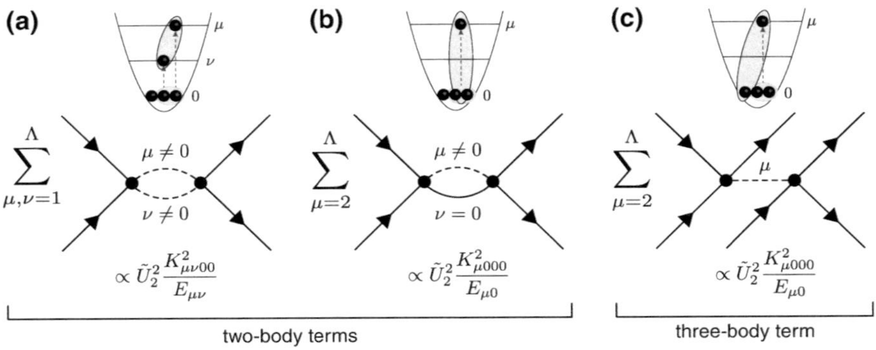

Fig. 7.6 Second-order diagrams relevant to the generation of effective two- and three-body interactions. **a**, **b** Diagrams of this type give rise to two-body interactions within the second order shift. Their contributions are canceled by introducing the counter term proportional to an adjustable constant A (see Eq. 7.38) for proper renormalization. **c** This diagram exemplarily shows a process generating three-body interactions. Two atoms in the lowest vibrational state $\mu = 0$ collide promoting one atom to a higher-lying state $\mu \neq 0$ that subsequently collides with a distinct third atom in state $\mu = 0$. An effective three-body interaction between the three distinct incoming atoms is created, mediated by a virtual transition to the $\mu \neq 0$ intermediate state. The diagrams are adapted from Johnson et al. [12]

where the indices μ and ν are used as a short hand notation for the indices of the three-dimensional single-particle orbitals introduced above. The occupation numbers of states with n atoms are expressed by

$$\begin{aligned} |n; 00\rangle &= |n, 0, \ldots, 0\rangle \\ |n; \mu\mu\rangle &= |n-2, 0, \ldots, 2_\mu, \ldots, 0\rangle \\ |n; \mu\nu\rangle &= |n-2, 0, \ldots, 1_\mu, \ldots, 1_\nu, \ldots, 0\rangle. \end{aligned} \tag{7.39}$$

The sum over the indices μ and ν terminates at a high-energy cutoff Λ which excludes highly excited orbitals from the calculation. $K_{\mu\nu00}$ denotes the overlap integral of the orbitals involved in the virtual processes and $s_{\mu\nu}$ takes the value 4 (1) for the case $\mu = \nu$ ($\mu \neq \nu$), respectively. The parameter A is introduced to ensure proper renormalization of the perturbation theory. It must be chosen such that $E_n^{(2)}$ does not generate two-body energies in addition to $E_n^{(1)}$. Therefore the condition $E_2^{(2)} = 0$ must be fulfilled and the contributions of all two-body diagrams that are generated by the first term of Eq. 7.38 (examples shown in Fig. 7.6a, b) are canceled in the renormalization process. The remaining three-body processes (example shown in Fig. 7.6c) give rise to effective three-body interactions

$$E_n^{(2)} = \frac{\tilde{U}_3}{6} n(n-1)(n-2), \tag{7.40}$$

where $\tilde{U}_3$ denotes the corresponding characteristic energy scale.

7.3 Experimental Realization

The conditions at a single site of an optical lattice are ideally suited to observe the quantum phase revival dynamics that is outlined above. First, the interaction energy U/h is typically on the order of some kilohertz leading to a revival time $t_{\text{rev}} = h/U$ on the order of 1 ms. This time scale is long enough to be easily resolved experimentally and still short enough to avoid disruption or dephasing due to loss processes or technical fluctuations. Second, the collapse time $t_{\text{coll}} = t_{\text{rev}}/\sqrt{\bar{n}}$ is slightly shorter than the revival time because typical mean fillings are $\bar{n} \gtrsim 1$. This ensures that the collapse is neither too fast for experimental detection nor too slow. In the latter case, the matter wave field might not have fully collapsed before the onset of the revival. Third, the quantum phase revival dynamics is realized on several thousand sites in parallel which generates strong experimental signatures as discussed in the previous section.

However, the parallel implementation of the dynamics on many sites requires a homogeneous lattice across the atomic cloud. On the one hand, it is crucial that the depth of the optical lattice is identical at all occupied sites such that the level spacings between the on-site orbitals do not differ from site to site. Otherwise, the revival times would not be well-defined across the lattice and the dynamics would wash out. On the other hand, mutual energy offsets between the lattice sites must be avoided as these can lead to rapid relative dephasing. Previous experiments were restricted to the observation of few revivals only, because the dynamics was realized in red-detuned optical lattices. The underlying harmonic confinement in those configurations led to rapid mutual dephasing between the sites which severely damped the observed dynamics after few revivals.

In this section, I report on the observation of quantum revival dynamics over more than 40 cycles. The remarkable increase of coherence time has been enabled by using a blue-detuned optical lattice and simultaneous compensation of the underlying anticonfinement by a copropagating red-detuned dipole trap (for details see Sect. 5.4). This improvement is a qualitative change that allows for the observation of novel physics. High spectral resolution, which is gained by the detection of long quantum revival traces, enables precise measurements of individual Fock state energies E_n^{MO} on an absolute energy scale. Our data unambiguously reveals characteristic Fock state energies that cannot be described within the single-orbital framework that is also employed in the context of the Bose-Hubbard model. Using the measured Fock state energies, we can precisely infer the strength of the effective multi-body interactions (see Sect. 7.2.4).

7.3.1 Preparing an Array of Miniature BECs

The experiment begins with the preparation of a pure Bose-Einstein condensate of ^{87}Rb atoms in the $|F, m_F\rangle = |1, +1\rangle$ hyperfine state. Away from any Feshbach

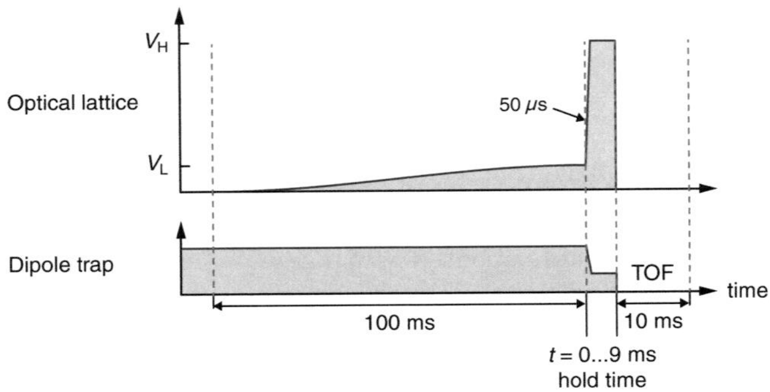

Fig. 7.7 Experimental sequence as described in the text

resonances atoms in this state interact repulsively; the background scattering length is quoted in the literature as $a_s \approx 102\ a_0$ with deviations on the per cent level [34, 35]. The atom number is adjusted for different data sets between 1.2×10^5 and 3.5×10^5 with shot-to-shot variations of $\pm 10\,\%$ or better. The atom cloud is held in a pancake-shaped crossed optical dipole trap operating at a wavelength of $\lambda_{\text{dip}} = 1030\,\text{nm}$. The trap frequencies are $\omega_z = 2\pi \times 130\,\text{Hz}$ in the direction of gravity and $\omega_\perp = 2\pi \times 32\,\text{Hz}$ in the orthogonal plane. Subsequently, a three-dimensional lattice of simple cubic type is superimposed and adiabatically ramped up within 100 ms to a depth V_L ranging between 3 E_{rec} and 13 E_{rec}, below the Mott insulator transition that is expected slightly below 15 E_{rec} for our parameters. The lattice is created by a laser operating at a blue-detuned wavelength of $\lambda_{\text{lat}} = 738\,\text{nm}$ as described in detail in Sect. 5.4. For the lattice depths V_L the many-body ground state of the system is a superfluid. Therefore, we expect the quantum states at the individual lattice sites to range from coherent states for shallow lattice depths to highly number-squeezed states for deeper lattices [21, 63–65].

After adiabatic loading, we suddenly increase the lattice depth from V_L to V_H, ranging from 25 E_{rec} to 41 E_{rec}. This suppresses the tunnel coupling, freezes out the atom number distribution at each lattice site and effectively creates an array of miniature Bose-Einstein condensates (see Fig. 7.7). The jump is performed within 50 μs, which is slow enough to avoid the population of higher lattice bands, but fast in comparison with tunneling in the first band. The nonadiabatic change of parameters abruptly exposes the system to a new Hamiltonian and the many-body state prepared at V_L is no longer the ground state or even an eigenstate. Therefore nonequilibrium dynamics arise. The quantum phase evolution on the individual lattice sites starts, governed by the Fock state energies of Eq. 7.36.

Quantum phase revivals have been studied in previous experiments [21–23]. However, these were limited to following the dynamics for short traces only, with typically four to five revival cycles. In those cases, the measurements were performed in a red-detuned optical lattice where a global harmonic confinement led to rapid relative dephasing of lattice sites. In these experiments, the different detunings of the optical lattice (blue-detuned) and the dipole trap (red-detuned) with respect to the atomic resonances ($\lambda_{D1} = 795$ nm and $\lambda_{D2} = 780$ nm) allow us to change the underlying harmonic confinement during the experimental sequence. Simultaneously with the jump to the lattice depth V_H, we reduce the dipole trap to cancel the harmonic confinement, creating a homogeneous lattice system in the horizontal plane (see also Sect. 5.4). After letting the system evolve for a hold time t, all trapping potentials are switched off simultaneously and an absorption image of the matter wave interference pattern is recorded after $t_{TOF} = 10$ ms time-of-flight expansion. As a measure of the ensemble averaged squared modulus of the matter wave amplitude $|\langle\hat{a}\rangle(t)|^2$ (see Sect. 4.2.2), we determine the visibility of the interference pattern for each image [66] according to

$$\mathcal{V} = \frac{N_{max} - N_{min}}{N_{max} + N_{min}}. \tag{7.41}$$

Here, N_{max} denotes the total atom number in the four boxes around the first order interference peaks and N_{min} the total atom number in boxes that are rotated by 45° around the central peak. The width of the boxes is chosen as 15 % of $2\hbar k_{lat} t_{TOF}/m$, which is the distance between the center and the first order diffraction peaks ($k_{lat} = 2\pi/\lambda_{lat}$).

7.3.2 Observation of Multi-Orbital Quantum Revivals

Figure 7.8 shows a typical time trace of the multi-orbital phase evolution. It displays about 40 revivals, while the coherence time allows to record the dynamics for even longer times. On top of a fast series of collapses and revivals, a slower modulation of the envelope is observed that indicates a beat between different frequencies in the system. Due to the large number of observed revivals, it is possible to analyze the spectral content of the trace using a numerical Fourier transform. Technical details on the Fourier analysis are provided in Appendix D.

In the resulting spectrum, five distinct frequency components of orders U/h and $2U/h$ can be clearly identified (Fig. 7.9). The smallest peaks at about 3.4 and 7.1 kHz originate from sites occupied by up to four atoms, because they correspond to the energy terms $E_4 - 2E_3 + E_2$ and $E_4 - E_3 - E_2$ with spectral weights of $\sqrt{12}\, c_2 |c_3|^2 c_4^*$ and $\sqrt{8}\, c_1^* c_2 c_3 c_4^*$, respectively. It is interesting to note that quantum revival spectroscopy can help to detect Fock states with very small probability amplitudes c_n. Their signal can be "amplified" through the interference with Fock states $|n-1\rangle$ and $|n-2\rangle$ with larger amplitudes c_{n-1} and c_{n-2}. This can be deduced, for example, from Eq. 7.26. The residual damping in the time trace most probably stems from residual harmonic confinement along the vertical z-axis, residual tunneling [67] via higher bands, and the finite extension of the atomic ensemble. The latter leads to the sampling of a slightly inhomogeneous distribution of lattice depths as a result of the Gaussian shape of the lattice laser beams.

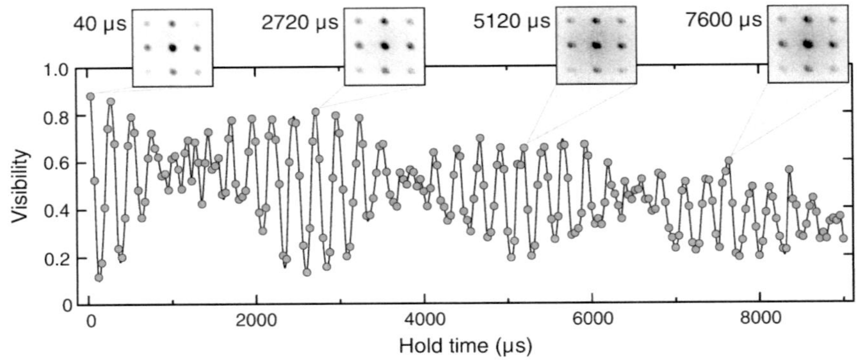

Fig. 7.8 Multi-orbital quantum revivals of a number-squeezed state. Collapse and revival dynamics of number-squeezed superposition states in a deep optical lattice. A Bose-Einstein condensate of about $1.9(3) \times 10^5$ ^{87}Rb atoms was adiabatically loaded into a $V_{\mathrm{L}} = 8\ E_{\mathrm{rec}}$ lattice within 100 ms. Quantum phase evolution was induced by a nonadiabatic jump to a $V_{\mathrm{H}} = 41(1)\ E_{\mathrm{rec}}$ deep lattice, while the superposition states with finite number fluctuations are preserved. The sample has an ensemble-averaged mean atom number of $\langle \bar{n} \rangle \approx 1.0$ and a central mean atom number of $\bar{n}_c \approx 2.5$. Simultaneously with the lattice jump, the underlying harmonic confinement is instantaneously minimized to maximize the coherence time. The quantum phase dynamics show a beat signal resulting from coherent multi-body interactions. Each data point corresponds to a single run of the experiment. The *solid line* interpolates the data and serves as a guide to the eye

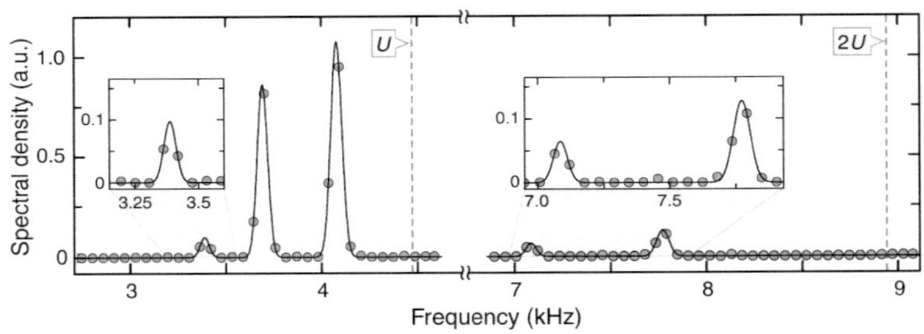

Fig. 7.9 Spectral analysis of the time trace in Fig. 7.8 reveals the contributing frequencies. The *solid line* shows Gaussian fits to the peaks. *Grey dashed lines* display the frequencies corresponding to the single-orbital interaction energies U and $2U$ at a lattice depth of 41 E_{rec}

As discussed in Sect. 4.2.2, the relation between $\mathcal{V}$ and $|\langle \hat{a} \rangle (t)|^2$ is monotonic but not exactly linear. Depending on the amount of nonlinearity, one might expect mixing between the spectral contributions of $|\langle \hat{a} \rangle (t)|^2$, when the data is evaluated using the visibility. Indeed, we observe additional frequencies in the spectra (see below) which we assign to nonlinear spectral mixing. However, those frequencies are suppressed by a factor of fifty compared to the dominant spectral features.

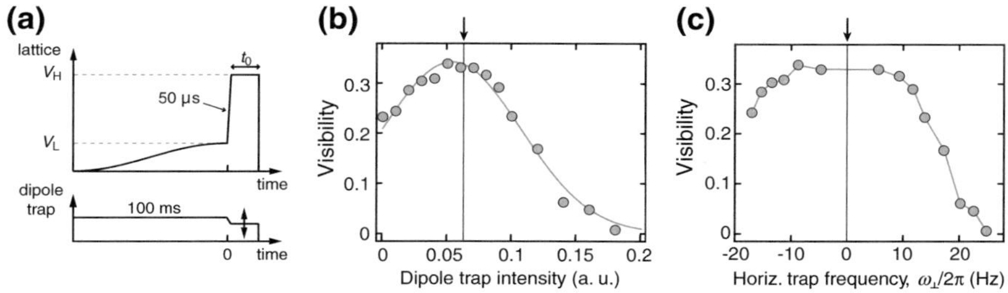

Fig. 7.10 Minimization of underlying confinement for quantum phase revival dynamics. **a** Picking a revival at a hold time t_0 of about 2.5 ms, we vary the intensity of the dipole trap laser and **b** monitor the change of the visibility at the revival, identifying a clear maximum. **c** Converting the dipole trap intensity to trap frequency using the model outlined in Sect. 5.3, we find that maximal coherence time is observed at the expected point of vanishing harmonic confinement $\omega_\perp = 0$ (arrow)

7.3.3 Minimization of the Harmonic Confinement

In order to obtain time traces with the long coherence time demonstrated in Fig. 7.8, it is critical to minimize the harmonic confinement in the system. Otherwise rapid relative dephasing between lattice sites restricts the measurement to only few revival cycles [21–23]. Based on the accurate model of our combined lattice plus dipole trap potential (Sect. 5.4), we can directly "dial-in" the desired trapping potential and cancel the harmonic confinement in the horizontal plane. To ensure that this theoretical potential does indeed yield the largest coherence times, we perform an independent cross check. We typically choose a phase revival at an intermediate hold time of about $t_0 \approx 2.5$ ms. At this point we monitor the visibility of the interference pattern as a function of the dipole trap intensity (see Fig. 7.10). From these data we determine the maximal visibility and find that the corresponding dipole trap intensity complies with the model of our trapping potential. In the experiments of this section it is crucial to minimize the harmonic confinement, but it is demonstrated in Sect. 7.5 that exotic condensate states can be created via quantum phase evolution in the presence of underlying trapping potentials.

7.3.4 Precision Measurement of Fock State Energies and Multi-Body Interactions

In order to measure the dependence of the multi-orbital Fock state energies E_n^{MO} on lattice depth, we have recorded several long collapse and revival traces with V_H between 25 E_{rec} and 41 E_{rec} as shown in Fig. 7.11. Owing to long coherence times and high spectral resolution, the Fourier transform of the traces reveals sets of frequencies of orders U/h and $2U/h$ with a relative statistical uncertainty on the level of few per cent. The frequencies increase monotonically with lattice depth. Already for the energy of the Fock state with two atoms we observe the stunning relevance

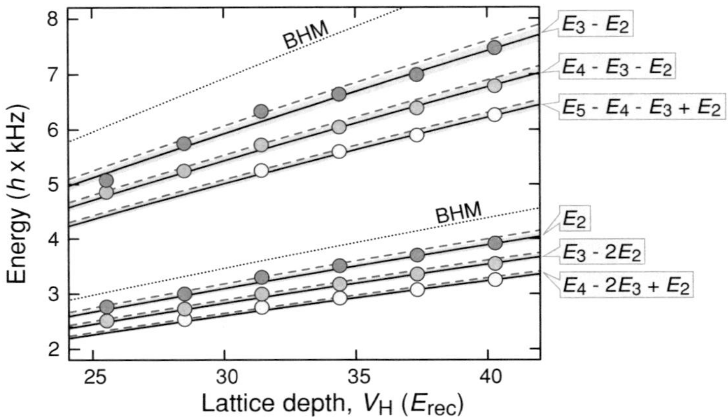

Fig. 7.11 Quantum phase revival energies in a multi-orbital system. Long collapse and revival traces have been recorded under identical loading conditions ($V_L = 8\ E_{rec}$), but variable lattice depth V_H during phase evolution. The contributing frequencies f_{mn} of orders U/h and $2U/h$ have been derived using numerical Fourier transform and are plotted as energies $h \times f_{mn}$. The typical experimental uncertainty is $\pm 50\ h \times$Hz. The shading of the data points reflects the relative spectral weight (the darker the higher). The *solid lines* with *grey shading* indicate the theoretically expected energies. They have been calculated using exact diagonalization (see Sect. 7.2.4) with an on-site basis set of 4^3 orbitals at an s-wave scattering length of $a_s = 102(2)\ a_0$, where a_0 is the Bohr radius. A calculation using a smaller basis set with 3^3 orbitals yields slightly higher energies (*dashed lines*). The *dotted lines* show the single-orbital interaction energies U and $2U$ that have been calculated using Wannier orbitals, which are commonly used in the Bose-Hubbard model (BHM). At low lattice depth we can only resolve the strongest peaks owing to smaller peak spacings

of multi-orbital effects in optical lattices: The measured energy E_2 is 10(1) % lower than predicted by a simple single-orbital Bose-Hubbard model.[7]

We compare the experimental data to the theoretical results of an exact diagonalization of the multi-orbital system at a single lattice site. The calculations have been performed by D.S. Lühmann at the University of Hamburg. According to the description in Sect. 7.2.4, we use hard boundary conditions and a contact interaction. The first assumption is justified because the on-site wavefunctions essentially do not extent to neighboring lattice sites at the high lattice depths used in the experiment. The second assumption is a subtle issue in the context of renormalization, where care must be taken in the use of the contact interaction [12, 68, 69]. Therefore and due to the uncertainties in the published values for the scattering length a_s (see above) [34, 35], we use a_s as an adjustable parameter to match our theory to the experimentally measured value of E_2. Based on the resulting best match, our theory has predictive power for the Fock states with higher atom number. When the diagonalization is performed on a Hilbert space with 4^3 single-particle orbitals, corresponding to the four lowest lying lattice bands in three dimensions, we find the best match for

[7] For this calculation we have taken Wannier orbitals as the on-site ground state wavefunction, which is commonly done within the Bose-Hubbard model.

E_2 at $a_s = 102(2)\,a_0$. This value also leads to remarkable agreement for all other measured energies suggesting applicability and high accuracy of this multi-orbital theory (Fig. 7.11, black solid lines).

It is a natural question whether the inclusion of higher-lying orbitals improves the accuracy of the calculation. While from a purely theoretical viewpoint the answer might seem to be affirmative, there are experimental reasons for this being not the case. A cutoff at around $s = 4$ seems reasonable since only the lowest four orbitals are classically bound, that is their energy is lower than the lattice depth. While within Bloch theory there is no fundamental difference between classically bound and unbound states, the band structure calculation in Fig. 2.8 shows that lattice bands with energies larger than the lattice depth essentially have a parabolic dispersion relation, similar to a free particle. Such states are prone to fast tunneling. Admixtures of these states would quickly be lost from our system for two reasons: First, the lattice does not have underlying confinement in the horizontal plane (see schematic potential landscape in Fig. 2.7) and, second, is also exposed to a linear gravitational gradient in the vertical direction.[8] Therefore, high-lying orbitals are likely to form fast loss channels and do not contribute to wavefunction renormalization.

Using the experimental and theoretical data on the multi-orbital Fock state energies E_n^{MO} (Fig. 7.12a), we can iteratively derive the strength of effective multi-body interactions using Eq. 7.36). The relevant relations read $U_2 = E_2^{\mathrm{MO}}$, $U_3 = E_3^{\mathrm{MO}} - 3U_2$, and $U_4 = E_4^{\mathrm{MO}} - 6U_2 - 4U_3$. Results are shown in Fig. 7.12b. We observe that the effective three-body interaction is attractive and the measured values agree well with the results obtained from exact diagonalization. Given the fact that exact diagonalization is solely based on a two-body contact interaction, this agreement indicates that intrinsic, direct three-body interactions are negligible for our experimental parameters. The energy scale of the measured effective four-body interaction strengths is as low as $h \times 100\,\mathrm{Hz}$ and also in good agreement with theory, if the experimental uncertainties are taken into account. Altogether, our data shows that the expansion of Fock state energies in terms of multi-body interactions converges quickly. This offers the possibility to efficiently incorporate interaction effects of multi-orbital physics into refined effective single-band lattice Hamiltonians.

7.3.5 Observation of Atom Number Statistics

In addition to measuring coherent interatomic interaction strengths on an absolute energy scale, quantum revival spectroscopy can be used to reveal the changes of the on-site number statistics for different many-body quantum states [26]. The spectral weights of the detected frequencies carry information on the probability

[8] In a recent experiment by Bakr et al. [70], which uses a similar optical lattice setup, atoms are selectively excited to higher-lying lattice bands. The fast loss in classically unbound bands is confirmed by monitoring the population of individual lattice sites using single site resolved imaging.

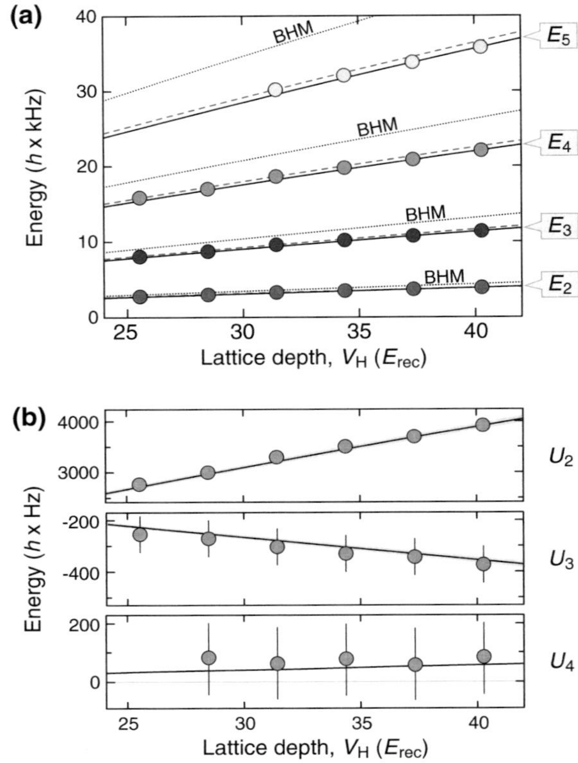

Fig. 7.12 Fock state energies and effective multi-body interactions in experiment (*points*) and theory (*lines*). **a** Multi-orbital Fock state energies as derived from both experimental and theoretical data in Fig. 7.11. The *solid* (*dashed*) lines show the multi-orbital Fock state energies E_n^{MO} as calculated by exact diagonalization on a Hilbert space with 4^3 (3^3) single-particle orbitals. The corresponding energies of the Bose-Hubbard model (BHM, *dotted lines*) lie considerably higher. **b** Effective two-body (*top*), three-body (*middle*) and four-body (*bottom*) interaction strengths as derived from experiment and theory, assuming $a_s = 102(2)$ a_0 and 4^3 contributing single-particle orbitals. The error bars correspond to one standard deviation

amplitudes of the Fock states c_n (Eq. 7.26). To demonstrate this, we adiabatically prepare three-dimensional arrays of coherent ($V_\mathrm{L} = 3\ E_\mathrm{rec}$) to highly number-squeezed states ($V_\mathrm{L} = 13\ E_\mathrm{rec}$) close to the Mott insulator transition around 15 E_rec. For the differently prepared states, we use an identical setting of $V_\mathrm{H} = 40$ E_rec to record the quantum phase revivals (Fig. 7.13). Thus, quantum revivals are employed as a detection sequence for the different many-body quantum states.

As V_L is increased from 3 to 11 E_rec, the time traces evolve from seemingly irregular oscillations into a clear beat signal because fewer frequencies contribute. This can be observed in the corresponding spectra displayed in Fig. 7.13b and is more quantitatively shown in Fig. 7.14. The narrowing of the spectra reflects a decrease in the variance of the atom number distribution. This is dominantly caused by number squeezing, which also leads to smaller peak amplitudes in the Fourier spectra. Additionally, stronger interactions in deeper lattices induce a reduction of the average on-site density, which also entails a smaller variance. We attribute the surprisingly pronounced dynamics at $V_\mathrm{L} = 13\ E_\mathrm{rec}$, in proximity of the Mott insulator transition, to the presence of up to three superfluid layers of the emergent Mott shells with fillings $n = 1, 2$ and 3.

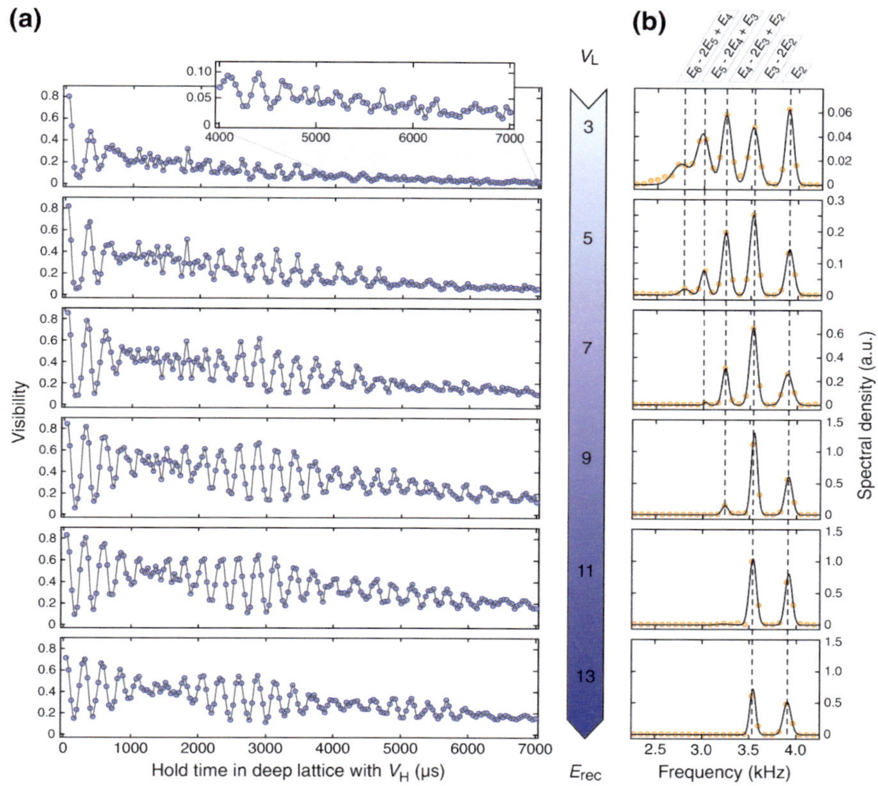

Fig. 7.13 Influence of the global atom number statistics on quantum revival spectra when the Mott insulator transition is approached. **a** Multi-orbital quantum revivals in a deep lattice ($V_H = 40$ E_{rec}) after adiabatic loading of 3.3(3) $\times$ 10^5 ^{87}Rb atoms into lattices with depths ranging from $V_L = 3$ E_{rec} to 13 E_{rec}. The mean atom number of the individual traces differed by as little as $\pm 1\,\%$. The coherence time for shallow lattices seems significantly reduced, however, the visibility reliably shows dynamics down to the per cent level (inset). **b** The corresponding Fourier spectra reveal frequency contributions from Fock states containing up to six atoms. The peak positions agree with the theoretical predictions obtained from exact diagonalization (*dashed vertical lines*) and are independent of V_L. Number squeezing manifests itself both in reduced peak amplitudes and in a narrowing of the spectra for increasing V_L. The *solid lines* show Gaussian fits to the peaks

7.4 Excursion: Detecting a Schrödinger Cat in the Quantum Noise

In Sect. 7.1, we have discussed the quantum collapse and revival dynamics of an ideal coherent state of interacting atoms. The amplitude of the matter wave field is maximally collapsed at half of the revival time $t_{rev}/2 = h/(2U)$. At this instant of time the quantum state of the system corresponds to a superposition of two quasi-coherent states which is strikingly illustrated by the Q-function (Sect. 7.1.4 and

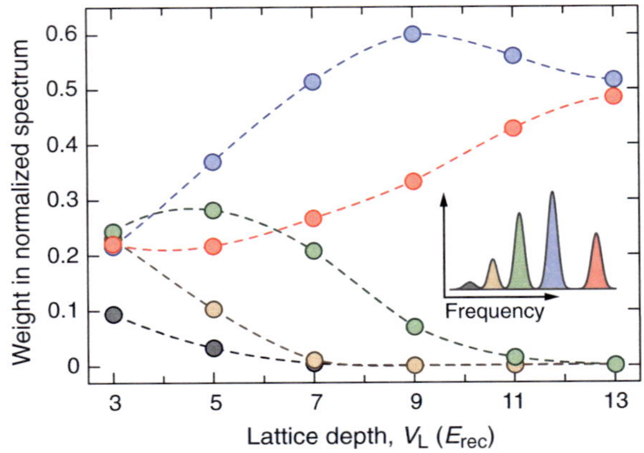

Fig. 7.14 Relative weight of frequencies in the spectra for different lattice depths V_L (see Fig. 7.13). For each spectrum the area under the individual peaks has been determined (see inset) and normalized to the area under all peaks of order U/h. The evaluation shows that the spectrum narrows down to the frequencies E_2/h and $(E_3 - 2E_2)/h$ with the theoretical weights $c_0 c_1^2 c_2$ and $c_1 c_2^2 c_3$, respectively

Fig. 7.2). Because two macroscopically occupied quantum states states are involved, it is justified to call the quantum state at maximal collapse a Schrödinger cat state [25, 71]. The experimental detection of the Schrödinger cat is challenging as it does not show any specific signatures in the momentum distribution observed in time-of-flight images. However, in this excursion we show that correlations in the quantum noise of experimental images might allow its identification.[9]

7.4.1 *The Quantum State at Maximal Collapse: A Schrödinger Cat*

In this excursion we refer to the time dependent coherent state 7.10 to keep the treatment specific. In principle, all calculations can also be carried out with the general superposition state $|\psi(t)\rangle = \sum_{n=0}^{\infty} c_n e^{-iE_n t/\hbar}|n\rangle$ that also comprises number-squeezed states. At maximal collapse the coherent state takes the form

$$|\psi_\alpha(t_{\text{rev}}/2)\rangle = |\psi_\alpha^{\text{coll}}\rangle = e^{-|\alpha|^2/2} \sum_{n=0}^{\infty} \frac{\alpha^n}{\sqrt{n!}} e^{-i\pi n(n-1)/2}|n\rangle. \tag{7.42}$$

The phase factor $\exp(-i\phi)$ with $\phi = \pi n(n-1)/2$ assumes the value 1 for $n = 4m$ and $n = 4m + 1$ and the value -1 for $n = 4m + 2$ and $n = 4m + 3$, where m is an integer number including zero. Therefore, we can decompose the sum into components with a relative phase shift of π:

[9] The ideas for this proposal have been sparked in discussions with Immanuel Bloch and Eugene Demler.

$$
\begin{aligned}
|\psi_\alpha^{\mathrm{coll}}\rangle = & \; e^{-|\alpha|^2/2} \sum_{m=0}^{\infty} \frac{\alpha^{4m}}{\sqrt{(4m)!}} |4m\rangle + \frac{\alpha^{4m+1}}{\sqrt{(4m+1)!}} |4m+1\rangle \\
& - e^{-|\alpha|^2/2} \sum_{m=0}^{\infty} \frac{\alpha^{4m+2}}{\sqrt{(4m+2)!}} |4m+2\rangle + \frac{\alpha^{4m+3}}{\sqrt{(4m+3)!}} |4m+3\rangle
\end{aligned} \tag{7.43}
$$

It is interesting to note that this state is not an eigenstate of the annihilation operator, but of the annihilation operator applied twice,

$$
\hat{a}\hat{a}|\psi_\alpha^{\mathrm{coll}}\rangle = -\alpha^2 |\psi_\alpha^{\mathrm{coll}}\rangle, \tag{7.44}
$$

which has crucial importance for the discussion of noise correlations at maximum collapse in the next section. From the analysis of the Q-function we know that $|\psi_\alpha^{\mathrm{coll}}\rangle$ corresponds to the superposition of two quasi-coherent states with a relative phase shift of π. Therefore, we approximate $|\psi_\alpha^{\mathrm{coll}}\rangle$ by the superposition

$$
|\psi_\alpha^{\mathrm{coll}}\rangle \approx \frac{1}{\sqrt{2}}(|i\alpha\rangle + |-i\alpha\rangle) \equiv |\mathrm{cat}\rangle, \tag{7.45}
$$

where the states $|i\alpha\rangle$ and $|-i\alpha\rangle$ are taken to be actual coherent states for simplicity. Using this state the matter wave field amplitude vanishes exactly

$$
\langle \mathrm{cat}|\hat{a}|\mathrm{cat}\rangle = \frac{i}{2}\left(\alpha + \alpha \langle -i\alpha|i\alpha\rangle - \alpha \langle i\alpha|-i\alpha\rangle - \alpha\right) = 0, \tag{7.46}
$$

while $\langle \psi_\alpha^{\mathrm{coll}}|\hat{a}|\psi_\alpha^{\mathrm{coll}}\rangle \gtrsim 0$ generally has a small, but non-vanishing value. However, the Schrödinger cat state shares the important property

$$
\hat{a}\hat{a}|\mathrm{cat}\rangle = -\alpha^2 |\mathrm{cat}\rangle \tag{7.47}
$$

with $|\psi_\alpha^{\mathrm{coll}}\rangle$. In the following analysis, we restrict ourselves to using $|\mathrm{cat}\rangle$ as it simplifies the calculations. At the end of the discussion, we will specify the changes that may occur when the accurate state $|\psi_\alpha^{\mathrm{coll}}\rangle$ is used.

7.4.2 Detection via Noise Correlations

In an optical lattice, the cat state discussed above will ideally form at each lattice site. Therefore, we assume the global many-body state to have the form

$$
|\Psi^{\mathrm{coll}}\rangle = \prod_i |\mathrm{cat}\rangle_i. \tag{7.48}
$$

We assume the system to have M lattice sites that are filled with N atoms in total, resulting in a mean on-site filling $\bar{n} = N/M$.

At the time of the collapse, the density in momentum space $n(\mathbf{k}) = \langle \hat{n}(\mathbf{k}) \rangle$, which is observed in time-of-flight imaging, does not show a direct signature of the Schrödinger cat state. Based on Eq. 7.24 we expect a density $n(\mathbf{k}, t_{\text{rev}}/2) = |w(\mathbf{k})|^2 M\bar{n}$. In order to identify a unique feature of the Schrödinger cat, we analyze the correlated density in the following. It is theoretically described by

$$\begin{aligned} \langle \hat{n}(\mathbf{k})\hat{n}(\mathbf{k}') \rangle &= \langle \hat{\psi}^\dagger(\mathbf{k})\hat{\psi}(\mathbf{k})\hat{\psi}^\dagger(\mathbf{k}')\hat{\psi}(\mathbf{k}') \rangle \\ &= \delta(\mathbf{k}-\mathbf{k}')\langle \hat{\psi}^\dagger(\mathbf{k})\hat{\psi}(\mathbf{k}) \rangle + \langle \hat{\psi}^\dagger(\mathbf{k})\hat{\psi}^\dagger(\mathbf{k}')\hat{\psi}(\mathbf{k})\hat{\psi}(\mathbf{k}') \rangle, \end{aligned} \tag{7.49}$$

where the commutation relation $[\hat{\psi}(\mathbf{k}), \hat{\psi}^\dagger(\mathbf{k}')] = \delta(\mathbf{k}-\mathbf{k}')$ is applied in the last step to achieve normal order. The term $\langle \hat{\psi}^\dagger(\mathbf{k})\hat{\psi}(\mathbf{k}) \rangle$ creates an autocorrelation peak which does not contain any information on the structure of the many-body quantum state. Therefore, it will be omitted. Using the transformation $\hat{\psi}(\mathbf{k}) = w(\mathbf{k}) \sum_i \hat{a}_i e^{-i\mathbf{k}\cdot\mathbf{r}_i}$, the correlated density can be expressed in terms of on-site operators $\hat{a}_i$:

$$\langle \hat{\psi}^\dagger(\mathbf{k})\hat{\psi}^\dagger(\mathbf{k}')\hat{\psi}(\mathbf{k})\hat{\psi}(\mathbf{k}') \rangle = |w(\mathbf{k})|^2 |w(\mathbf{k}')|^2 \sum_{ijkl} e^{-i\mathbf{k}\cdot(\mathbf{r}_j-\mathbf{r}_i)} e^{-i\mathbf{k}'\cdot(\mathbf{r}_l-\mathbf{r}_k)} \langle \hat{a}_i^\dagger \hat{a}_k^\dagger \hat{a}_j \hat{a}_l \rangle. \tag{7.50}$$

Taking the expectation value within the many-body Schrödinger cat state 7.48, a straightforward calculation yields the result

$$\langle \hat{a}_i^\dagger \hat{a}_k^\dagger \hat{a}_j \hat{a}_l \rangle = (\alpha_i^* \alpha_j)^2 \delta_{ik}\delta_{jl} + \bar{n}_i \bar{n}_k \delta_{il}\delta_{kj} + \bar{n}_i \bar{n}_k \delta_{ij}\delta_{kl} \tag{7.51}$$

and the correlated density of the many-body Schrödinger cat takes the form

$$\begin{aligned} &\langle \hat{\psi}^\dagger(\mathbf{k})\hat{\psi}^\dagger(\mathbf{k}')\hat{\psi}(\mathbf{k})\hat{\psi}(\mathbf{k}') \rangle = \\ &\quad |w(\mathbf{k})|^2 |w(\mathbf{k}')|^2 \left(\sum_{ij} e^{-i(\mathbf{k}+\mathbf{k}')\cdot(\mathbf{r}_j-\mathbf{r}_i)} (\alpha_i^* \alpha_j)^2 + \sum_{ik} e^{-i(\mathbf{k}-\mathbf{k}')\cdot(\mathbf{r}_k-\mathbf{r}_i)} \bar{n}_i \bar{n}_k + \sum_{ik} \bar{n}_i \bar{n}_k \right). \end{aligned} \tag{7.52}$$

To understand the meaning of this formula, it is crucial to note that the positions of the lattice sites form a regular, simple cubic array according to $\mathbf{r}_i = (\lambda_{\text{lat}}/2)\mathbf{n}_i$, where $\mathbf{n}_i \in \mathbb{Z}^3$ is a three-dimensional vector with integer valued entries.

The second sum of Eq. 7.52 gives rise to the typical correlation signals that have been observed in experiments analyzing a Mott insulator of bosons [72, 73]. The experimental observable for these correlations is commonly defined as

$$C(\mathbf{d}) = \frac{\int d\mathbf{k} \langle \hat{n}(\mathbf{k})\hat{n}(\mathbf{k}-\mathbf{d}) \rangle}{\int d\mathbf{k} \langle \hat{n}(\mathbf{k}) \rangle \langle \hat{n}(\mathbf{k}-\mathbf{d}) \rangle} - 1, \tag{7.53}$$

where $\mathbf{d} = \mathbf{k} - \mathbf{k}'$ defines the difference of the momenta $\mathbf{k}$ and $\mathbf{k}'$, which corresponds to the distance of pixels in a time-of-flight image. Assuming identical and real valued coherent state amplitudes $\alpha_i = \alpha = \alpha^*$ at each lattice site, we obtain

$$\begin{aligned} C(\mathbf{d}) &= \frac{1}{M^2} \frac{\int d\mathbf{k} |w(\mathbf{k})|^2 |w(\mathbf{k}-\mathbf{d})|^2 \sum_{ij} e^{-i(2\mathbf{k}-\mathbf{d})(\mathbf{r}_j - \mathbf{r}_i)}}{\int d\mathbf{k} |w(\mathbf{k})|^2 |w(\mathbf{k}-\mathbf{d})|^2} + \frac{1}{M^2} \sum_{ik} e^{-i\mathbf{d}\cdot(\mathbf{r}_k - \mathbf{r}_i)} \\ &\approx \frac{1}{M^2} \sum_{ik} e^{-i\mathbf{d}\cdot(\mathbf{r}_k - \mathbf{r}_i)}. \end{aligned} \tag{7.54}$$

The second line holds, because the rapid oscillations in the numerator average out the first term to zero. This means the Schrödinger cat state is expected to show the same correlations that are also observed for a bosonic Mott insulator [73], which therefore do not provide a distinctive feature.

However, the unique property of the cat state is captured by the first term of Eq. 7.52. This term can be detected by defining an anticorrelation function, in which the role of $\mathbf{k}'$ is taken by $-\mathbf{k}'$. As an appropriate experimental observable we propose

$$AC(\mathbf{d}) = \frac{\int d\mathbf{k} \langle \hat{n}(\mathbf{k}) \hat{n}(\mathbf{d}-\mathbf{k}) \rangle}{\int d\mathbf{k} \langle \hat{n}(\mathbf{k}) \rangle \langle \hat{n}(\mathbf{d}-\mathbf{k}) \rangle} - 1, \tag{7.55}$$

which should allow to detect the exotic $\mathbf{k}/-\mathbf{k}$ correlations of the Schrödinger cat state according to

$$\begin{aligned} AC(\mathbf{d}) &= \frac{1}{M^2} \sum_{ij} e^{-i\mathbf{d}(\mathbf{r}_j - \mathbf{r}_i)} + \frac{1}{M^2} \frac{\int d\mathbf{k} |w(\mathbf{k})|^2 |w(\mathbf{k}-\mathbf{d})|^2 \sum_{ik} e^{-i(2\mathbf{k}-\mathbf{d})\cdot(\mathbf{r}_k - \mathbf{r}_i)}}{\int d\mathbf{k} |w(\mathbf{k})|^2 |w(\mathbf{k}-\mathbf{d})|^2} \\ &\approx \frac{1}{M^2} \sum_{ij} e^{-i\mathbf{d}\cdot(\mathbf{r}_j - \mathbf{r}_i)}. \end{aligned} \tag{7.56}$$

The anticorrelations defined here are expected to vanish for a bosonic Mott insulator. We further note that a superfluid state is expected to neither show correlations $C(\mathbf{d})$ nor anticorrelations $AC(\mathbf{d})$.

The application of the anticorrelation function 7.55 on an actual experimental image necessitates a delicate procedure. To obtain $AC(\mathbf{d})$, the image $n(\mathbf{k})$ needs to be correlated with its point reflected counterpart $n(-\mathbf{k})$ displaced by a relative distance $\mathbf{d}$. The point reflection must be centered on $\mathbf{k} = 0$. In principle, the center of the image can be determined by fitting the maximum of the envelope originating from the Wannier function $|w(\mathbf{k})|^2$. Nevertheless, it constitutes a critical step of the scheme because complications can arise from finite imaging resolution or pixel size. It is important to note that each experimental image $n(\mathbf{k})$ merely is a single realization of the atomic density. Due to the atom shot noise present in the image, it does not correspond to the expectation values used in the definitions 7.53 and 7.55. However, the quantum mechanical expectation values can be derived by appropriately averaging over an ensemble of independently acquired experimental images as described and argued in detail in [72–74].

Finally, we note that the anticorrelations, that have been derived for an idealized cat state 7.45 to simplify the calculation, are expected to emerge in the same way for the exact state at maximum quantum collapse $|\psi_\alpha^{\rm coll}\rangle$. In the above derivation, the $\mathbf{k}/-\mathbf{k}$ correlations arise because the cat state fulfills $\hat{a}\hat{a}|\text{cat}\rangle = -\alpha^2|\text{cat}\rangle$, while $\hat{a}|\text{cat}\rangle \neq \alpha|\text{cat}\rangle$ at the same time. Similarly, the exact maximally collapsed state has the properties $\hat{a}\hat{a}|\psi_\alpha^{\rm coll}\rangle = -\alpha^2|\psi_\alpha^{\rm coll}\rangle$ and $\hat{a}|\psi_\alpha^{\rm coll}\rangle \neq \alpha|\psi_\alpha^{\rm coll}\rangle$. Nevertheless, it remains an interesting open question whether the Schrödinger cat can survive when the full multinomial wavefunction is used as the many-body quantum state instead of the Gutzwiller-type product ansatz. This topic is currently being theoretically investigated [40], (P. Zoller, private communication, 2011).

7.5 Excursion: Atom Optical Transformation of BECs in an Optical Lattice

In this brief section, we show that interesting physics can also arise when collapse and revival dynamics happens in the presence of an underlying confinement. It turns out that a global potential can be used as an atom optical element that imprints a locally varying phase on the matter wave field, similar to the effect of a lens or a tailored phase mask on a light wave. For example, this can enable the transformation of an initial condensate state into an exotic condensate at $\mathbf{q} = 0$ with a non-trivial momentum distribution [75, 76].

In a theoretical treatment, an underlying potential can generally be incorporated by assigning an energy offset ϵ_i to each lattice site i. Accordingly, the energy of a Fock state $|n\rangle_i$ at this lattice site reads $E_n^{(i)} + \epsilon_i n$, and the global many-body state given in Eq. 7.19 for a lattice with M occupied sites takes the more general form

$$|\Psi(t)\rangle = \prod_{i=1}^{M} |\psi_i(t)\rangle \quad \text{with} \quad |\psi_i(t)\rangle = \sum_{n=0}^{\infty} c_n^{(i)} e^{i(E_n^{(i)}+\epsilon_i n)t/\hbar} |n\rangle_i. \tag{7.57}$$

Employing the approximations used in Sect. 7.2.2 ($E_n^{(i)} = E_n$ and $c_n^{(i)} = c_n$ for all occupied lattice sites i), a straightforward calculation yields the modified time dependent momentum distribution as a generalization of Eq. 7.24

$$n(\mathbf{k}, t) = |w(\mathbf{k})|^2 \left(M\bar{n} + |\langle\hat{a}\rangle(t)|^2 \sum_{i\neq j} e^{-i\mathbf{k}\cdot(\mathbf{r}_j-\mathbf{r}_i)} e^{-i(\epsilon_j-\epsilon_i)t/\hbar} \right). \tag{7.58}$$

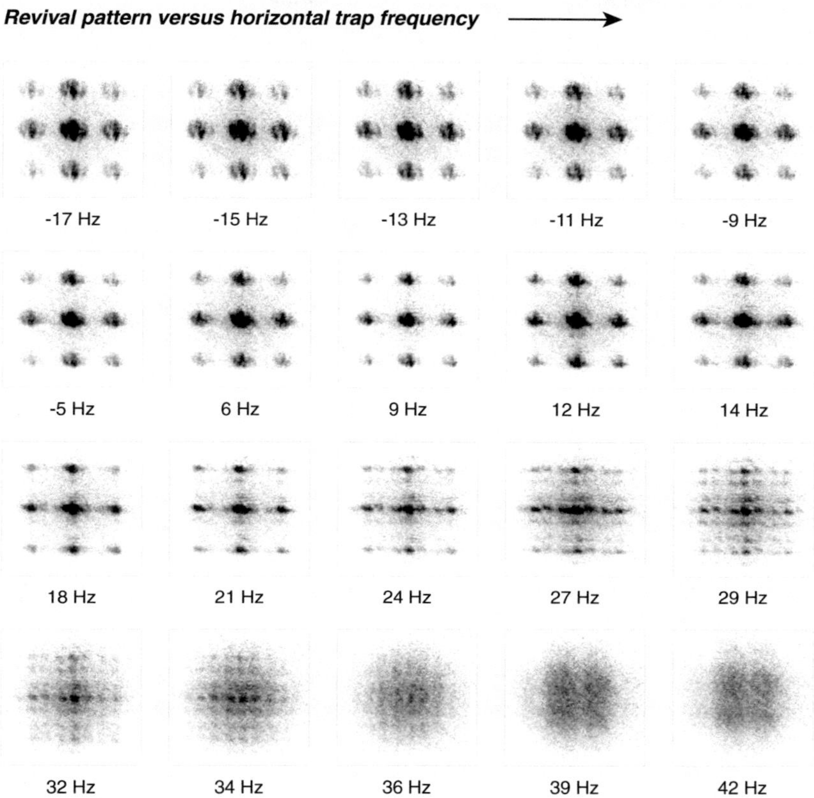

Fig. 7.15 Time evolution originating from underlying harmonic confinement; the displayed images are the raw data that entered the evaluation in Fig. 7.10. A fixed evolution time t_0 of about 2.5 ms was chosen coinciding with a strong quantum phase revival in a $V_{\mathrm{H}} = 40\,E_{\mathrm{rec}}$ deep lattice. The horizontal trap frequency $\omega_{\perp}/(2\pi)$ was increased starting from slightly negative values, corresponding to an anticonfining potential. At vanishing confinement, the regular momentum pattern of a superfluid state appears, while more complex structures of the momentum distribution arise for increasing confinement, indicating fractional rephasing, that is higher-order Talbot fringes

The quantum phase revival dynamics contained in $|\langle \hat{a} \rangle(t)|^2$ are given by Eq. 7.26, while an additional phase term $e^{-i(\epsilon_j - \epsilon_i)t/\hbar}$ complements the interference term $e^{-i\mathbf{k}\cdot(\mathbf{r}_j - \mathbf{r}_i)}$.

For the case of isotropic harmonic confinement, the local energy offset is proportional to the squared distance of the lattice site from the trap center according to $\epsilon_i = m\omega^2 \mathbf{r}_i^2/2$, where ω denotes the trap frequency of the global underlying confinement, $\mathbf{r}_i = (\lambda_{\mathrm{lat}}/2)\mathbf{n}_i$ is the position of lattice site i and $\mathbf{n}_i$ is a three-dimensional vector with integer entries. Additionally, the local phase is a linear function of the evolution time t. Expressed in the language of optics, this setting is analogous to a grating that is illuminated by a coherent light wave and directly followed by a thin lens. A thin lens also creates a phase shift that depends quadratically on the distance from the lens center [77].

In the matter wave case, the "focal length" of this lens is inversely proportional to the evolution time t. However, at times when $(\epsilon_j - \epsilon_i)t$ is an integer multiple of 2π the lens does not have an observable effect because all phase terms have the same value. For a typical harmonic potential with $\omega \approx 2\pi \times 100\,\text{Hz}$, such a complete rephasing in the external potential occurs after about 100 ms, while at certain intermediate times already fractional rephasing is expected [76].[10] By choosing a larger trap frequency ω, this time scale can be significantly shortened. Generally speaking, the time evolution due to an underlying potential is a coherent process which induces dephasing and subsequent rephasing, but not necessarily decoherence. Practically, however, the time scale for rephasing may be so long that decoherence due to technical reasons or atom loss precludes its observation.

We have observed the onset of the evolution that originates from the harmonic confinement, as displayed in Fig. 7.15. In these data, a fixed observation time is chosen and the energy offsets ϵ_i are varied via the external potential. Within the given evolution time rephasing would be expected for large enough compression. However, we merely observe indications for fractional rephasings, which occur at smaller compressions. At larger compression the coherence of the evolution is limited, probably due to anharmonicities of the trapping potential and a slight variation of the lattice depth across the extent of the atomic cloud. Additionally, effects arising from finite time-of-flight expansion may play an important role [47].

Going beyond the case of harmonic confinement, Eq. 7.58 suggests an interesting possibility to manipulate the frozen-out condensate state in a deep optical lattice. Recent experiments have shown that the engineering and projection of almost arbitrary potential landscapes with resolutions on the order of the lattice spacing are within technical reach [81–83]. Using such experimental capabilities, it may be possible to transform the initial quantum state into exotic nonequilibrium condensate states [76] by appropriately engineering the local energy offsets ϵ_i. We note that the quantum phase revival term $|\langle\hat{a}\rangle(t)|^2$ does not play an active role in this transformation. The transformation is solely based on the evolution in the external potential. However, the additional collapse and revival dynamics cannot be avoided as the lattice has to be deep during the evolution in the external potential in order to avoid a spatial redistribution of atoms. The quantum phase revivals effectively discretize the observation of the evolution into time steps of t_{rev} as only the fully revived state can form a condensate.

[10] Please note the similarity to the Talbot effect [78], that is the self-imaging of waves in the near-field (Fresnel) diffraction pattern of a grating [77, 79]. The fractional rephasing is analogous to higher-order Talbot fringes [80].

7.6 Conclusion and Outlook

The conclusion of this chapter is two-fold: First, a new type of spectroscopy has been developed that allows for the precise measurement of the energy of atom number states on an absolute energy scale. Second, the application of this novel technique has revealed that the energies of Fock states on the sites of an optical lattice can generally not be understood within a single-orbital framework. Instead multi-orbital effects give rise to effective multi-body interactions.

The novel method of quantum revival spectroscopy is deeply rooted in the laws of quantum mechanics. By making use of the delocalization of atoms in an optical lattice and nonequilibrium quantum dynamics induced by an interaction quench, the individual energies of atom number states are converted into frequencies. These are observed in the resulting collapse and revival dynamics. In order to distinguish the individual frequencies, a high spectral resolution is needed which requires a long coherence time of the dynamics. The latter is experimentally achieved through the accurate control and cancellation of the underlying confining potential using the combination of a blue-detuned optical lattice and a red-detuned dipole trap. Quantum revival dynamics in a multi-orbital system not only distinguishes individual Fock states by their energies [50], but additionally allows to obtain information on the atom number statistics based on the spectral weight of the measured frequencies. In recently developed experimental setups with single-site and single-atom resolution [81, 82, 84, 85], the atom number statistics in an optical lattice can be accessed more directly. However, quantum revival spectroscopy forms a unique probe of delocalization and quantum superposition that can prove exceptionally helpful in the identification and analysis of complex, delocalized quantum phases [86].

Our precision measurement has strikingly revealed the importance of multi-orbital physics in optical lattices. While the single-orbital Bose-Hubbard model predicts energies that are integer multiples of the interaction parameter U, we have observed characteristic, renormalized energies for each Fock state as interactions promote particles to excited orbitals. The comparison of our experimental data to an exact diagonalization including excited orbitals has yielded excellent agreement. In a field theoretical approach, the energies of Fock states can be expressed by an effective single-band Hamiltonian that includes a series of multi-body interactions as higher-order corrections to the usual two-body contact interaction. We have extracted the strength of the effective three- and four-body interactions from our data, showing that the higher-order contributions are significant in leading orders and converge quickly at high orders. It is an intriguing prospect to use the multi-body interactions demonstrated here for the simulation of effective field theories. In particular, by making use of a Feshbach resonance, it might be possible to study the breakdown of the low energy limit of a field theory when interactions get strong or to test the Gell-Mann and Low theorem describing the evolution of eigenstates when adiabatically switching from noninteracting to interacting eigenstates [87], (P. R. Johnson, E. Tiesinga, private communication, 2010). Taking a glance beyond single lattice sites, very recent studies have shown that multi-body interactions can have significant impact on the many-body physics of lattice quantum gases [56, 88]; multi-body interactions might help in realizing novel strongly correlated quantum phases, for example with topological order [5] or exotic ground-state properties [6]. Finally, quantum phase revivals may not only be used for the detection of effective, but also intrinsic multi-body interactions, which might be present or enhanced in the vicinity of Feshbach or Efimov resonances [89].

The methods and measurements presented in this chapter make an important contribution to a detailed understanding of the Hamiltonians that are actually realized in optical lattice systems. The precise knowledge of interactions is a crucial input for the comparison of lattice-based quantum simulators with many-body quantum theory. It is important in quantum computation schemes that implement gate operations via coherent interatomic collisions [90, 91], as well as in experiments that work towards the interaction-induced dynamical creation of spin-squeezed or Schrödinger cat states with ultracold atoms [64, 92]. The renormalization of lattice parameters due to interatomic interactions has been found to have a crucial role in complex quantum systems such as Bose-Fermi mixtures [93]. We will discuss this case in Chaps. 8 and 9. Besides understanding atom number dependent interaction parameters, it is equally important to elucidate the renormalization of the tunnel parameters in future investigations [56, 94, 95].

References

1. A. Auerbach, *Interacting Electrons and Quantum Magnetism* (Springer, New York, 1994)
2. D. Jaksch, P. Zoller, The cold atom Hubbard toolbox. Ann. Phys. **315**, 52 (2005)
3. M. Lewenstein, A. Sanpera, V. Ahufinger, B. Damski, A. Sen, U. Sen, Ultracold atomic gases in optical lattices: mimicking condensed matter physics and beyond. Adv. Phys. **56**, 243 (2007)
4. K.R.A. Hazzard, E.J. Mueller, On-site correlations in optical lattices: band mixing to coupled quantum hall puddles. Phys. Rev. A **81**, 031602 (2010)
5. B. Paredes, T. Keilmann, J.I. Cirac, Pfaffian-like ground state for three-body hard-core bosons in one-dimensional lattices. Phys. Rev. A **75**, 053611 (2007)
6. B. Capogrosso-Sansone, S. Wessel, H.P. Büchler, P. Zoller, G. Pupillo, Phase diagram of one-dimensional hard-core bosons with three-body interactions. Phys. Rev. B **79**, 020503 (2009)
7. T. Kraemer, M. Mark, P. Waldburger, J.G. Danzl, C. Chin, B. Engeser, A.D. Lange, K. Pilch, A. Jaakkola, H.C. Nägerl, R. Grimm, Evidence for Efimov quantum states in an ultracold gas of caesium atoms. Nature **440**, 315 (2006)
8. S. Knoop, F. Ferlaino, M. Mark, M. Berninger, H. Schobel, H.C. Nägerl, R. Grimm, Observation of an Efimov-like trimer resonance in ultracold atom-dimer scattering. Nat. Phys. **5**, 227 (2009)
9. M. Zaccanti, B. Deissler, C. D'Errico, M. Fattori, M. Jona-Lasinio, S. Müller, G. Roati, M. Inguscio, G. Modugno, Observation of an Efimov spectrum in an atomic system. Nat. Phys. **5**, 586 (2009)
10. F. Ferlaino, S. Knoop, M. Berninger, W. Harm, J.P. D'Incao, H.-C. Nägerl, R. Grimm, Evidence for universal four-body states tied to an Efimov trimer. Phys. Rev. Lett. **102**, 140401 (2009)
11. T. Lompe, T.B. Ottenstein, F. Serwane, A.N. Wenz, G. Zürn, S. Jochim, Radio-frequency association of Efimov trimers. Science **330**, 940 (2010)
12. P.R. Johnson, E. Tiesinga, J.V. Porto, C.J. Williams, Effective three-body interactions of neutral bosons in optical lattices. New J. Phys. **11**, 093022 (2009)
13. I. Bloch, J. Dalibard, W. Zwerger, Many-body physics with ultracold gases. Rev. Mod. Phys. **80**, 885 (2008)

14. B. Yurke, D. Stoler, Generating quantum mechanical superpositions of macroscopically distinguishable states via amplitude dispersion. Phys. Rev. Lett. **57**, 13 (1986)
15. E.M. Wright, D.F. Walls, J.C. Garrison, Collapses and revivals of Bose-Einstein condensates formed in small atomic samples. Phys. Rev. Lett. **77**, 2158 (1996)
16. E.M. Wright, T. Wong, M.J. Collett, S.M. Tan, D.F. Walls, Collapses and revivals in the interference between two Bose-Einstein condensates formed in small atomic samples. Phys. Rev. A **56**, 591 (1997)
17. D.F. Walls, M.J. Collett, T. Wong, S.M. Tan, E.M. Wright, Phase dynamics and Bose-broken symmetry in atomic Bose-Einstein condensates. Philos. Trans. R. Soc. Lond. Ser. A **355**, 2393 (1997)
18. G.J. Milburn, J. Corney, E.M. Wright, D.F. Walls, Quantum dynamics of an atomic Bose-Einstein condensate in a double-well potential. Phys. Rev. A **55**, 4318 (1997)
19. A. Imamoḡlu, M. Lewenstein, L. You, Inhibition of coherence in trapped Bose-Einstein condensates. Phys. Rev. Lett. **78**, 2511 (1997)
20. Y. Castin, J. Dalibard, Relative phase of two Bose-Einstein condensates. Phys. Rev. A **55**, 4330 (1997)
21. M. Greiner, O. Mandel, T. Hänsch, I. Bloch, Collapse and revival of the matter wave field of a Bose-Einstein condensate. Nature **419**, 51 (2002)
22. M. Anderlini, J. Sebby-Strabley, J. Kruse, J.V. Porto, W.D. Phillips, Controlled atom dynamics in a double-well optical lattice. J. Phys. B **39**, S199 (2006)
23. J. Sebby-Strabley, B.L. Brown, M. Anderlini, P.J. Lee, W.D. Phillips, J.V. Porto, P.R. Johnson, Preparing and probing atomic number states with an atom interferometer. Phys. Rev. Lett. **98**, 200405 (2007)
24. M. Brune, F. Schmidt-Kaler, A. Maali, J. Dreyer, E. Hagley, J.M. Raimond, S. Haroche, Quantum rabi oscillation: a direct test of field quantization in a cavity. Phys. Rev. Lett. **76**, 1800 (1996)
25. S. Haroche, J.-M. Raimond, *Exploring the Quantum: Atoms, Cavities, and Photons (Oxford Graduate Texts)*, 1st edn. (Oxford University Press, USA, 2006)
26. E. Tiesinga, P.R. Johnson, Collapse and revival dynamics of number-squeezed superfluids of ultracold atoms in optical lattices. Phys. Rev. A **83**, 063609 (2011)
27. R.J. Glauber, Coherent and incoherent states of the radiation field. Phys. Rev. **131**, 2766 (1963)
28. P. G. J. van Dongen, *Quantentheorie für Fortgeschrittene*. Lecture Notes, Johannes Gutenberg-Universität Mainz (2004/2005)
29. D.M. Meekhof, C. Monroe, B.E. King, W.M. Itano, D.J. Wineland, Generation of nonclassical motional states of a trapped atom. Phys. Rev. Lett. **76**, 1796 (1996)
30. R. Loudon, *The Quantum Theory of Light*, (Oxford Science Publications) 3rd edn. (OUP, Oxford, 2000)
31. M.R. Andrews, C.G. Townsend, H.-J. Miesner, D.S. Durfee, D.M. Kurn, W. Ketterle, Observation of interference between two Bose condensates. Science **275**, 637 (1997)
32. S. Will, *Atom Optical Experiments with Ultracold Sodium Atoms*, Master's thesis, Johannes Gutenberg-Universität Mainz/Massachusetts Institute of Technology, (2006)
33. G.-B. Jo, Y. Shin, S. Will, T.A. Pasquini, M. Saba, W. Ketterle, D.E. Pritchard, M. Vengalattore, M. Prentiss, Long phase coherence time and number squeezing of two Bose-Einstein condensates on an atom chip. Phys. Rev. Lett. **98**, 030407 (2007)
34. N.N. Klausen, J.L. Bohn, C.H. Greene, Nature of spinor Bose-Einstein condensates in rubidium. Phys. Rev. A **64**, 053602 (2001)
35. E.G.M. van Kempen, S.J.J.M.F. Kokkelmans, D.J. Heinzen, B.J. Verhaar, Interisotope determination of ultracold rubidium interactions from three high-precision experiments. Phys. Rev. Lett. **88**, 093201 (2002)
36. D. Jaksch, C. Bruder, J.I. Cirac, C.W. Gardiner, P. Zoller, Cold bosonic atoms in optical lattices. Phys. Rev. Lett. **81**, 3108 (1998)
37. D. Jaksch, *Bose-Einstein condensation and applications*. Ph.D. Thesis, Leopold-Franzens-Universität Innsburck, 1999
38. M. Scully, S. Zubairy, *Quantum Optics* (Cambridge University Press, Cambridge, 1997)

39. W. Zwerger, Mott-Hubbard transition of cold atoms in optical lattices. J. Opt. B **5**, S9 (2003)
40. K. Pawłowski, K. Rzążewski, Background atoms and decoherence in optical lattices. Phys. Rev. A **81**, 013620 (2010)
41. J. Schachenmayer, A.J. Daley, P. Zoller, Atomic matter-wave revivals with definite atom number in an optical lattice. Phys. Rev. A **83**, 043614 (2011)
42. C. Orzel, A.K. Tuchman, M.L. Fenselau, M. Yasuda, M.A. Kasevich, Squeezed states in a Bose-Einstein condensate. Science **291**, 2386 (2001)
43. M. Greiner, O. Mandel, T. Esslinger, T. Hänsch, I. Bloch, Quantum phase transition from a superfluid to a Mott insulator in a gas of ultracold atoms. Nature **415**, 39 (2002)
44. D.S. Rokhsar, B.G. Kotliar, Gutzwiller projection for bosons. Phys. Rev. B **44**, 10328 (1991)
45. M. Greiner, *Ultracold quantum gases in three-dimensional optical lattice potentials*. Ph.D. Thesis, LMU München, 2003
46. D.-S. Lühmann, *Multiorbital physics in optical lattices*. Ph.D. Thesis, Universität Hamburg, 2009
47. F. Gerbier, S. Trotzky, S. Fölling, U. Schnorrberger, J.D. Thompson, A. Widera, I. Bloch, L. Pollet, M. Troyer, B. Capogrosso-Sansone, N.V. Prokof'ev, B.V. Svistunov, Expansion of a quantum gas released from an optical lattice. Phys. Rev. Lett. **101**, 155303 (2008)
48. P. G. J. van Dongen, *Quantenmechanik I*. Lecture Notes, Johannes Gutenberg-Universität Mainz (2005/2006)
49. O.E. Alon, A.I. Streltsov, L.S. Cederbaum, Zoo of quantum phases and excitations of cold bosonic atoms in optical lattices. Phys. Rev. Lett. **95**, 030405 (2005)
50. G.K. Campbell, J. Mun, M. Boyd, P. Medley, A.E. Leanhardt, L.G. Marcassa, D.E. Pritchard, W. Ketterle, Imaging the Mott insulator shells by using atomic clock shifts. Science **313**, 649 (2006)
51. J. Li, Y. Yu, A. Dudarev, Q. Niu, Interaction broadening of Wannier functions and Mott transitions in atomic BEC. New J. Phys. **8**, 154 (2006)
52. D.-S. Lühmann, K. Bongs, K. Sengstock, D. Pfannkuche, Self-trapping of bosons and fermions in optical lattices. Phys. Rev. Lett. **101**, 050402 (2008)
53. J. Larson, A. Collin, J.-P. Martikainen, Multiband bosons in optical lattices. Phys. Rev. A **79**, 033603 (2009)
54. H.P. Büchler, Microscopic derivation of Hubbard parameters for cold atomic gases. Phys. Rev. Lett. **104**, 090402 (2010)
55. A. Mering, M. Fleischhauer, Multiband and nonlinear hopping corrections to the three-dimensional Bose-Fermi-Hubbard model. Phys. Rev. A **83**, 063630 (2011)
56. O. Dutta, A. Eckardt, P. Hauke, B. Malomed, M. Lewenstein, Bose-Hubbard model with occupation-dependent parameters. New J. Phys. **13**, 023019 (2011)
57. L. Cao, I. Brouzos, S. Zöllner, P. Schmelcher, Interaction-driven interband tunneling of bosons in the triple well. New J. Phys. **13**, 033032 (2011)
58. A. Szabo, N. S. Ostlund, *Modern Quantum Chemistry—Introduction to Advanced Electronic Structure Theory*, 1st edn. (revised edn.) (McGraw-Hill Publishing Company, New York, 1989)
59. S. Will, T. Best, U. Schneider, L. Hackermüller, D.-S. Lühmann, I. Bloch, Time-resolved observation of coherent multi-body interactions in quantum phase revivals. Nature **465**, 197 (2010)
60. A. Muryshev, G.V. Shlyapnikov, W. Ertmer, K. Sengstock, M. Lewenstein, Dynamics of dark solitons in elongated Bose-Einstein condensates. Phys. Rev. Lett. **89**, 110401 (2002)
61. I.E. Mazets, T. Schumm, J. Schmiedmayer, Breakdown of integrability in a quasi-1D ultracold bosonic gas. Phys. Rev. Lett. **100**, 210403 (2008)
62. D.-S. Lühmann, K. Bongs, D. Pfannkuche, Excitation spectrum of Mott shells in optical lattices. J. Phys. B **42**, 145305 (2009)
63. F. Gerbier, S. Fölling, A. Widera, O. Mandel, I. Bloch, Probing number squeezing of ultracold atoms across the superfluid-Mott insulator transition. Phys. Rev. Lett. **96**, 090401 (2006)
64. J. Estéve, C. Gross, A. Weller, S. Giovanazzi, M.K. Oberthaler, Squeezing and entanglement in a Bose-Einstein condensate. Nature **455**, 1216 (2008)

65. P. Cheinet, S. Trotzky, M. Feld, U. Schnorrberger, M. Moreno-Cardoner, S. Fölling, I. Bloch, Counting atoms using interaction blockade in an optical superlattice. Phys. Rev. Lett. **101**, 090404 (2008)
66. F. Gerbier, A. Widera, S. Fölling, O. Mandel, T. Gericke, I. Bloch, Phase coherence of an atomic Mott insulator. Phys. Rev. Lett. **95**, 050404 (2005)
67. U.R. Fischer, R. Schützhold, Tunneling-induced damping of phase coherence revivals in deep optical lattices. Phys. Rev. A **78**, 061603 (2008)
68. K. Huang, *Statistical Mechanics*, 2nd edn. (Wiley, New York, 1987)
69. T. Busch, B.-G. Englert, K. Rzażewski, M. Wilkens, Two cold atoms in a harmonic trap. Found. Phys. **28**, 549 (1998)
70. W.S. Bakr, P.M. Preiss, M.E. Tai, R. Ma, J. Simon, M. Greiner, Orbital excitation blockade and algorithmic cooling in quantum gases. Nature **480**, 500 (2011)
71. M.W. Jack, Effect of atom loss on collapse and revivals of phase coherence in small atomic samples. Phys. Rev. A **67**, 043612 (2003)
72. E. Altman, E. Demler, M.D. Lukin, Probing many-body states of ultracold atoms via noise correlations. Phys. Rev. A **70**, 013603 (2004)
73. S. Fölling, F. Gerbier, A. Widera, O. Mandel, T. Gericke, I. Bloch, Spatial quantum noise interferometry in expanding ultracold atom clouds. Nature **434**, 481 (2005)
74. S. Fölling, *Probing strongly correlated states of ultracold atoms in optical lattices*. Ph.D. Thesis, Johannes Gutenberg-Universität Mainz, 2008
75. M. Gustavsson, E. Haller, M.J. Mark, J.G. Danzl, R. Hart, A.J. Daley, H.-C. Nägerl, Interference of interacting matter waves. New J. Phys. **12**, 065029 (2010)
76. M. Buchhold, U. Bissbort, S. Will, W. Hofstetter, Creating exotic condensates via quantum-phase-revival dynamics in engineered lattice potentials. Phys. Rev. A **84**, 023631 (2011)
77. B. Saleh, M. Teich, *Fundamentals of Photonics* (Wiley-Interscience, New York, 1991)
78. M.J. Mark, E. Haller, J.G. Danzl, K. Lauber, M. Gustavsson, H.-C. Nägerl, Demonstration of the temporal matter-wave Talbot effect for trapped matter waves. New J. Phys. **13**, 085008 (2011)
79. M.S. Chapman, C.R. Ekstrom, T.D. Hammond, J. Schmiedmayer, B.E. Tannian, S. Wehinger, D.E. Pritchard, Near-field imaging of atom diffraction gratings: the atomic Talbot effect. Phys. Rev. A **51**, R14 (1995)
80. S. Nowak, C. Kurtsiefer, T. Pfau, C. David, High-order Talbot fringes for atomic matter waves. Opt. Lett. **22**, 1430 (1997)
81. W. Bakr, J. Gillen, A. Peng, S. Fölling, M. Greiner, A quantum gas microscope for detecting single atoms in a Hubbard-regime optical lattice. Nature **462**, 74 (2009)
82. J. Sherson, C. Weitenberg, M. Endres, M. Cheneau, I. Bloch, S. Kuhr, Single-atom-resolved fluorescence imaging of an atomic Mott insulator. Nature **467**, 68 (2010)
83. B. Zimmermann, T. Müller, J. Meineke, T. Esslinger, H. Moritz, High-resolution imaging of ultracold fermions in microscopically tailored optical potentials. New J. Phys. **13**, 043007 (2010)
84. W.S. Bakr, A. Peng, M.E. Tai, R. Ma, J. Simon, J.I. Gillen, S. Fölling, L. Pollet, M. Greiner, Probing the superfluid-to-Mott insulator transition at the single-atom level. Science **329**, 547 (2010)
85. C. Weitenberg, M. Endres, J. Sherson, M. Cheneau, P. Schauß, T. Fukuhara, I. Bloch, S. Kuhr, Single-spin addressing in an atomic Mott insulator. Nature **471**, 319 (2011)
86. M. Foss-Feig, A.M. Rey, Phase diagram of the bosonic Kondo-Hubbard model. Phys. Rev. A **84**, 053619 (2011)
87. A. Fetter, J. Walecka, *Quantum Theory of Many-Particle Systems* (Dover Publications, New York, 2003)
88. S. Pielawa, T. Kitagawa, E. Berg, S. Sachdev, Correlated phases of bosons in tilted frustrated lattices. Phys. Rev. B **83**, 205135 (2011)
89. C. Chin, R. Grimm, P. Julienne, E. Tiesinga, Feshbach resonances in ultracold gases. Rev. Mod. Phys. **82**, 1225 (2010)

90. A. A Sørensen, L. M. Duan, J. I. Cirac, P. Zoller, Many-particle entanglement with Bose-Einstein condensates. Nature **409**, 63 (2001)
91. O. Mandel, M. Greiner, A. Widera, T. Rom, T. Hänsch, I. Bloch, Controlled collisions for multi-particle entanglement of optically trapped atoms. Nature **425**, 937 (2003)
92. C. Gross, T. Zibold, E. Nicklas, J. Esteve, M.K. Oberthaler, Nonlinear atom interferometer surpasses classical precision limit. Nature **464**, 1165 (2010)
93. T. Best, S. Will, U. Schneider, L. Hackermüller, D. van Oosten, D.-S. Lühmann, I. Bloch, Role of interactions in ^{87}Rb-^{40}K Bose-Fermi mixtures in a 3D optical lattice. Phys. Rev. Lett. **102**, 030408 (2009)
94. D.-S. Lühmann, O. Jürgensen, K. Sengstock, Multi-orbital and density-induced tunneling of bosons in optical lattices. New J. Phys. **14**, 033021 (2012)
95. U. Bissbort, F. Deuretzbacher, W. Hofstetter, *Effective multibody-induced tunneling and interactions in the Bose-Hubbard model of the lowest dressed band of an optical lattice*. Phys. Rev. A **86**, 023617 (2012)

Chapter 8
Interacting Mixtures of Bosons and Fermions in Optical Lattice Potentials

In the context of real solid state materials, it is a crucial question to what extent impurities influence quantum phases and phase transitions. This question is so important, because impurities can hardly be avoided in real systems and have an impact on many-body quantum phases that is complex and often not well understood. For example, high-temperature superconductivity is observed in materials that are intentionally doped with certain impurity atoms, but their exact role in the mechanism responsible for superconductivity is still not settled [1]. Experimental input on the effect of impurities is highly sought-after and ultracold atoms in optical lattice potentials seem to be an ideal model system, offering the possibility to introduce impurities, for example interacting fermions, into an otherwise almost defect free quantum system, for example a bosonic Mott insulator. When Bose-Fermi mixtures in optical lattice potentials became experimentally available, pioneering investigations addressed the coherence properties of the bosons in the presence of interacting fermions [2, 3]. Those studies raised the question, in which way the fermions influence the superfluid to Mott insulator transition that had so far only been observed in purely bosonic systems [4]. It was found that in the presence of fermions the Mott insulating phase was entered already at lower lattice depth, in contrast to the common theoretical expectations [5, 6]. However, the interpretation of the data in the early experiments [2, 3] remained vague, because the interaction between the bosonic and fermionic atoms was fixed at the fairly large attractive background scattering length of $a_{\mathrm{BF}} \approx -189\ a_0$ [7]. The only tunable parameter was the fermionic filling.

In this chapter I report on a refined study on the role of interspecies interactions in a quantum degenerate mixture of bosonic ^{87}Rb and fermionic ^{40}K in a 3D optical lattice [8]. For attractive interspecies interactions a pronounced modification of the bosonic coherence properties is identified, which appears as a marked shift of the superfluid to Mott insulator transition towards lower lattice depths. However, for repulsive interactions the coherence of the bosonic component is essentially unaffected, being compatible with phase separation that already sets in at rather weak repulsion. The theoretical analysis reveals that our data for attractive interactions can be qualitatively understood in terms of Bose-Hubbard parameters that

S. Will, *From Atom Optics to Quantum Simulation*, Springer Theses,
DOI: 10.1007/978-3-642-33633-1_8,

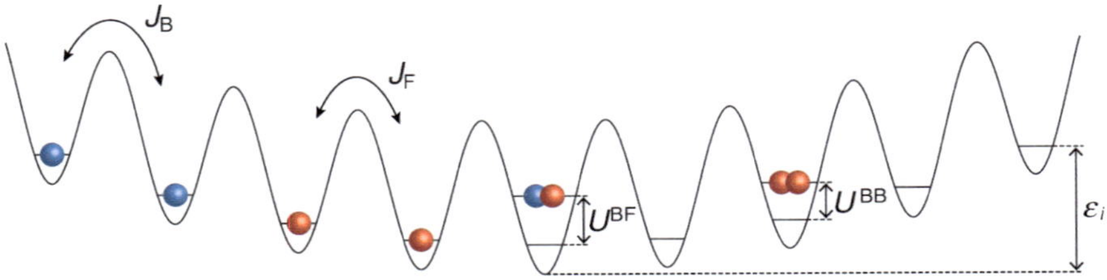

Fig. 8.1 Illustration of the matrix elements of the Bose-Fermi Hubbard model (see text)

are renormalized by interspecies interactions owing to the presence of higher lattice bands. The indirect evidence for renormalized Hubbard parameters obtained here has motivated the measurements reported in Chap. 9 that provide a quantitative confirmation. A thorough discussion of the experiments and results presented in this chapter is also given in the PhD thesis of Thorsten Best [9]. Here the most important aspects are summarized in view of Chap. 9.

8.1 Theoretical Framework: Bose-Fermi Hubbard Model

We consider a mixture of bosonic ^{87}Rb and fermionic ^{40}K atoms loaded in an optical lattice with both species being prepared in a single spin state. Interactions among the spin polarized fermions are suppressed, while interactions among the bosons and between bosons and fermions must be considered.

It is convenient to think of this system within the framework of the single-band Bose-Fermi Hubbard model [10–12]:

$$\hat{H} = -J^{\mathrm{B}} \sum_{\langle i,j \rangle} \hat{a}_i^\dagger \hat{a}_j - J^{\mathrm{F}} \sum_{\langle i,j \rangle} \hat{c}_i^\dagger \hat{c}_j + \frac{U^{\mathrm{BB}}}{2} \sum_i \hat{n}_i(\hat{n}_i - 1) + U^{\mathrm{BF}} \sum_i \hat{n}_i \hat{m}_i, \quad (8.1)$$

where $\hat{a}_i$ ($\hat{c}_i$) denotes the bosonic (fermionic) annihilation operator and $n_i \equiv \hat{a}_i^\dagger \hat{a}_i$ ($m_i \equiv \hat{c}_i^\dagger \hat{c}_i$) the bosonic (fermionic) atom number operators, respectively. The bosonic (fermionic) tunneling parameters J^{B} (J^{F}) and the Bose-Bose interaction energy U^{BB} are defined as discussed in Chap. 3 (see Fig. 8.1). The Bose-Fermi interaction energy is given by

$$U^{\mathrm{BF}} = \frac{2\pi\hbar^2 a_{\mathrm{BF}}}{\mu_{\mathrm{BF}}} \int d^3r \, |w_{\mathrm{B}}(\mathbf{r})|^2 \, |w_{\mathrm{F}}(\mathbf{r})|^2, \quad (8.2)$$

where a_{BF} is the interspecies s-wave scattering length, $\mu_{\mathrm{BF}} = m_{\mathrm{B}} m_{\mathrm{F}}/(m_{\mathrm{B}} + m_{\mathrm{F}})$ the reduced mass of the colliding boson-fermion pair and $w_{\mathrm{B}}(\mathbf{r})$ ($w_{\mathrm{F}}(\mathbf{r})$) the bosonic (fermionic) Wannier function at a lattice site. Generally, also energy offsets ϵ_i^{B} and

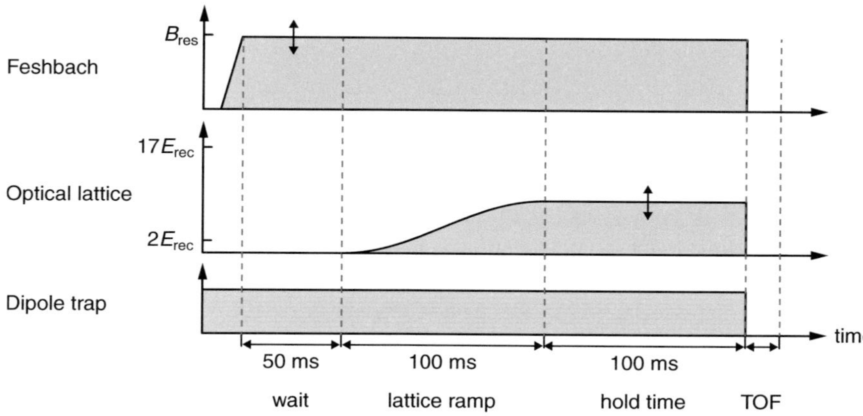

Fig. 8.2 Experimental sequence to adiabatically load a Bose-Fermi mixture into a 3D optical lattice. Measurements are performed at various magnetic fields, corresponding to various interspecies scattering length a_{BF}, and lattice depths ranging from 2 to 17 E_{rec}

ϵ_i^F of each lattice site i (for example, due to an external trapping potential) must be taken into account, but are omitted here for simplicity.

Based on the single-band model (8.1) a variety of quantum phases have been predicted, including charge density waves, spin-density waves [13], polaronic quasiparticles [14, 15] and perturbed Mott insulating states [16, 17]. In limited parameter regimes even more exotic phases are expected, such as superfluids formed by the correlated or anticorrelated flow of composite quasiparticles (for example a fermion plus a bosonic hole) [18] and supersolid phases [19, 20]. The prospect of realizing a supersolid phase with ultracold atoms is particularly exciting, as the possible observation of supersolidity in cold Helium systems is still heavily debated [21, 22]. Extensions of the above Hamiltonian that consider a two-component fermionic spin-mixture instead of spin-polarized fermions, have led to the prediction of enhanced fermion-fermion interactions mediated by Bose-Fermi interactions, which in turn may give rise to superfluidity at high transition temperatures [23]. The direct experimental observation of mediated interactions is reported in Chap. 9 [24].

For single component lattice quantum gases, such as the ones discussed in Chaps. 6 and 7, the description in terms of a single-band Hubbard model can capture the main physical effects in broad parameter ranges of experimental relevance [25]. In contrast, the experimental studies on Bose-Fermi mixtures conducted so far [2, 3] and in particular the data presented in this chapter [8] strongly suggest, that physics beyond the single-band Bose-Fermi Hubbard model must be considered in a realistic description. The interplay of two atomic species enhances the roles of multi-band [26] and finite temperature effects [27] as well as physical loss channels. Numerous theoretical investigations are directed towards a better understanding of these effects [5, 28–32].

8.2 Experimental Realization

The experimental sequence aims at the adiabatic preparation of an interacting mixture of bosonic ^{87}Rb and fermionic ^{40}K in a three-dimensional optical lattice (see Fig. 8.2). Our starting point is a quantum degenerate mixture of $N_{\mathrm{B}} = 4 \times 10^5$ rubidium and up to $N_{\mathrm{F}} = 3 \times 10^5$ potassium atoms in their respective hyperfine ground states $|F, m_F\rangle = |1, +1\rangle$ and $|9/2, -9/2\rangle$, which is created by the experimental sequence outlined in Chap. 5. The mixture is stored in a crossed dipole trap ($\lambda_{\mathrm{dip}} =$ 1030 nm) formed by elliptical laser beams to provide tight vertical confinement for good interspecies overlap in the presence of gravity (see Sect. 5.3).

After the preparation of the mixture, a homogeneous magnetic field is used to address the interspecies Feshbach resonance at $B_{\mathrm{res}} = 546.75$ G, which has a width of 3.1 G [7, 33] (see Sect. 2.3.5). This allows us to tune the interspecies scattering length a_{BF} between $-170\ a_0$ and $+800\ a_0$ below the resonance and between $-800\ a_0$ and $-200\ a_0$ above the resonance. The accuracy of the interspecies scattering length a_{BF} is determined by the magnetic field stability and strongly varies depending on the distance from the resonance position. In the parameter regimes of the measurements presented here, we reach an accuracy of $\pm 20\ a_0$ close to and $\pm 0.1\ a_0$ far away from the resonance.

After 50 ms of settling time for the magnetic field, a 3D optical lattice is slowly ramped up within 100 ms to final depths ranging between $V_{\mathrm{lat}} = 2$ to 17 E_{rec}. Different from all other experiments reported in this thesis, the lattice wavelength used here is $\lambda_{\mathrm{lat}} = 755$ nm. At this "magic" wavelength the lattice depth, measured in units of the respective recoil energies $E_{\mathrm{rec}}^{\mathrm{B,F}} = h^2/(2m_{\mathrm{B},\ F}\lambda_{\mathrm{lat}}^2)$, is equal for both species (see Sect. 5.4). Therefore the shape of the on-site Wannier functions is identical for ^{87}Rb and ^{40}K, which ensures maximum overlap of the wavefunctions, while the tunneling rate of the bosons is slower compared to the fermions by a factor $m_{\mathrm{F}}/m_{\mathrm{B}}$, which is the mass ratio. The harmonic anti-confinement of the blue-detuned optical lattice is compensated using the dipole trap, which is kept at a constant intensity irrespective of the lattice depth. Therefore the effective horizontal and vertical trap frequencies slightly depend on V_{lat}: At 2 E_{rec} they are $2\pi \times (32, 113)$ Hz for ^{87}Rb and $2\pi \times (44, 194)$ Hz for ^{40}K; at 17 E_{rec} they are $2\pi \times (28, 112)$ Hz for ^{87}Rb and $2\pi \times (31, 191)$ Hz for ^{40}K.

The mixture is held in the lattice for 100 ms, which is long compared to the tunneling time for all lattice depths used in the measurements and therefore should ensure the formation of a fully equilibrated sample. Then all traps as well as the magnetic field are instantaneously switched off and both the bosonic and the fermionic atom clouds undergo 18 ms of time-of-flight (TOF) expansion. The interference pattern of the ^{87}Rb atoms is recorded using absorption imaging.[1] From these images the interference contrast is extracted in terms of the visibility [34],

---

[1] Also the fermionic clouds are imaged to monitor the atom number. However, apart from counting the fermion number those images are not further evaluated.

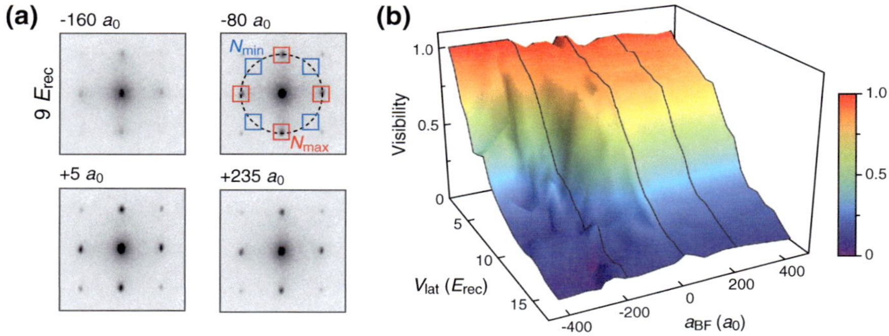

Fig. 8.3 **a** Time-of-flight images of the bosonic component of an interacting Bose-Fermi mixture in a $V_{\text{lat}} = 9\,E_{\text{rec}}$ deep lattice. The influence of interspecies interactions on the bosonic interference pattern is shown in four exemplary images ranging from strong attraction to strong repulsion. The visibility (see Eq. 8.3) is used to quantitatively measure the contrast of the patterns. **b** The visibility of the bosonic component in mixtures with about $N_{\text{B}} = 4 \times 10^5$ ^{87}Rb and $N_{\text{F}} = 3 \times 10^5$ ^{40}K atoms is shown as a function of the lattice depth V_{lat} and the interspecies scattering length a_{BF}. The lattice depth in units of the respective recoil energy is identical for the bosonic and the fermionic component ($\lambda_{\text{lat}} = 755\,\text{nm}$)

$$\mathcal{V} = \frac{N_{\text{max}} - N_{\text{min}}}{N_{\text{max}} + N_{\text{min}}}, \tag{8.3}$$

where N_{max} denotes the total atom number in the four boxes around the first-order interference peaks and N_{min} the total atom number in the boxes that are rotated by 45° (see Fig. 8.3a). The visibility is a robust experimental measure for macroscopic phase coherence and is discussed in detail in Chap. 4. In addition, we determine the condensate fraction [35]: After fitting and subtracting a broad Gaussian background, which corresponds to the Fourier transform of the on-site wavefunctions, the eight first-order diffraction peaks as well as the central peak are fitted by anisotropic two-dimensional bimodal Gaussians. We define the condensate fraction as the number of atoms in the narrow feature of the bimodal Gaussians divided by the total atom number.

8.2.1 Analyzing the Visibility

We probe the coherence of the bosonic component at various lattice depths and several interspecies interactions. Three fermion numbers are chosen ranging from low ($N_{\text{F}} \approx 0.25 N_{\text{B}}$) over intermediate ($N_{\text{F}} \approx 0.5 N_{\text{B}}$) to high ($N_{\text{F}} \approx 0.75 N_{\text{B}}$) in order to identify possible effects of the fermionic filling. The visibility analysis for the complete data set at high N_{F} is shown in Fig. 8.3b. Cuts of this plot at several fixed lattice depths are displayed in Fig. 8.4 to illustrate the details.

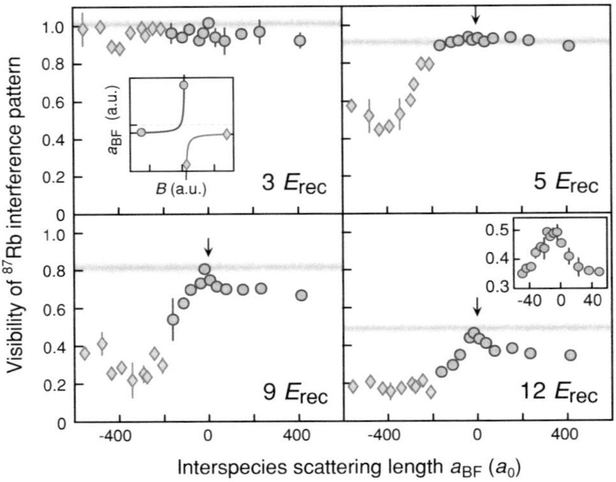

Fig. 8.4 Visibility versus interspecies scattering length at four lattice depths ranging from 3 E_{rec} to 12 E_{rec} for high fermion number ($N_{\mathrm{F}} = 3 \times 10^5$). The cuts are taken from the full data set shown in Fig. 8.3b. At lattice depths larger than 3 E_{rec} the visibility shows a maximum at a position, which is consistent with $a_{\mathrm{BF}} \approx 0$ (indicated by the *arrow*) given the experimental uncertainties. The *diamonds* (*circles*) indicate points measured *above* (*below*) resonance, as shown in the upper inset. The *gray* shaded horizontal lines represent the visibility in a purely bosonic ^{87}Rb cloud measured for the same experimental parameters

For shallow lattices of less than 3 E_{rec} we find a high visibility, which stays at an almost constant level independent of the interspecies interaction strength. At $a_{\mathrm{BF}} \approx 0$ we observe a monotonic decay of the visibility versus lattice depth compatible with the superfluid to Mott insulator transition in a purely bosonic ^{87}Rb sample [34]. This is quantitatively supported in Fig. 8.4 by comparison to the *gray* shaded lines and shows that the fermion cloud can experimentally be tuned to full transparency for the bosons.

For small values of a_{BF} between about $\pm 30\ a_0$ we observe a symmetrical decrease of visibility that appears to be centered around $a_{\mathrm{BF}} \approx 0$ within our measurement accuracy (lower inset of Fig. 8.4). Initially, it has been suggested that this symmetry could be explained by a particle-hole transformation of the fermionic operators $\hat{c}_i \rightarrow (-1)^i \hat{c}_i^\dagger$, which changes the sign of the Bose-Fermi interaction $U^{\mathrm{BF}} \rightarrow -U^{\mathrm{BF}}$ in the Bose-Fermi Hubbard model (8.1). However, strictly speaking particle-hole symmetry only exists at half fermionic filling in a homogeneous system [13]. Given the experimental robustness of the feature being rather insensitive to changes of the fermion number, particle-hole symmetry can probably not explain our observation. Recent theoretical studies employing Hartree-Fock mean field theory [31] and dynamical mean field theory (DMFT) [32] strongly suggest that an intricate interplay between interspecies interactions, finite temperature and underlying harmonic confinement explains the observed behavior in the regime of small a_{BF}. These studies show that Bose-Fermi interactions lead to a redistribution of atoms in the underlying harmonic trap similar to the situation without lattice discussed in Sect. 2.1.5. The redistribution goes along with an effective reduction of the size of the available Hilbert space in close analogy to our observations in attractively interacting Fermi-Fermi mixtures in a 3D optical lattice (see Sect. 6.3 and [36]). When bosons and fermions interact attractively, they have a strong tendency to occupy the same lattice sites, while avoiding each other for repulsive interactions. Assuming that the total entropy of the Bose-Fermi system stays at a constant value during the lattice loading, the reduced Hilbert space induces an increase of the absolute temperature, that entails a significant reduction of the bosonic coherence.

Towards both sides of the symmetry peak we observe a further decay of visibility, which is significantly stronger for intermediate attractive than for comparable repulsive scattering lengths. Therefore, we conjecture a fundamental difference in the underlying mechanisms on either side. The behavior near the attractive background scattering length $a_{BG} = -189\ a_0$ is compatible with previous experimental observations [2, 3]. At even stronger attractions ($a_{BF} \lesssim -300\ a_0$), we find a significant loss of ^{87}Rb atoms, which interestingly is accompanied by an increase of visibility. This regime is discussed in the next section in more detail.

For strong repulsion, the visibility remains almost constant on a high level, slightly below the maximal value for $a_{BF} \approx 0$. This indicates that bosons and fermions do not occupy the same lattice sites in this regime. On the one hand, for moderately repulsive interactions this behavior would be compatible with a local separation of the species as it would occur in anti-correlated mixed phases [37] or in a supersolid [19, 20]. On the other hand, it could simply hint at global phase separation [13, 32]. It is hard to distinguish between the two scenarios experimentally, as only simultaneous high resolution images of the local bosonic and fermionic occupation would provide a definite answer. Nevertheless, under the current conditions global phase separation seems to be the more likely scenario, being promoted by the slightly different trap shapes for the two species in the presence of gravity (see Sect. 5.3.1). For $a_{BF} > 400\ a_0$ we observe very strong atom losses that are essentially independent of the hold time, suggesting that they occur early during lattice loading. In this regime the mixture is held closely below the Feshbach resonance, where the highest lying molecular level has a very low binding energy (see Sect. 2.3). Heteronuclear three-body processes are likely to efficiently populate the molecular state forming a strong loss channel.

It is an important question, whether the Bose-Fermi systems prepared here are in thermal equilibrium. We have checked this by probing the reversibility of the initial thermodynamical change of state. To this end we slowly tune the interspecies scattering length to zero after the mixture has been prepared at a certain value of a_{BF}. Provided that enough time is given for re-equilibration, we detect at least partial reversibility of the initial visibility decrease in the range $-200\ a_0 < a_{BF} < 400\ a_0$, which coincides with the regime of no or minor atom losses. This suggests that the reduction of coherence in the given range of interactions is caused by an adiabatic change of state, a reversible redistribution of atoms during lattice loading.

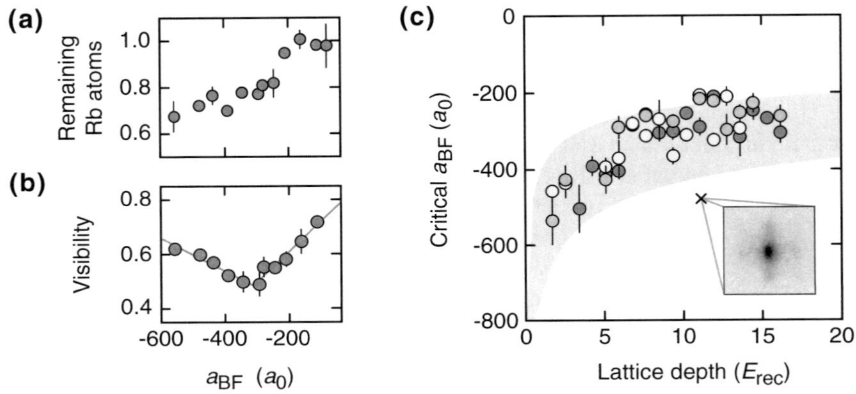

Fig. 8.5 Critical interspecies scattering lengths for the loss dominated regime at attractive interactions. The experimentally observed loss feature of ^{87}Rb atoms (**a**) is accompanied by a simultaneous minimum in the visibility versus interspecies interaction (**b**), shown here for $V_{lat} = 9\ E_{rec}$ and $N_F = 2 \times 10^5$ fermions. (**c**) The critical interspecies interactions obtained from bilinear fits to the visibility data are shown as a function of lattice depth. *Darker* colors indicate higher fermion numbers N_F. The *gray* shaded area shows a theoretical estimate of the scattering length, at which bosonic three-body losses become relevant on the scale of the experimental hold time (see main text). The inset shows a typical time-of-flight image in the loss-dominated regime with increased visibility ($V_{lat} = 11\ E_{rec}$ and $a_{BF} = -480\ a_0$)

Based on the above observations, we qualitatively identify five distinct regimes: First, lossless coexistence in shallow lattices for all scattering lengths. For deeper lattices there are further regimes characterized by very strong interspecies attraction, intermediate attraction, weak interaction of either sign and strong repulsion. This classification also holds for the low and intermediate values of N_F, although the symmetric feature around $a_{BF} \approx 0$ is most pronounced for the highest ^{40}K numbers.

8.2.2 Attractive Interactions and On-Site Collapse

As described above, we observe significant losses of ^{87}Rb accompanied by an increase of visibility for interspecies attraction beyond a critical value as shown in Fig. 8.5a, b. However, we note that this increase of visibility does not go along with sharp diffraction peaks in the time-of-flight images that would be expected for an actual condensate (see inset of Fig. 8.5c). Nevertheless, we use the minimum in the visibility to obtain the critical value for the onset of the loss regime. A bilinear model is fitted to the relevant part of the visibility profiles to determine the minimum quantitatively. In Fig. 8.5c the resulting critical scattering lengths are plotted versus lattice depth, including data for all fermion numbers N_F. The critical values appear to be independent of N_F, which suggests that the visibility increase at strong attractions is dominantly governed by the on-site physics.

The enhancement of losses is probably caused by the accumulation of bosonic and fermionic density at the lattice sites. On the one hand, the attractive interspecies interactions enforce an increased bosonic occupation on sites that are occupied by a fermion. On the other hand, interaction-induced narrowing of the bosonic and fermionic on-site wavefunctions due to the admixture of higher bands gives rise to an increased peak density. This is in analogy to the modification of on-site wavefunctions observed in Chap. 7 for purely bosonic samples. A theoretical treatment of the Bose-Fermi case with attractive interactions is provided in [26]. These two effects are likely to lead to a significant increase in the three-body loss rate.

8.2.2.1 Variational Model

We employ a variational harmonic oscillator model (see Chap. 9 for a detailed description) to estimate the losses that are caused by the enhanced peak density of the on-site wavefunctions. Within this model we minimize the total on-site energy with respect to the widths σ_B and σ_F of the bosonic and fermionic Gaussian density profiles for given lattice depth V_lat, occupation numbers and interspecies interaction a_BF. We find that for all experimentally relevant boson numbers ($0 \leq n \leq 5$), the on-site wavefunctions collapse when a sufficiently strong attraction a_BF is chosen; the "optimized" widths σ_B and σ_F approach zero. However, already the density increase in the vicinity of this collapse gives rise to an enhancement of the bosonic three-body loss rate $\dot{N}_3$ due to the $\dot{N}_3 \propto \sigma_\mathrm{B}^{-6}$ scaling. As a practical assumption, we can consider a site lost when $\dot{N}_3\tau \gg 1$, where $\tau = 100$ ms is the experimental hold time in the lattice. In bulk measurements the three-body loss coefficient for ^{87}Rb has been measured to lie in the range $5.8(1.9) \times 10^{-30}$ cm^6/s [38]. Assuming that this value is also applicable in the tightly confining potential of a single lattice well, occupations with more than three bosons are expected to be lost on time scales faster than our experimental hold time even in the absence of fermions. On the other hand, we find a ratio of lost ^{87}Rb to ^{40}K atoms between three and four, which suggests that three-body losses of two bosons together with one fermion do not play a major role in our system in accordance with the observations in reference [2]. Consequently, we focus our attention on sites occupied by three bosons and one fermion. For this case we use our variational model to derive the critical scattering lengths for bosonic three-body losses at which $\dot{N}_3\tau = 1$. The results agree surprisingly well with the experimentally observed onset of the loss dominated regime (see Fig. 8.5c).

We conclude that for strong attraction the dominant process is the loss of highly occupied sites due to the interaction-induced enhancement of the bosonic on-site density [26, 39]. The reason for the associated increase of visibility may be found in the removal of strongly localized atoms on highly occupied sites from the system. This might cause an enhanced bosonic mobility due to defects and local incommensurability. The incoherent background in the time-of-flight images is simply reduced, giving rise to a better contrast of the interference pattern.

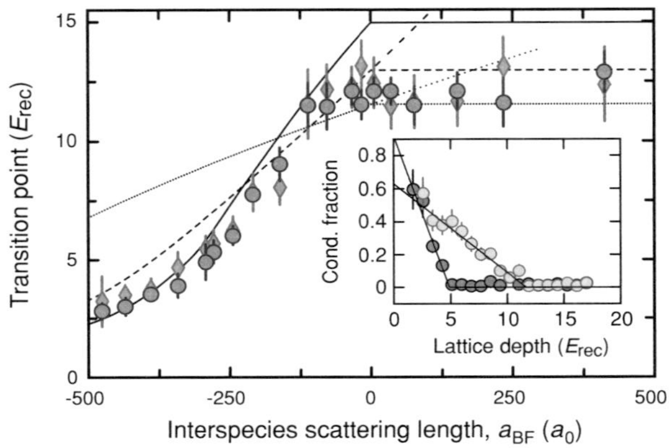

Fig. 8.6 Shift of the Mott insulator transition as a function of interspecies interaction. The *diamonds* (*circles*) represent the experimental runs at $N_F = 2 \times 10^5 = N_B/2$ ($N_F = 3 \times 10^5 = 3N_B/4$). The error bars include the fit uncertainty and the average over typically two experimental runs. The *lines* show theoretical results for the transition point as obtained by the variational model for $n = 1$ (*dotted line*) as well as exact diagonalization for Mott shells with $n = 1$ (*dashed line*) and $n = 2$ (*solid line*) bosons. For $a_{BF} > 0$ the *solid lines* correspond to a phase separation scenario, while the *dashed lines* would correspond to a scenario, in which bosons and fermions occupy the same sites. The inset shows the behavior of the condensate fraction at $a_{BF} = -295\ a_0$ (*black circles*) and $a_{BF} = +235\ a_0$ (*gray circles*) at $N_F = 2 \times 10^5$. The *black solid lines* show bilinear fits to the data. The kink determines the respective critical lattice depth V_{lat}^{crit} corresponding to the transition point

8.2.3 Measuring the Shift of the Mott Transition

In order to investigate the stability of bosonic superfluidity in the mixture, we determine the condensate fraction of the recorded time-of-flight images as described above. Compared to the visibility, the condensate fraction allows for a clearer quantitative determination of the transition point between the superfluid and the Mott insulating phase showing an abrupt increase. For all scattering lengths a_{BF} and all fermion numbers, we find the condensate fraction to decay monotonically towards zero when the lattice depth is increased. The critical lattice depth V_{lat}^{crit} at which the condensate fraction vanishes is regarded as the transition point. It is the kink position of a bilinear fit model as illustrated in the inset of Fig. 8.6.

The critical lattice depth depends sensitively on the interspecies interaction strength, while a difference between medium and high fermion numbers $N_F = N_B/2$ and $N_F = 3N_B/4$ cannot be discerned (see Fig. 8.6).[2] The transition point shows a significant shift by up to 10 E_{rec} towards lower lattice depths for strong attractive interactions. In this regime, we expect the bosons to move on top of an essentially homogeneous fermion background with filling one, corresponding to a band

[2] For low $N_F = N_B/4$ the assumption of a homogeneous fermion filling is probably not valid. Here, the decrease of the condensate fraction shows a more complex behavior that cannot be captured by our bilinear fit model.

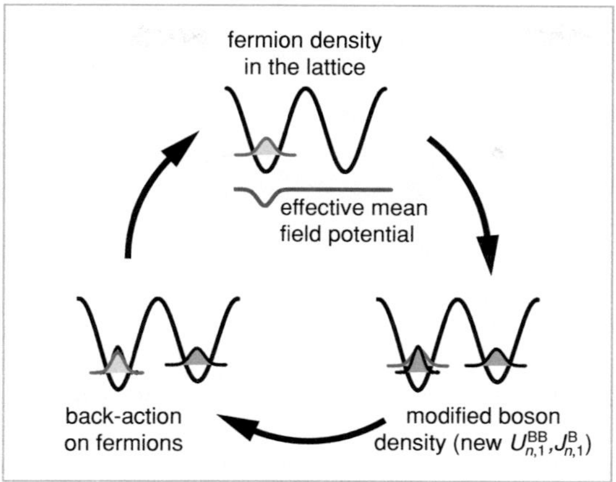

Fig. 8.7 Self-consistent calculation of the deformation of on-site orbitals induced by attractive interactions. Starting with a single fermion at a lattice site, an added boson feels an attractive mean-field potential. This narrows the bosonic on-site wavefunction compared to the noninteracting case. In turn, also the fermionic wavefunction becomes narrower. The wavefunction deformation eventually gives rise to renormalized interaction strengths $U^{BB}_{n,1}$ and tunneling parameters $J^{B}_{n,1}$. The procedure is repeated until convergence, see also Ref. [28]

insulator. This scenario is confirmed by recent theoretical investigations [32] and supported by the experimental observation of enhanced fermionic filling at attractive interspecies interactions [24] (see Chap. 9). The underlying carpet of fermions forms an attractive potential for the bosons that adds to the depth of the lattice potential. This effective deepening of the lattice sites modifies the bosonic tunneling J^{B} and the Bose-Bose interaction strength U^{BB} promoting the superfluid to Mott insulator transition towards lower lattice depths.

Figure 8.6 shows that the transition shift via renormalized Hubbard parameters is quantitatively supported by numerical calculations that have been carried out by Dirk-Sören Lühmann at the University of Hamburg [8, 40]. The exact diagonalization of an on-site multi-orbital system filled with a fermion ($m = 1$) and n bosons yields modified on-site wavefunctions that are used to calculate the renormalized interaction energy (see Fig. 8.7). The corresponding renormalized tunneling parameter is deduced from a band structure calculation at the interaction-induced effective lattice depth experienced by the bosons. It is known from accurate Monte-Carlo simulations that the Mott transition of the $n = 1$ ($n = 2$) shell in a homogeneous three-dimensional lattice happens at a critical ratio $U^{BB}/J^{B} = 29.36$ ($U^{BB}/J^{B} = 49.86$) [41]. Using this result, the theoretical critical lattice depth for a given interspecies interaction is obtained when the renormalized Hubbard parameters reach the critical ratio U^{BB}/J^{B} for the Mott transition. The such derived critical lattice depths are shown in Fig. 8.6 both for the $n = 1$ (green line) and the $n = 2$ (red line) shell.

The general agreement between theoretical and experimental data is remarkably good. At around $a_{BF} \approx 0$ the experimental data lies slightly below the theoreti-

Fig. 8.8 Double well model illustrating the accumulation of bosons on a homogeneous fermionic background. *Black* (*gray*) balls indicate bosons (fermions)

cal prediction for $n = 1$, which indicates a systematic deviation between the kink position of the bilinear fit and the theoretical transition point. Furthermore, finite size effects and other experimental uncertainties (for example, the calibration of the lattice depth) might play a role. For interspecies scattering lengths in the range $-250\ a_0 < a_{\mathrm{BF}} < 0\ a_0$ the experimental data tends to agree better with the $n = 1$ theory (green solid line), while for stronger attraction exceptional agreement with the theory for $n = 2$ is found (red solid line). This brings us to the conjecture that the bosonic filling smoothly changes from one to two, when the interspecies attraction is increased. A double well model system with renormalized interaction parameters can be used to illustrate the interaction-induced redistribution of bosons on top of a homogeneous fermionic background. The accumulation process depicted in Fig. 8.8 can occur, if

$$U_{2,1}^{\mathrm{BB}} + 2U_2^{\mathrm{BF}} - 2U_1^{\mathrm{BF}} < 0, \tag{8.4}$$

where $U_{n,m}^{\mathrm{BB}}$ and U_n^{BF} denote the renormalized Bose-Bose and Bose-Fermi interaction strengths for n bosons and m fermions. We note that this process is forbidden within the single-band Bose-Fermi Hubbard model, where the left hand side equals to U^{BB}, independent of the filling and larger than zero for repulsive Bose-Bose interactions. In contrast, exact diagonalization of the multi-orbital system predicts that the redistribution is energetically favored for $a_{\mathrm{BF}} < -200\ a_0$ almost independent of the lattice depth [40]. This is consistent with the data of Fig. 8.6, where the experiment follows the $n = 2$ theory below $a_{\mathrm{BF}} \approx -250\ a_0$.

For repulsive interspecies interactions the model outlined here predicts a shift of the superfluid to Mott insulator transition towards deeper lattices, given the bosons move on a homogeneous background of fermions with filling one. However, analogous to the constant visibility for positive scattering lengths a_{BF} observed in Fig. 8.4, a variation of the transition point as a function of the interspecies interactions is not observed on the repulsive side. This again indicates that bosons and fermions do not occupy the same lattice sites in this regime. Recent studies on Bose-Fermi mixtures of ytterbium atoms provide similar evidence for a strong tendency to phase separation for repulsive interspecies interactions [42].

8.3 Conclusions and Outlook

In this chapter we have analyzed the coherence properties of the bosonic component in a Bose-Fermi mixture of ^{87}Rb and ^{40}K in an optical lattice. For attractive interspecies interactions we have found a significant shift of the bosonic superfluid to Mott insulator transition towards lower lattice depth. The remarkable agreement of the data with a numerical exact diagonalization of a multi-orbital system suggests that this shift can be attributed to renormalized tunneling and interaction parameters of an effective Hubbard model. On the side of repulsive interspecies interactions, we interpret the absence of a shift of the transition point as a signature for local or global phase separation between the bosonic and fermionic components. Additionally, we have observed that the fermions are fully transparent to the bosons at $a_{\mathrm{BF}} = 0$, which is a promising starting point for tunable impurities [37, 43, 44]. In the vicinity (about $\pm 30\ a_0$) of vanishing interspecies interactions, we find a symmetric decay of visibility in the bosonic interference patterns. Recent theoretical investigations have explained this phenomenon in terms of an entropy redistribution arising from the interplay of interspecies interactions, finite temperature and finite system size [31, 32].

It is an important message of this chapter that the application of a single-band Hamiltonian is not sufficient to deduce and understand the complex many-body system of interacting bosons and fermions in an optical lattice. Interactions bring multiband effects into play, which can lead to a significant renormalization of interaction and tunneling parameters [26, 30, 45]. Additionally, the finite entropy situation in this closed quantum system can give rise to adiabatic heating or cooling effects [5, 27, 29] as well as an involved redistribution of atoms. The detailed investigation of these effects will be a worthwhile challenge and help to identify regimes, in which intriguing quantum phases, such as charge density waves [14, 20], composite particles [12] and supersolidity [19, 20, 46–48], can be realized and observed. In the next chapter we make use of quantum revival spectroscopy (see Chap. 7) to explore interaction effects of few-body Bose-Fermi systems on individual lattice sites.

References

1. P.A. Lee, N. Nagaosa, X.-G. Wen, Doping a Mott insulator: physics of high-temperature superconductivity. Rev. Mod. Phys. **78**, 17 (2006)
2. K. Günter, T. Stöferle, H. Moritz, M. Köhl, T. Esslinger, Bose-Fermi mixtures in a three-dimensional optical lattice. Phys. Rev. Lett. **96**, 180402 (2006)
3. S. Ospelkaus, C. Ospelkaus, O. Wille, M. Succo, P. Ernst, K. Sengstock, K. Bongs, Localization of bosonic atoms by fermionic impurities in a three-dimensional optical lattice. Phys. Rev. Lett. **96**, 180403 (2006)
4. M. Greiner, O. Mandel, T. Esslinger, T. Hänsch, I. Bloch, Quantum phase transition from a superfluid to a Mott insulator in a gas of ultracold atoms. Nature **415**, 39 (2002)
5. L. Pollet, C. Kollath, U. Schollwöck, M. Troyer, Mixture of bosonic and spin-polarized fermionic atoms in an optical lattice. Phys. Rev. A **77**, 023608 (2008)

6. R.M. Lutchyn, S. Tewari, S. Das Sarma, Boson Hubbard model with weakly coupled fermions. Phys. Rev. B **78**, 220504 (2008)
7. A. Simoni, M. Zaccanti, C. D'Errico, M. Fattori, G. Roati, M. Inguscio, G. Modugno, Near-threshold model for ultracold KRb dimers from interisotope Feshbach spectroscopy. Phys. Rev. A **77**, 052705 (2008)
8. T. Best, S. Will, U. Schneider, L. Hackermüller, D. van Oosten, D.-S. Lühmann, I. Bloch, Role of interactions in ^{87}Rb-^{40}K Bose-Fermi mixtures in a 3D optical lattice. Phys. Rev. Lett. **102**, 030408 (2009)
9. T. Best, *Interacting Bose-Fermi mixtures in optical lattices*. Ph.D. thesis, Universität Mainz, 2011
10. A. Albus, F. Illuminati, J. Eisert, Mixtures of bosonic and fermionic atoms in optical lattices. Phys. Rev. A **68**, 023606 (2003)
11. A.B. Kuklov, B.V. Svistunov, Counterflow superfluidity of two-species ultracold atoms in a commensurate optical lattice. Phys. Rev. Lett. **90**, 100401 (2003)
12. M. Lewenstein, L. Santos, M.A. Baranov, H. Fehrmann, Atomic Bose-Fermi mixtures in an optical lattice. Phys. Rev. Lett. **92**, 050401 (2004)
13. L. Pollet, M. Troyer, K. Van Houcke, S.M.A. Rombouts, Phase diagram of Bose-Fermi mixtures in one-dimensional optical lattices. Phys. Rev. Lett. **96**, 190402 (2006)
14. L. Mathey, D.-W. Wang, W. Hofstetter, M.D. Lukin, E. Demler, Luttinger liquid of polarons in one-dimensional Boson-Fermion mixtures. Phys. Rev. Lett. **93**, 120404 (2004)
15. A. Privitera, W. Hofstetter, Polaronic slowing of fermionic impurities in lattice Bose-Fermi mixtures. Phys. Rev. A **82**, 063614 (2010)
16. M. Cramer, J. Eisert, F. Illuminati, Inhomogeneous atomic Bose-Fermi mixtures in cubic lattices. Phys. Rev. Lett. **93**, 190405 (2004)
17. C.N. Varney, V.G. Rousseau, R.T. Scalettar, Quantum Monte Carlo study of the visibility of one-dimensional Bose-Fermi mixtures. Phys. Rev. A **77**, 041608 (2008)
18. A. Zujev, A. Baldwin, R.T. Scalettar, V.G. Rousseau, P.J.H. Denteneer, M. Rigol, Superfluid and Mott-insulator phases of one-dimensional Bose-Fermi mixtures. Phys. Rev. A **78**, 033619 (2008)
19. H.P. Büchler, G. Blatter, Supersolid versus phase separation in atomic Bose-Fermi mixtures. Phys. Rev Lett. **91**, 130404 (2003)
20. I. Titvinidze, M. Snoek, W. Hofstetter, Supersolid Bose-Fermi mixtures in optical lattices. Phys. Rev. Lett. **100**, 100401 (2008)
21. E. Kim, M.H.W. Chan, Probable observation of a supersolid helium phase. Nature **427**, 225 (2004)
22. S. Balibar, The enigma of supersolidity. Nature **464**, 176 (2010)
23. F. Illuminati, A. Albus, High-temperature atomic superfluidity in lattice Bose-Fermi mixtures. Phys. Rev. Lett. **93**, 090406 (2004)
24. S. Will, T. Best, S. Braun, U. Schneider, I. Bloch, Coherent interaction of a single fermion with a small bosonic field. Phys. Rev. Lett. **106**, 115305 (2011)
25. D. Jaksch, P. Zoller, The cold atom Hubbard toolbox. Ann. Phys. **315**, 52 (2005)
26. D.-S. Lühmann, K. Bongs, K. Sengstock, D. Pfannkuche, Self-trapping of Bosons and Fermions in optical lattices. Phys. Rev. Lett. **101**, 050402 (2008)
27. M. Cramer, S. Ospelkaus, C. Ospelkaus, K. Bongs, K. Sengstock, J. Eisert, Do mixtures of bosonic and fermionic atoms adiabatically heat up in optical lattices? Phys. Rev. Lett. **100**, 140409 (2008)
28. D.-S. Lühmann, K. Bongs, D. Pfannkuche, Excitation spectrum of Mott shells in optical lattices. J. Phys. B **42**, 145305 (2009)
29. L. Pollet, C. Kollath, K.V. Houcke, M. Troyer, Temperature changes when adiabatically ramping up an optical lattice. New J. Phys. **10**, 065001 (2008)
30. S. Tewari, R.M. Lutchyn, S. Das Sarma, Effects of fermions on the superfluid-insulator phase diagram of the Bose-Hubbard model. Phys. Rev. B **80**, 054511 (2009)
31. M. Cramer, Interaction-dependent temperature effects in Bose-Fermi mixtures in optical lattices. Phys. Rev. Lett. **106**, 215302 (2011)

32. M. Snoek, I. Titvinidze, I. Bloch, W. Hofstetter, Effect of interactions on harmonically confined Bose-Fermi mixtures in optical lattices. Phys. Rev. Lett. **106**, 155301 (2011)
33. C. Klempt, T. Henninger, O. Topic, J. Will, W. Ertmer, E. Tiemann, J. Arlt, ^{40}K-^{87}Rb Feshbach resonances: modeling the interatomic potential. Phys. Rev. A **76**, 020701 (2007)
34. F. Gerbier, A. Widera, S. Fölling, O. Mandel, T. Gericke, I. Bloch, Phase coherence of an atomic Mott insulator. Phys. Rev. Lett. **95**, 050404 (2005)
35. I.B. Spielman, W.D. Phillips, J.V. Porto, Condensate fraction in a 2D Bose gas measured across the Mott-insulator transition. Phys. Rev. Lett. **100**, 120402 (2008)
36. L. Hackermüller, U. Schneider, M. Moreno-Cardoner, T. Kitagawa, T. Best, S. Will, E. Demler, E. Altman, I. Bloch, B. Paredes, Anomalous expansion of attractively interacting fermionic atoms in an optical lattice. Science **327**, 1621 (2010)
37. A. Mering, M. Fleischhauer, One-dimensional Bose-Fermi-Hubbard model in the heavy-fermion limit. Phys. Rev. A **77**, 023601 (2008)
38. E.A. Burt, R.W. Ghrist, C.J. Myatt, M.J. Holland, E.A. Cornell, C.E. Wieman, Coherence, correlations, and collisions: what one learns about Bose-Einstein condensates from their decay. Phys. Rev. Lett. **79**, 337 (1997)
39. M. Bruderer, W. Bao, D. Jaksch, Self-trapping of impurities in Bose-Einstein condensates: strong attractive and repulsive coupling. Europhys. Lett. **82**, 30004 (2008)
40. D.-S. Lühmann, *Multiorbital Physics in Optical Lattices*. Ph.D. Thesis, Universität Hamburg, 2009
41. B. Capogrosso-Sansone, N.V. Prokof'ev, B.V. Svistunov, Phase diagram and thermodynamics of the three-dimensional Bose-Hubbard model. Phys. Rev. B **75**, 134302 (2007)
42. S. Sugawa, K. Inaba, S. Taie, R. Yamazaki, M. Yamashita, Y. Takahashi, Interaction and filling induced quantum phases of dual Mott insulators of bosons and fermions. Nat. Phys. **7**, 642 (2011)
43. U. Gavish, Y. Castin, Matter-wave localization in disordered cold atom lattices. Phys. Rev. Lett. **95**, 020401 (2005)
44. K.V. Krutitsky, M. Thorwart, R. Egger, R. Graham, Ultracold bosons in lattices with binary disorder. Phys. Rev. A **77**, 053609 (2008)
45. R.M. Lutchyn, S. Tewari, S. Das Sarma, Loss of superfluidity by fermions in the boson Hubbard model on an optical lattice. Phys. Rev. A **79**, 011606 (2009)
46. R. Roth, K. Burnett, Quantum phases of atomic boson-fermion mixtures in optical lattices. Phys. Rev. A **69**, 021601 (2004)
47. I. Titvinidze, M. Snoek, W. Hofstetter, Generalized dynamical mean-field theory for Bose-Fermi mixtures in optical lattices. Phys. Rev. B **79**, 144506 (2009)
48. P.P. Orth, D.L. Bergman, K. Le Hur, Supersolidity of cold-atom Bose-Fermi mixtures in optical lattices. Phys. Rev. A **80**, 023624 (2009)

Chapter 9
Coherent Interaction of a Single Fermion with a Small Bosonic Field

Multi-component systems play a central role in quantum many-body physics. From interacting atoms and photons to electrons and phonons, the interplay of interactions in binary mixtures gives rise to intriguing quantum phenomena such as superradiance, BCS superfluidity or polaron physics [1–4]. Recently, the problem of impurities embedded in an external quantum environment has also shifted into the focus of ultracold atom experiments. For example, fermionic spin impurities in a Fermi sea have lead to the observation of a Fermi polaron [5, 6] and the interactions between a single ion and a Bose-Einstein condensate have been studied [7, 8]. When such impurity systems are scaled down to the few-body regime, they can share important properties with models for atomic nuclei [9].

In this Chapter, I present the experimental study of an elementary few-body system consisting of a single fermionic atom and a coherent field of bosonic atoms. So far, research on Bose-Fermi mixtures in optical lattices has mainly focussed on the coherence properties of the global quantum many-body state [10–13] (see Chap. 8). In the present chapter, the local properties of miniature Bose-Fermi systems on individual lattice sites are investigated. Using quantum revival spectroscopy (see Chap. 7), the absolute strengths of intra- and interspecies interactions are precisely measured as a function of the interspecies scattering length, tuned by means of a Feshbach resonance (see Sect. 2.3). Already moderate Bose-Fermi interactions give rise to notable changes of the on-site wavefunctions, which are observed as modifications of the Bose-Bose interaction energy [14–19]. This is a direct confirmation of the renormalization of Hubbard parameters that is indirectly deduced in Chap. 8. Furthermore, in the context of our investigations, a novel method is devised that allows to selectively infer the mean fermionic filling within the volume, in which bosons and fermions overlap.

S. Will, *From Atom Optics to Quantum Simulation*, Springer Theses,
DOI: 10.1007/978-3-642-33633-1_9, © Springer-Verlag Berlin Heidelberg 2013

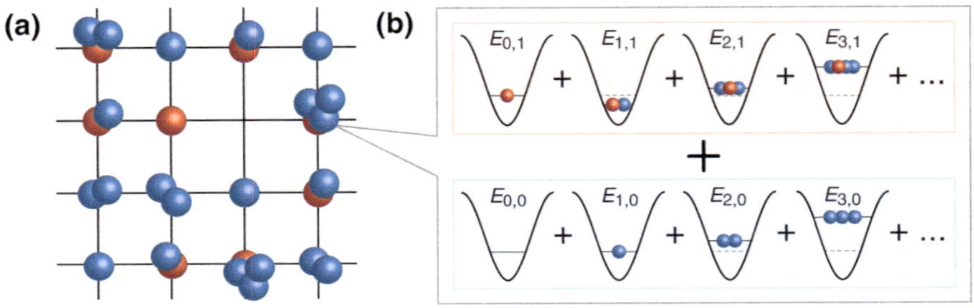

Fig. 9.1 Few-body on-site states of a delocalized Bose-Fermi mixture. **a** In a shallow optical lattice both the bosonic and the fermionic species are delocalized and show atom number fluctuations. The cartoon picture displays a possible result of a projection measurement assuming an average fermionic filling of $\bar{m} = 0.5$. *Blue* (*red*) balls indicate bosons (fermions). **b** The local quantum state at a lattice site corresponds to coherent superpositions of bosonic atom number states *with* and *without* a fermion with eigenenergies $E_{n,m}$ (see main text)

9.1 Theoretical Model

We start with a basic theoretical model for the quantum state relevant to this chapter and its dynamics (see Appendix E for additional details). We consider a delocalized Bose-Fermi mixture in a shallow optical lattice. Then, the lattice depth is rapidly increased, suppressing both bosonic and fermionic tunneling ($J^{\mathrm{B}}, J^{\mathrm{F}} \rightarrow 0$) and freezing out the delocalized atom distributions of bosons and fermions. In this setting, the eigenstates at a lattice site are given by product atom number states $|n\rangle|m\rangle$, containing n bosons (where n is an integer number) and m fermions (where m is either 0 or 1) (see Fig. 9.1). We denote the respective eigenenergies by $E_{n,m}$. For a delocalized mixture, both the bosonic and fermionic component show number fluctuations and the corresponding on-site quantum states can be described as a coherent superposition of bosonic atom number states both in the absence ($m = 0$) and the presence ($m = 1$) of a fermion. We model the time evolution at a lattice site as a superposition of phase evolutions with and without a fermion, which are governed by the eigenenergies $E_{n,0}$ and $E_{n,1}$,

$$|\psi_{\mathrm{BF}}(t)\rangle = \sum_{n=0}^{\infty} c_n e^{-iE_{n,0}t/\hbar}|n\rangle|0\rangle + d_n e^{-iE_{n,1}t/\hbar}|n\rangle|1\rangle. \tag{9.1}$$

Here, c_n and d_n denote the probability amplitudes of finding n bosons without ($m = 0$) and with ($m = 1$) a fermion, respectively.

Within the single-band Bose-Fermi Hubbard model [10] (compare Sect. 8.1) the eigenenergies are given by

$$E_{n,m} = \frac{U^{\mathrm{BB}}}{2} n(n-1) + U^{\mathrm{BF}}\, n\, m. \tag{9.2}$$

Here, the Bose-Bose interaction energy, $U^{\mathrm{BB}} \propto a_{\mathrm{BB}} \int d^3r|\phi_{\mathrm{B}}(\mathbf{r})|^4$, and the Bose-Fermi interaction energy, $U^{\mathrm{BF}} \propto a_{\mathrm{BF}} \int d^3r|\phi_{\mathrm{B}}(\mathbf{r})|^2|\phi_{\mathrm{F}}(\mathbf{r})|^2$, are independent of the bosonic and fermionic atom numbers n and m since the model is restricted to the lowest lattice band; a_{BB} (a_{BF}) denotes the intraspecies (interspecies) scattering length and $\phi_{\mathrm{B}}(\mathbf{r})$ ($\phi_{\mathrm{F}}(\mathbf{r})$) the bosonic (fermionic) ground state orbital at a lattice site. While essential features of the resulting quantum dynamics are captured by this single-orbital model, the experimentally observed dynamics contain signatures that can only be explained within a multi-orbital approach. Here, the interaction-induced deformation of on-site wavefunctions gives rise to interaction strengths $U^{\mathrm{BB}}_{n,m}$ and U^{BF}_n that explicitly depend on the number of bosons and fermions [14–19].

9.1.1 Quantum Phase Revivals in a Bose-Fermi System

The interactions between the bosonic field and the fermion are encoded in the dynamics of the on-site wavefunction $|\psi_{\mathrm{BF}}(t)\rangle$. Quantum revival spectroscopy (see Chap. 7) allows to probe the corresponding interaction energies by sampling the time-dependent coherence of the bosonic component. In an idealized picture (see Sect. 7.2.2), the bosonic coherence is proportional to $|\langle\psi_{\mathrm{BF}}(t)|\hat{a}|\psi_{\mathrm{BF}}(t)\rangle|^2 \equiv |\langle\hat{a}\rangle(t)|^2$, where $\hat{a}$ denotes the annihilation operator for a boson at a lattice site. The time evolution of the coherence is governed by the interference of the two dynamical evolutions *with* and *without* a fermion:

$$|\langle\hat{a}\rangle(t)|^2 = \left|\sum_{n=0}^{\infty} C_n(t) + D_n(t)\right|^2, \tag{9.3}$$

where the purely bosonic contribution enters as

$$C_n(t) = \sqrt{n+1}\, c_n^* c_{n+1}\, e^{-i(E_{n+1,0}-E_{n,0})t/\hbar}$$

and the Bose-Fermi interactions are contained in the term

$$D_n(t) = \sqrt{n+1}\, d_n^* d_{n+1}\, e^{-i(E_{n+1,1}-E_{n,1})t/\hbar}.$$

Computing Eq. 9.3 further, three distinct sets of spectral contributions are identified:

$$\sum_{n,\tilde{n}=0}^{\infty} C_{\tilde{n}}^*(t)C_n(t) \quad\rightarrow\quad E^{\mathrm{BB}}_{n,\tilde{n};0} = E_{n+1,0} - E_{n,0} - E_{\tilde{n}+1,0} + E_{\tilde{n},0} \tag{9.4}$$

$$\sum_{n,\tilde{n}=0}^{\infty} D_{\tilde{n}}^*(t)D_n(t) \quad\rightarrow\quad E^{\mathrm{BB}}_{n,\tilde{n};1} = E_{n+1,1} - E_{n,1} - E_{\tilde{n}+1,1} + E_{\tilde{n},1} \tag{9.5}$$

$$\sum_{n,\tilde{n}=0}^{\infty} D_{\tilde{n}}^{*}(t) C_n(t) \quad \rightarrow \quad E_{n,\tilde{n}}^{\mathrm{BF}} = E_{n+1,0} - E_{n,0} - E_{\tilde{n}+1,1} + E_{\tilde{n},1} \tag{9.6}$$

The first set of energies (9.4) corresponds to the purely bosonic contributions of orders U^{BB}, $2U^{\mathrm{BB}}$, and so on, *without* the influence of a fermion. It exactly represents the spectral features of the bosonic quantum phase revivals discussed in Chap. 7. The second set (9.5) yields the bosonic contributions of orders U^{BB}, $2U^{\mathrm{BB}}$, and so on, *with* the influence of a fermion. The third set (9.6) corresponds to mixed contributions of orders $|U^{\mathrm{BF}}|$, $|U^{\mathrm{BB}}+U^{\mathrm{BF}}|$, $|U^{\mathrm{BB}}-U^{\mathrm{BF}}|$,$|2U^{\mathrm{BB}}+U^{\mathrm{BF}}|$, $|2U^{\mathrm{BB}}-U^{\mathrm{BF}}|$, and so on, which appear as new strong features in the experimental data. It is important to note that quantum revival spectroscopy reveals *absolute* interaction energies, but does not distinguish the sign of the interactions. However, whether Bose-Fermi interactions are attractive or repulsive, can indirectly be determined from the interaction-induced deformation of the bosonic on-site wavefunction, which we discuss below in detail (see Sect. 9.2.3).

9.1.2 Coherent State and Single-Orbital Approximation

In general, the collapse and revival dynamics of the coherence as captured by Eq. 9.3 can be very complex. However, we obtain a simplified analytic expression when the influence of the Bose-Bose and the Bose-Fermi interaction on the bosonic atom number statistics is neglected, which is justified in the limit of rather small interactions. Such conditions motivate the use of coherent states for the bosonic field (compare Chap. 7), whose probability amplitudes are scaled by the mean fermionic atom number $\bar{m}$ according to $c_n = \sqrt{1-\bar{m}}\, e^{-|\alpha|^2/2}\alpha^n/\sqrt{n!}$ and $d_n = \sqrt{\bar{m}}\, e^{-|\alpha|^2/2}\alpha^n/\sqrt{n!}$, where $\alpha = \sqrt{\bar{n}}e^{i\phi}$ denotes the complex field amplitude with the mean bosonic atom number $\bar{n}$ and initial phase ϕ. With the additional assumption of single-orbital eigenenergies following Eq. 9.2, we obtain the quantum phase evolution (for the explicit derivation see E.1)

$$|\langle\hat{a}\rangle(t)|^2/\bar{n} = e^{2\bar{n}(\cos(U^{\mathrm{BB}}t/\hbar)-1)} \times \{1 - 2\bar{m}(1-\bar{m})[1-\cos(U^{\mathrm{BF}}t/\hbar)]\}. \tag{9.7}$$

Figure 9.2 illustrates these idealized dynamics for several mean fermionic fillings $\bar{m}$. The suppression of revivals is a striking signature for the interference between the Bose-Bose and the Bose-Fermi phase evolution. Particularly, for the case $U^{\mathrm{BF}}/U^{\mathrm{BB}} = z + 0.5$, where z is an integer number, the suppression factor s has a simple relation to the mean fermionic filling,

$$s = (1-2\bar{m})^2, \tag{9.8}$$

shown in the right panel of Fig. 9.2.

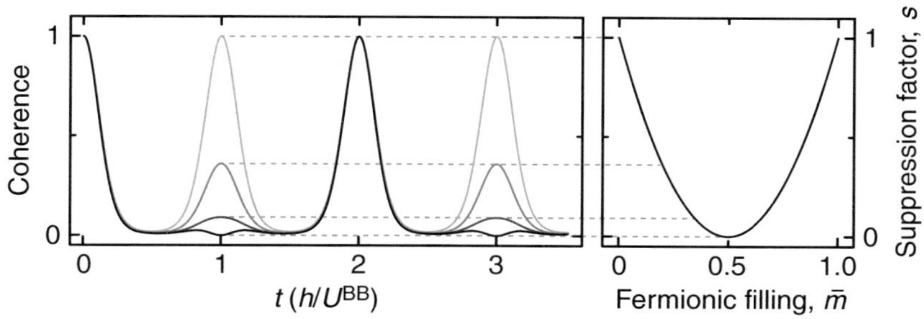

Fig. 9.2 Quantum phase revivals in a few-body Bose-Fermi system. Suppression of odd quantum phase revivals in a Bose-Fermi mixture at $|U^{\text{BF}}/U^{\text{BB}}| = 0.5$ for variable fermionic fillings: $\bar{m} = 0$, 0.2, 0.35 and 0.5 (*solid lines*, *darker color* for larger $\bar{m}$). The *right* panel displays the suppression factor s as a function of the filling $\bar{m}$

9.2 Experimental Realization

The experiment is performed with a quantum degenerate mixture of $1.7(3) \times 10^5$ bosonic ^{87}Rb and $2.1(4) \times 10^5$ fermionic ^{40}K atoms in their respective hyperfine ground states $|F, m_F\rangle = |1, +1\rangle$ and $|9/2, -9/2\rangle$. The Bose-Einstein condensate (BEC) and the Fermi cloud are held in a pancake-shaped dipole trap ($\lambda_{\text{dip}} = 1030$ nm) with tight vertical confinement to ensure good overlap of the two species in the presence of gravity. The BEC is quasi-pure and the Fermi cloud has a temperature of typically $T/T_{\text{F}} = 0.2$, where T_{F} is the Fermi temperature.

The Feshbach resonance at $B_{\text{res}} = 546.75(6)$ G (see Sect. 2.3.5) is used to tune the interspecies scattering length a_{BF} between $-161.2(1)\ a_0$ and $+134(19)\ a_0$, where a_0 is the Bohr radius. We rely on the parametrization of the resonance by Simoni et al. [20] quoting a background scattering length of $a_{\text{BG}} = -189\ a_0$ and a width of $\Delta = -3.1$ G. In addition, Ref. [20] provides both an experimental and a theoretical value for the zero crossing of the scattering length at 543.3(5) G and 543.66(8) G, respectively. Using our data (see below), we have independently determined the magnetic field at vanishing interspecies interactions. Our measurements yield 543.6(2) G in excellent agreement with the theoretical value, which supports the accuracy of this parametrization. The Bose-Bose intraspecies scattering length stays fixed at $a_{\text{BB}} = +102(2) a_0$ [21–23].

After 25 ms of settling time for the magnetic field, a three-dimensional (3D) optical lattice operating at $\lambda_{\text{lat}} = 738$ nm is adiabatically ramped up within 50 ms to a depth of $V_{\text{L}}^{\text{B}} = 5.2\ E_{\text{rec}}^{\text{B}}$, where $E_{\text{rec}}^{\text{B}} = h^2/(2m_{\text{B}}\lambda_{\text{lat}}^2)$ denotes the recoil energy for ^{87}Rb. The lattice depth for the fermions in units of the recoil energy for ^{40}K, $E_{\text{rec}}^{\text{F}} = h^2/(2m_{\text{F}}\lambda_{\text{lat}}^2)$, is lower by a factor of 1.50 ± 0.015 (see Sect. 5.4) such that $V_{\text{L}}^{\text{F}} = 3.5\ E_{\text{rec}}^{\text{F}}$. At this lattice depth and for all interspecies interactions used in the experiment, the bosons are expected to form a superfluid and the fermions are delocalized (see Chap. 8). The horizontal and vertical trapping frequencies of the dipole trap in the presence of the shallow lattice are $2\pi \times (25, 94)$ Hz for ^{87}Rb and $2\pi \times (36, 173)$ Hz for ^{40}K. (Fig. 9.3)

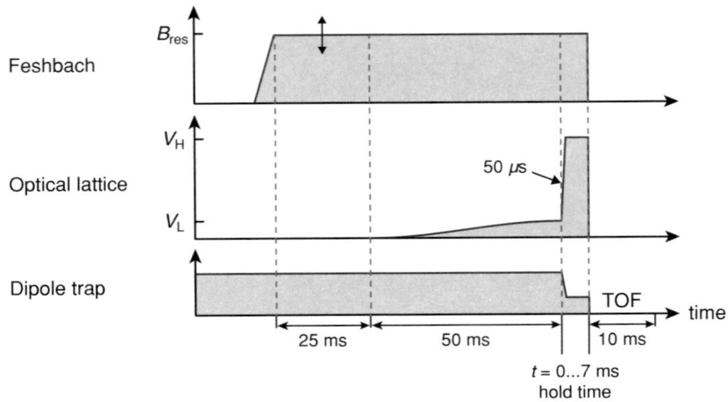

Fig. 9.3 Visualization of the experimental sequence. The steps of the sequence are timed to ensure adiabatic preparation, on the one hand, and minimal total duration to avoid heating or atom loss, on the other hand

Then, a 3D array of coherent bosonic fields with partial fermionic filling (compare Fig. 9.1) is created by rapidly increasing the lattice depth from V_L^B to $V_H^B = 28.2(3)$ E_{rec}^B (corresponding to $V_H^F = 18.8(2)$ E_{rec}^F) within 50 μs, being slow enough to avoid populating higher lattice bands, but fast in comparison to tunneling within the first band. The lattice jump suppresses the tunnel coupling, freezes out the atom distributions and initiates quantum phase evolution at each lattice site. In order to ensure a long coherence time, we reduce the dipole trap simultaneously with this jump to cancel the harmonic confinement in the horizontal plane, avoiding relative dephasing of lattice sites as detailed in Sect. 7.3.

We sample long time traces of the quantum phase evolution to achieve high spectral resolution, which provides us with detailed information on the interactions in the Bose-Fermi system. Rather low bosonic and fermionic filling are used in the measurements to limit the number of spectral contributions. In order to suppress three-body atom loss on sites with high occupation, we restrict ourselves to moderate interspecies interactions and do not use higher lattice depths V_H (compare Chap. 8).

After variable hold times t (up to 7 ms, in steps of 40 μs) all trapping potentials are switched off and the bosonic and fermionic clouds expand during 10 ms time-of-flight. Absorption images of the bosonic interference pattern as well as the diffracted fermion cloud are recorded. In this section we focus our data evaluation on the bosonic images, from which we derive the coherence $|\langle\hat{a}\rangle(t)|^2$ by evaluating the ratio of the summed atom numbers in the central, first- and second-order coherence peaks to the total atom number, N_{coh}/N_{tot} [24] (see Sect. 4.2.2). A brief, preliminary discussion of the fermion images is given at the end of this chapter (see Sect. 9.3).

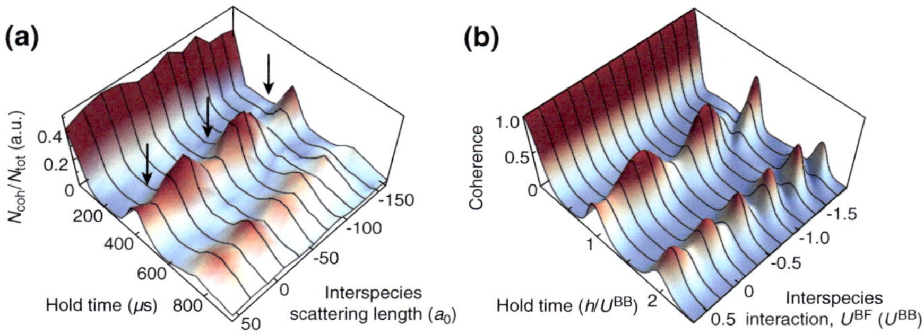

Fig. 9.4 Interference of Bose-Bose and Bose-Fermi phase dynamics. **a** The measured initial quantum phase revivals are modulated as a function of the interspecies scattering length a_{BF}. The intraspecies scattering length a_{BB} is fixed at $+102(2)\ a_0$. The first revival is suppressed at $a_{\mathrm{BF}} = -126(2), a_0, -40(3)a_0$ and $+41(3)a_0$, as marked by the *arrows*. *Black lines* indicate the traces recorded in the experiment. **b** The general structure of the experimental data is well captured by the simple model of Eq. 9.7. In the simulation the bosonic filling is kept fixed at $\bar{n} = 0.85$ and the fermionic filling is varied according to the results extracted from **a** (see also inset of Fig. 9.9). Additionally, a temporal exponential decay comparable to the experimental data ($\tau = 1.1$ ms) is applied

9.2.1 Measurement of the Mean Fermionic Filling

The experimental data reveals a modulation of the initial quantum phase revivals depending on the interspecies interaction strength, as shown in Fig. 9.4a. For the first revival, we detect three local minima with suppression factors $s = 0.57(3)$, $0.43(3)$ and $0.16(2)$ as the attraction is increased. The mean fermionic filling can be determined from the suppression factor by inverting Eq. 9.8: $\bar{m} = (1 \mp \sqrt{s})/2$, where the minus (plus) sign corresponds to $\bar{m} < 0.5$ ($\bar{m} > 0.5$). The resulting fermionic fillings read $\bar{m} = 0.12(1)$, $0.17(1)$ and $0.30(1)$, respectively. Note that in all cases the minus sign has been chosen, because fermionic fillings $\bar{m} > 0.5$ are not expected for our experimental parameters. Qualitatively, the dynamics of the first few revivals are remarkably well captured by the single-orbital coherent state model Eq. 9.7 as illustrated by the comparison of experimental data and theoretical simulation in Fig. 9.4. For short observation times multi-orbital effects, in particular the atom number dependence of U^{BB} and U^{BF}, cannot be resolved.

9.2.2 Direct Observation of Bose-Fermi Interactions

Quantitative information about the interactions in the Bose-Fermi few-body system is obtained by sampling long time traces of quantum phase revivals yielding high spectral resolution, as displayed in Fig. 9.5. The trace at almost vanishing interspecies attraction (a_{BF}= $-8(6)a_0$) and its Fourier spectrum (see Figs. 9.5a and 9.6a) show the

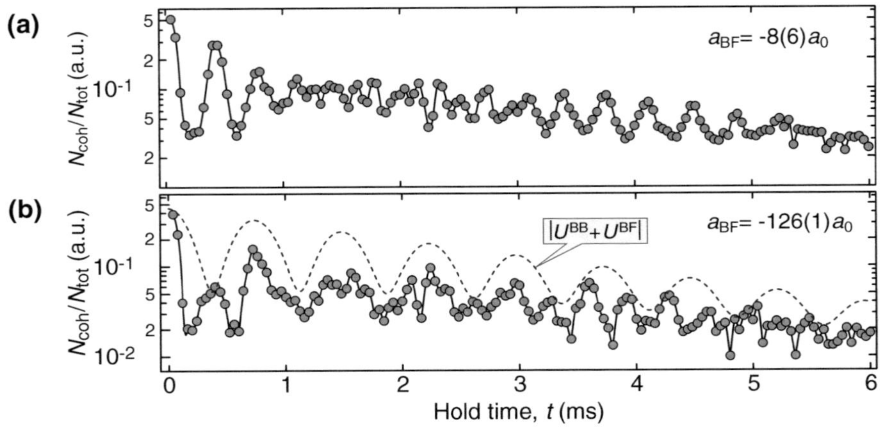

Fig. 9.5 High-resolution quantum revival traces of the Bose-Fermi system are shown for vanishing, **a**, and strong interspecies attraction, **b**. For the later case, one observes that every second revival is suppressed by an envelope (*gray dashed line*) that corresponds to the spectral components of order $|U^{BB} + U^{BF}|$. Each data point represents a single run of the experiment. Special care has been taken to ensure identical conditions for each run, particularly stable atom numbers. The *solid lines* interpolate the data and serve as a guide to the eye

signatures of effective multi-body interactions observed for a purely bosonic system (see Chap. 7).[1] However, at stronger interspecies attraction additional envelopes are observed to modulate the quantum collapse and revival dynamics (see Fig. 9.5b). The corresponding spectra (see Fig. 9.6) reveal new striking features: first, prominent additional peaks that arise from direct Bose-Fermi interactions of orders $|U^{BF}|$, $|U^{BB} + U^{BF}|$ and $|2U^{BB} + U^{BF}|$ and, second, a small, but significant upshift of the Bose-Bose interaction energies of orders U^{BB} and $2U^{BB}$.

The emerging Bose-Fermi features of orders $|U^{BF}|$, $|U^{BB} + U^{BF}|$ and $|2U^{BB} + U^{BF}|$ exhibit an almost linear dependence on a_{BF} (see Figs. 9.6 and 9.7). Inspecting these features more closely one observes that each of them consists of a comb of energies. This splitting arises from an explicit boson number dependence of the Bose-Bose ($U^{BB}_{n,m}$) and Bose-Fermi (U^{BF}_{n}) interaction strengths. How many of those spectral comb teeth can be expected? We will answer this question exemplarily for the energies of order $|U^{BF}|$.

The spectral contributions that explicitly contain Bose-Fermi interaction terms are generally given by the terms $E^{BF}_{n,\tilde{n}}$ as derived above (Eq. 9.6). The energies of order $|U^{BF}|$ are obtained for $n = \tilde{n}$, that is

$$E^{BF}_{n,n} = E_{n+1,0} - E_{n,0} - E_{n+1,1} + E_{n,1} \tag{9.9}$$

[1] For the technical details of the numerical Fourier analysis see Appendix D.

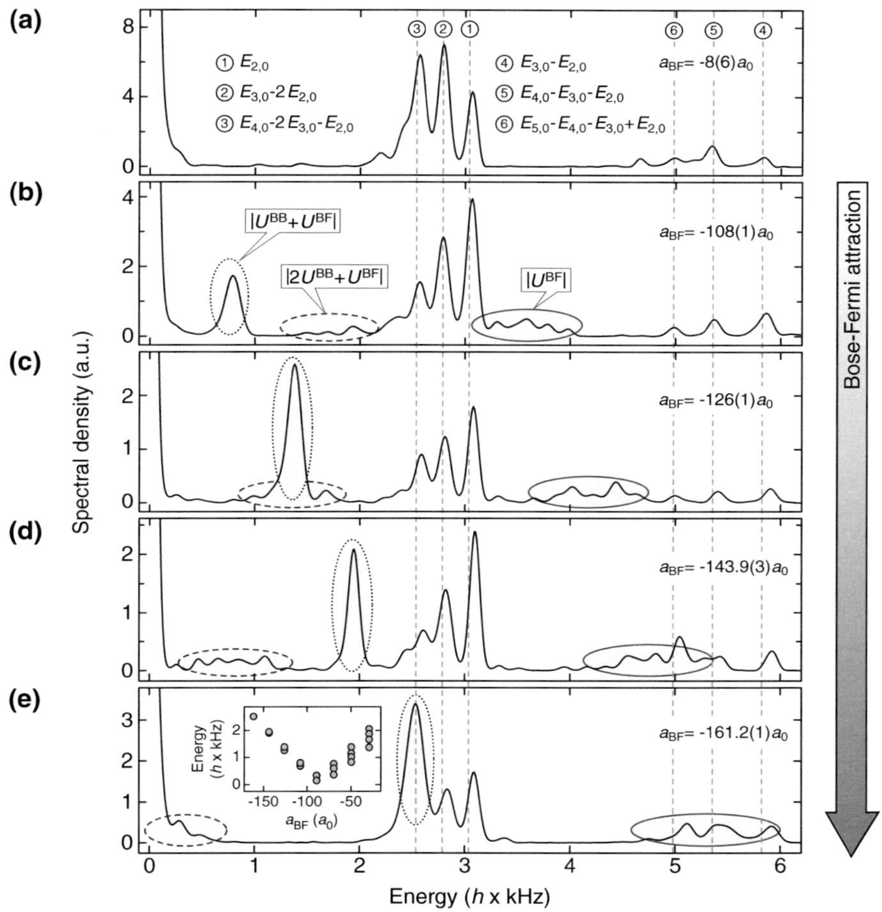

Fig. 9.6 Fourier spectra of time traces for vanishing, **a**, and linearly increasing Bose-Fermi attraction, **b–e**. The contributions involving direct Bose-Fermi interactions ($|U^{\mathrm{BF}}|$, $|U^{\mathrm{BB}} + U^{\mathrm{BF}}|$ and $|2U^{\mathrm{BB}} + U^{\mathrm{BF}}|$) are highlighted by *shaded ovals* (*solid*, *dotted* and *dashed lines*, respectively). The $|U^{\mathrm{BB}} + U^{\mathrm{BF}}|$ components are shown in the inset of panel **e** as a function of a_{BF}. *Dashed vertical lines* indicate the spectral contributions of orders U^{BB} (① to ③) and $2U^{\mathrm{BB}}$ (④ to ⑥) for a purely bosonic system or vanishing interspecies interactions [23]

with a spectral weight of $(n+1)c_n^* c_{n+1} d_n c_{n+1}^*$, where c_n, d_n are defined as in Eq. 9.1. Using Eq. 9.2 and considering an explicit atom number dependence of U^{BB} and U^{BF} we obtain for the first spectral contributions:

$$\begin{aligned} E_{0,0}^{\mathrm{BF}} &= U_1^{\mathrm{BF}} \\ E_{1,1}^{\mathrm{BF}} &= (U_{2,1}^{\mathrm{BB}} - U_{2,0}^{\mathrm{BB}}) + 2U_2^{\mathrm{BF}} - U_1^{\mathrm{BF}} \\ E_{2,2}^{\mathrm{BF}} &= 3(U_{3,1}^{\mathrm{BB}} - U_{3,0}^{\mathrm{BB}}) - (U_{2,1}^{\mathrm{BB}} - U_{2,0}^{\mathrm{BB}}) + 3U_3^{\mathrm{BF}} - 2U_2^{\mathrm{BF}} \end{aligned} \tag{9.10}$$

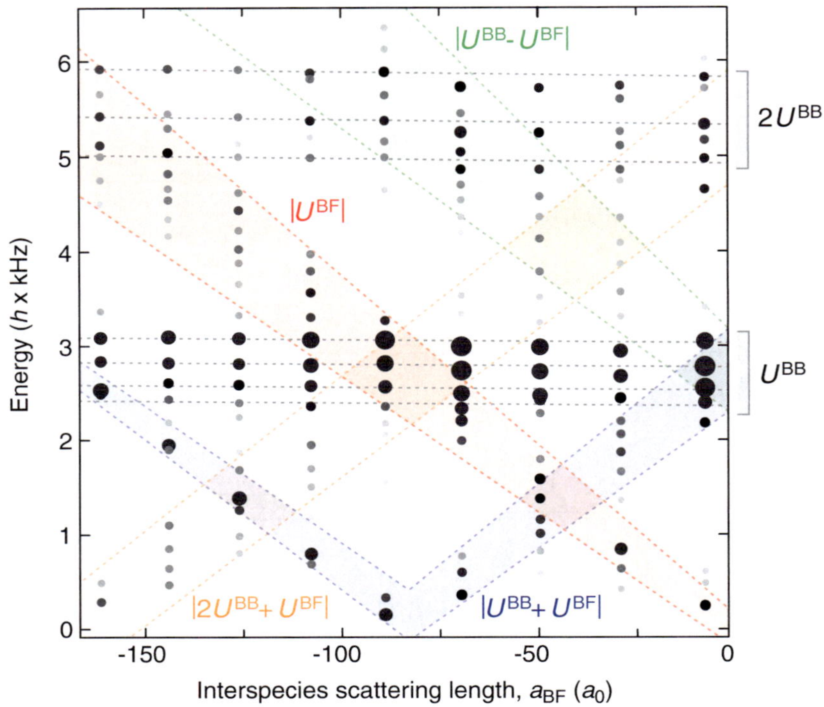

Fig. 9.7 Summary of the spectral features in a broad range of interspecies interactions. The peaks of nine Fourier spectra (including those of Fig. 9.6) are fitted by Gaussians. The resulting positions are shown as *bubbles*, whose size as well as *gray shading* indicate the peak height (the *larger* and *darker*, the *higher*). Spectral components arising from direct Bose-Fermi interactions are highlighted by *shaded* areas corresponding to energies of order $|U^{\mathrm{BF}}|$ (*red*), $|U^{\mathrm{BB}} - U^{\mathrm{BF}}|$ (*green*), $|U^{\mathrm{BB}} + U^{\mathrm{BF}}|$ (*blue*) and $|2U^{\mathrm{BB}} + U^{\mathrm{BF}}|$ (*orange*). Additionally, *black dashed lines* serve as guides to the eye for the energies of orders U^{BB} and $2U^{\mathrm{BB}}$. Note that the three data sets between $a_{\mathrm{BF}} = -75\, a_0$ and $-25\, a_0$ are measured at a slightly lower lattice depth $V_{\mathrm{H}} = 26.7(3)\ E^{\mathrm{B}}_{\mathrm{rec}}$. This explains the slight shift with respect to the other data

and so forth. Assuming that a bosonic on-site occupation of up to $n = 6$ is statistically significant (corresponding to the appearance of four peaks of order U^{BB}, see Chap. 7) we conclude from the above equation that up to six individual components of order $|U^{\mathrm{BF}}|$ are expected. A similar reasoning shows that under the same conditions five components of order $|U^{\mathrm{BB}} + U^{\mathrm{BF}}|$ and four of order $|2U^{\mathrm{BB}} + U^{\mathrm{BF}}|$ can be observed.

The separation between the individual components varies as a function of a_{BF}, which is shown in the inset of Fig. 9.6e for the energies of order $|U^{\mathrm{BB}} + U^{\mathrm{BF}}|$. The components narrow down for increasingly attractive interspecies interactions. In the range $-150a_0 \lesssim a_{\mathrm{BF}} \lesssim -100a_0$ they cannot be resolved anymore and appear as a single peak. This narrowing is caused by the interaction-induced changes of $U^{\mathrm{BB}}_{n,m}$ and U^{BF}_{n} that can compensate each other due to opposite signs. An accurate quantitative

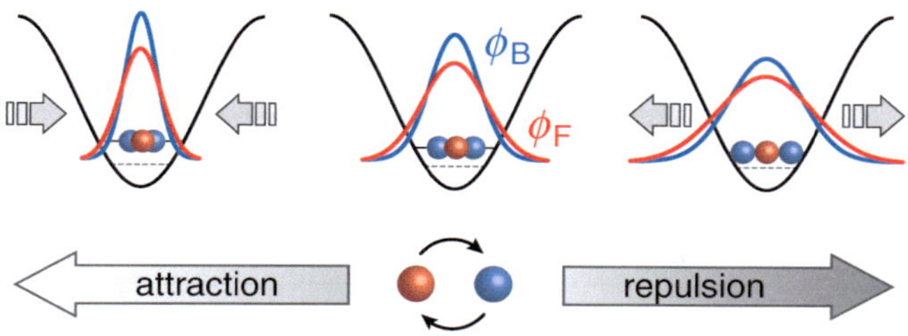

Fig. 9.8 The schematic shows that repulsive (attractive) interspecies interactions broaden (shrink) the on-site orbitals $\phi_B(\mathbf{r})$ and $\phi_F(\mathbf{r})$ and thereby affect the Bose-Bose interaction strengths $U_{n,1}^{BB} \propto a_{BB} \int d^3r\, |\phi_B(\mathbf{r})|^4$

prediction of this behavior would require modeling of interacting multi-body systems in a multi-orbital potential, which is a difficult problem and rarely treated in literature so far.

9.2.3 *Revealing Fermion-Mediated Bose-Bose Interactions*

The notable upshift of the interaction strengths of orders U^{BB} and $2U^{BB}$ observed in Fig. 9.6 for increasing attraction is induced by the presence of an interacting fermion. For the case of repulsive interspecies interactions a broadening of the on-site wavefunctions $\phi_B(\mathbf{r})$ and $\phi_F(\mathbf{r})$ is expected. Correspondingly the Bose-Bose interaction strength is effectively reduced, while the reverse happens for attractive interspecies interactions as illustrated in Fig. 9.8. These modifications are shown in detail in Fig. 9.9 for the highest spectral components of order U^{BB} and $2U^{BB}$ both for negative and positive interspecies scattering length.

We note that the present effect shares similarities with the atom number dependence of U^{BB} discussed for the purely bosonic system in Chap. 7. However, here the mechanism is more indirect. We observe a modification of the collision process among bosons, although the fermion does not take part in the actual collision. In this sense the change in the bosonic interaction strength is mediated by the presence of a fermion.

In order to analyze the data quantitatively, we consider the individual energy terms that give rise to the observed peaks. The highest energy of order U^{BB} is given by the superposition of the energy of two bosons without a fermion (see Eq. 9.4)

$$E_{1,0;0}^{BB} = E_{2,0} = U_{2,0}^{BB}, \tag{9.11}$$

and the respective energy in the presence of a fermion (see Eq. 9.5).

$$E_{1,0;1}^{BB} = E_{2,1} - 2E_{1,1} = U_{2,1}^{BB} + 2\left(U_2^{BF} - U_1^{BF}\right). \tag{9.12}$$

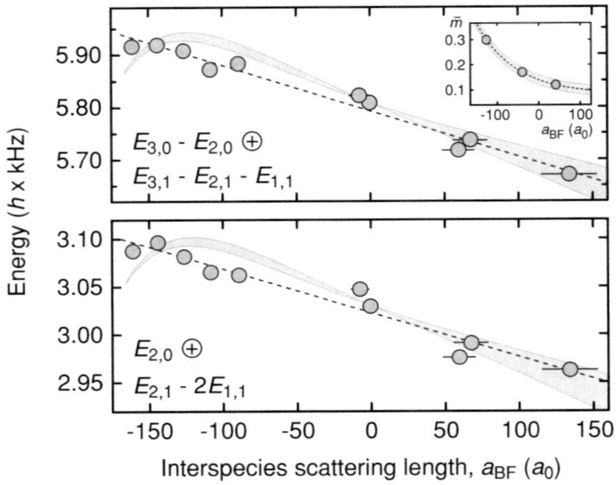

Fig. 9.9 Modification of the Bose-Bose interaction strength induced by an interacting fermion. The highest spectral contributions of order U^{BB} ($2U^{\text{BB}}$) are shown as a function of a_{BF} in the *lower* (*upper*) panel [Corresponding to the spectral features close to the *dashed line* ① (④) in Fig. 9.6]. The underlying energies $E_{2,0}$ and $E_{2,1} - 2E_{1,1}$ ($E_{3,0} - E_{2,0}$ and $E_{3,1} - E_{2,1} - E_{1,1}$) cannot be resolved individually appearing as a single superposition peak in the spectra. The *dashed line* shows a linear fit with a slope of $s_{2,1} = -0.46(4)\,\text{Hz}/a_0$ ($s_{3,1} = -0.87(5)\,\text{Hz}/a_0$). *Shaded* areas show the shift calculated within a variational model (see main text). It explicitly accounts for the measured values for the fermionic filling fitted by an exponential function, $\bar{m}(a_{\text{BF}})$ (inset). The measurement has been performed at a lattice depth $V_{\text{H}} = 28.2(3)\ E_{\text{rec}}^{\text{B}}$. The data at $a_{\text{BF}} = 0\,a_0$ have been obtained in a sample without fermions

For vanishing fermionic filling, the first energy (Eq. 9.11) is observed, while at unity fermionic filling solely the second energy (Eq. 9.12) would be detected. In our data, the fermionic filling lies in the range $0 < \bar{m} < 0.5$ as measured above. Therefore, the energies (Eqs. 9.11 and 9.12) are superimposed in the spectra and the individual peaks cannot be resolved due to a very small spacing. Taking into account the fermionic filling as a function of the interspecies scattering length, $\bar{m}(a_{\text{BF}})$, the effective position of the composite peak can be modeled as

$$\begin{aligned}[1 - \bar{m}(a_{\text{BF}})]\, E_{1,0;0}^{\text{BB}} + \bar{m}(a_{\text{BF}}) E_{1,0;1}^{\text{BB}} \\ = U_{2,0}^{\text{BB}} + \bar{m}(a_{\text{BF}}) \left[U_{2,1}^{\text{BB}} - U_{2,0}^{\text{BB}} + 2\left(U_2^{\text{BF}} - U_1^{\text{BF}}\right)\right].\end{aligned} \tag{9.13}$$

Not only pure Bose-Bose interaction strengths contribute to the signal, but also the difference in Bose-Fermi interaction energy comparing one to two bosons.

Similarly, the expressions for the highest energy of order $2U^{\text{BB}}$ is derived. The contribution without a fermion reads

$$E_{2,0;0}^{\text{BB}} = E_{3,0} - E_{2,0} = 3\, U_{3,0}^{\text{BB}} - U_{2,0}^{\text{BB}}, \tag{9.14}$$

and the analogous component in the presence of a fermion is given by

$$E^{\mathrm{BB}}_{2,0;1} = E_{3,1} - E_{2,1} - E_{1,1} = 3U^{\mathrm{BB}}_{3,1} + 3U^{\mathrm{BF}}_{3} - U^{\mathrm{BB}}_{2,1} - 2U^{\mathrm{BF}}_{2} - U^{\mathrm{BF}}_{1}. \quad (9.15)$$

The weighted superposition of the energies (Eqs. 9.14 and 9.15) reads

$$\begin{aligned}&[1 - \bar{m}(a_{\mathrm{BF}})]\, E^{\mathrm{BB}}_{2,0;0} + \bar{m}(a_{\mathrm{BF}}) E^{\mathrm{BB}}_{2,0;1} \qquad (9.16)\\ &= 3U^{\mathrm{BB}}_{3,0} - U^{\mathrm{BB}}_{2,0} + \bar{m}(a_{\mathrm{BF}}) \Big[3\left(U^{\mathrm{BB}}_{3,1} - U^{\mathrm{BB}}_{3,0}\right) - \left(U^{\mathrm{BB}}_{2,1} - U^{\mathrm{BB}}_{2,0}\right)\\ &+ 3U^{\mathrm{BF}}_{3} - 2U^{\mathrm{BF}}_{2} - U^{\mathrm{BF}}_{1}\Big].\end{aligned}$$

Using this derivation and the slopes $s_{2,1} = -0.46(4)$ Hz/a_0 and $s_{3,1} = -0.87(5)$ Hz/a_0 extracted for the lower and upper panel of Fig. 9.9, a conservative upper bound for the shifts $U^{\mathrm{BB}}_{2,1}$ and $U^{\mathrm{BB}}_{3,1}$ is derived. Assuming a constant fermionic filling $\bar{m} = 0.3$ and neglecting the shift arising from $2(U^{\mathrm{BF}}_{2} - U^{\mathrm{BF}}_{1})$ (see Eq. 9.13), we obtain $(\partial U^{\mathrm{BB}}_{2,1}/\partial a_{\mathrm{BF}})/h \approx s_{2,1}/\bar{m} = -1.5(2)$ Hz/a_0. Similarly, neglecting the shifts arising from $3U^{\mathrm{BF}}_{3} - 2U^{\mathrm{BF}}_{2} - U^{\mathrm{BF}}_{1}$ (see Eq. 9.16), we can approximate $(\partial U^{\mathrm{BB}}_{3,1}/\partial a_{\mathrm{BF}})/h \approx (s_{2,1} + s_{3,1})/(3\bar{m}) = -1.4(2)$ Hz/a_0. The variational approach introduced in the next section shows that the omission of the U^{BF} terms is reasonable for moderately attractive interactions in the range $-125a_0 \lesssim a_{\mathrm{BF}} \lesssim 0a_0$.

9.2.3.1 Variational Mean-Field Model

Precise modeling of the interaction energies in few-body systems is a demanding theoretical problem. A multi-orbital system for the bosons and the fermion would have to be solved, which goes beyond the scope of this work. Nevertheless, we aim at an approximate theory to validate the observed trend in the Bose-Bose interaction strengths (see Fig. 9.9). To this end, we employ a variational mean-field model, in which the on-site Hamiltonian is harmonically approximated. Generally, the model must be taken with care, because typical errors due to the harmonic approximation amount to more than 10 % when the *absolute* interaction strengths U^{BB} and U^{BF} are calculated (compare Fig. 3.2).[2] However, it serves well to determine the correct order of magnitude of the *shifts* of the Bose-Bose energies.

At first all energy contributions in the Eqs. 9.13 and 9.16 are calculated as a function of the interspecies scattering length a_{BF}. To this end we approximate the lattice site by an isotropic harmonic oscillator potential and assign a Gaussian wavefunction to the bosons and the fermion,

---

[2] For typical lattice depths used in experiments (below 50 E_{rec}) the harmonic approximation of a lattice site typically entails large errors, because the actual on-site wavefunction in a sinusoidal lattice deviates significantly from the Gaussian ground state wavefunction of a harmonic oscillator potential. Consequently, the interaction energy U and even more the tunneling coupling J deviate from their actual values (see Chap. 3).

$$\phi^{\mathrm{B}}_{\mathrm{harm}}(\mathbf{r};\sigma_{\mathrm{B}}) = \frac{1}{(\sqrt{\pi}\,\sigma_{\mathrm{B}})^{3/2}}\,e^{-\frac{r^2}{2\sigma_{\mathrm{B}}^2}} \tag{9.17}$$

and

$$\phi^{\mathrm{F}}_{\mathrm{harm}}(\mathbf{r};\sigma_{\mathrm{F}}) = \frac{1}{(\sqrt{\pi}\,\sigma_{\mathrm{F}})^{3/2}}\,e^{-\frac{r^2}{2\sigma_{\mathrm{F}}^2}}, \tag{9.18}$$

respectively. The full on-site Hamiltonian reads $\hat{H} = \hat{H}_{\mathrm{B}} + \hat{H}_{\mathrm{F}} + \hat{H}_{\mathrm{BF}}$ with the individual terms [25]

$$\hat{H}_{\mathrm{B}} = \int d^3r\,\hat{\psi}^{\dagger}_{\mathrm{B}}(\mathbf{r})\left[-\frac{\hbar^2}{2m_{\mathrm{B}}}\Delta + V_{\mathrm{B}}(\mathbf{r}) + \frac{g_{\mathrm{BB}}}{2}\hat{\psi}^{\dagger}_{\mathrm{B}}(\mathbf{r})\hat{\psi}_{\mathrm{B}}(\mathbf{r})\right]\hat{\psi}_{\mathrm{B}}(\mathbf{r}), \tag{9.19}$$

$$\hat{H}_{\mathrm{F}} = \int d^3r\,\hat{\psi}^{\dagger}_{\mathrm{F}}(\mathbf{r})\left[-\frac{\hbar^2}{2m_{\mathrm{F}}}\Delta + V_{\mathrm{F}}(\mathbf{r})\right]\hat{\psi}_{\mathrm{F}}(\mathbf{r}), \tag{9.20}$$

$$\hat{H}_{\mathrm{BF}} = g_{\mathrm{BF}}\int d^3r\,\hat{\psi}^{\dagger}_{\mathrm{B}}(\mathbf{r})\hat{\psi}^{\dagger}_{\mathrm{F}}(\mathbf{r})\hat{\psi}_{\mathrm{F}}(\mathbf{r})\hat{\psi}_{\mathrm{B}}(\mathbf{r}). \tag{9.21}$$

Here, $g_{\mathrm{BB}} = 4\pi\hbar^2 a_{\mathrm{BB}}/m_{\mathrm{B}}$ and $g_{\mathrm{BF}} = 2\pi\hbar^2 a_{\mathrm{BF}}/\mu$ are the Bose-Bose and the Bose-Fermi coupling strength, respectively, and $\mu = m_{\mathrm{B}}m_{\mathrm{F}}/(m_{\mathrm{B}} + m_{\mathrm{F}})$ is the reduced mass. The terms $V_{\mathrm{B}}(\mathbf{r}) = m_{\mathrm{B}}\omega_{\mathrm{B}}^2\mathbf{r}^2/2$ and $V_{\mathrm{F}}(\mathbf{r}) = m_{\mathrm{F}}\omega_{\mathrm{F}}^2\mathbf{r}^2/2$ denote the approximate harmonic on-site potential. The respective trap frequencies are given by $\omega_{\mathrm{B}} = 2\sqrt{s_{\mathrm{B}}}E^{\mathrm{B}}_{\mathrm{rec}}/\hbar$ and $\omega_{\mathrm{F}} = 2\sqrt{s_{\mathrm{F}}}E^{\mathrm{F}}_{\mathrm{rec}}/\hbar$, where s_{B} and s_{F} denote the lattice depth for the bosons and fermions in units of the respective recoil energy ($s_{\mathrm{B}} = 1.50 s_{\mathrm{F}}$ for our experimental parameters).

According to a mean-field ansatz, the field operators in the Hamiltonian $\hat{H}$ are replaced by a single mode of the field, given by the variational Gaussian wavefunctions

$$\hat{\psi}_{\mathrm{B}}(\mathbf{r}) \rightarrow \phi^{\mathrm{B}}_{\mathrm{harm}}(\mathbf{r};\sigma_{\mathrm{B}})\,\hat{a} \quad \text{and} \quad \hat{\psi}_{\mathrm{F}}(\mathbf{r}) \rightarrow \phi^{\mathrm{F}}_{\mathrm{harm}}(\mathbf{r};\sigma_{\mathrm{F}})\,\hat{c}. \tag{9.22}$$

Here, $\hat{a}$ and $\hat{c}$ are the bosonic and fermionic annihilation operators, respectively (compare Sect. 2.1). Taking the expectation value $\langle\hat{H}\rangle$ with respect to the general Bose-Fermi quantum state $|n\rangle|m\rangle$ with n bosons ($n \in \mathbb{N}$) and m (m either 0 or 1) fermions, the energy functional at a certain interspecies scattering length a_{BF} reads

$$\begin{aligned} &E(n, m, a_{\mathrm{BF}}; \sigma_{\mathrm{B}}, \sigma_{\mathrm{F}}) \\ &= n\cdot\frac{3}{4}\frac{\hbar^2}{m_{\mathrm{B}}}\frac{1}{\sigma_{\mathrm{B}}^2} + m\cdot\frac{3}{4}\frac{\hbar^2}{m_{\mathrm{F}}}\frac{1}{\sigma_{\mathrm{F}}^2} + n\cdot\frac{3}{4}m_{\mathrm{B}}\omega_{\mathrm{B}}^2\sigma_{\mathrm{B}}^2 + m\cdot\frac{3}{4}m_{\mathrm{F}}\omega_{\mathrm{F}}^2\sigma_{\mathrm{F}}^2 \\ &\quad + n(n-1)\cdot\frac{1}{\sqrt{2\pi}}\frac{\hbar^2 a_{\mathrm{BB}}}{m_{\mathrm{B}}}\frac{1}{\sigma_{\mathrm{B}}^3} + n\,m\cdot\frac{1}{\sqrt{2\pi}}\frac{\hbar^2 a_{\mathrm{BF}}}{\mu}\frac{1}{(\sigma_{\mathrm{B}}^2 + \sigma_{\mathrm{F}}^2)^{3/2}}, \end{aligned} \tag{9.23}$$

where the Gaussian widths σ_{B} and σ_{F} are used as variational parameters. The on-site trap frequencies are given by $\omega_{\mathrm{B}} = 2\sqrt{s_{\mathrm{B}}}E^{\mathrm{B}}_{\mathrm{rec}}/\hbar$ and $\omega_{\mathrm{F}} = 2\sqrt{s_{\mathrm{F}}}E^{\mathrm{F}}_{\mathrm{rec}}/\hbar$; s_{B} (s_{F})

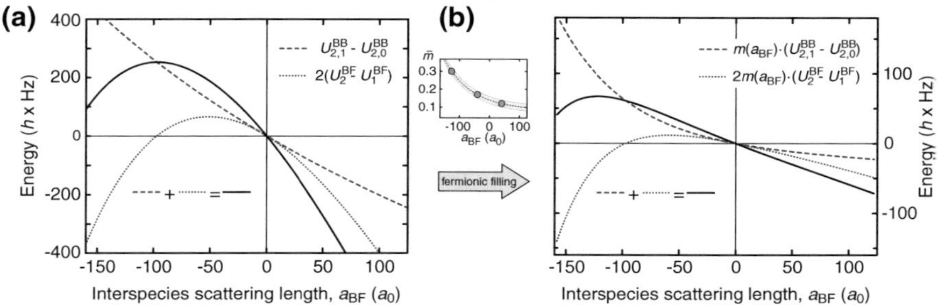

Fig. 9.10 Variational model for atom number dependent on-site interaction strengths $U_{n,m}^{\mathrm{BB}}$ and U_n^{BF}. **a** The contributions to the shift with respect to $U_{2,0}^{\mathrm{BB}}$ [see Eq. (9.13)] are displayed as a function of the interspecies scattering length a_{BF}: $U_{2,1}^{\mathrm{BB}} - U_{2,0}^{\mathrm{BB}}$ (*dashed*) and $2\left(U_2^{\mathrm{BF}} - U_1^{\mathrm{BF}}\right)$ (*dotted*) and their sum (*solid line*). **b** Weighing the shift with the interaction dependent fermionic filling $\bar{m}(a_{\mathrm{BF}})$ (as measured in the experiment) yields an almost linear trend (*solid line*) for $a_{\mathrm{BF}} \gtrsim -125\, a_0$

denotes the lattice depth for the bosonic (fermionic) species in units of the respective recoil energy and $s_{\mathrm{B}} = 1.50 \cdot s_{\mathrm{F}}$ for the parameters of the optical lattice used here (see above).

We numerically minimize Eq. 9.23 to obtain the optimized widths $\sigma_{\mathrm{B}}(n, m, a_{\mathrm{BF}})$ and $\sigma_{\mathrm{F}}(n, m, a_{\mathrm{BF}})$. The corresponding Bose-Bose and Bose-Fermi interaction energies per atom pair can be calculated according to

$$U_{n,m}^{\mathrm{BB}}(a_{\mathrm{BF}}) = \sqrt{\frac{2}{\pi}}\frac{\hbar^2 a_{\mathrm{BB}}}{m_{\mathrm{B}}}\frac{1}{\sigma_{\mathrm{B}}(n, m, a_{\mathrm{BF}})^3} \tag{9.24}$$

and

$$U_n^{\mathrm{BF}}(a_{\mathrm{BF}}) = \frac{1}{\sqrt{2\pi}}\frac{\hbar^2 a_{\mathrm{BF}}}{\mu}\frac{1}{\left[\sigma_{\mathrm{B}}(n, m, a_{\mathrm{BF}})^2 + \sigma_{\mathrm{F}}(n, m, a_{\mathrm{BF}})^2\right]^{3/2}}. \tag{9.25}$$

With these energies we can calculate the *shifts* of the Bose-Bose interaction energies, given by the squared brackets in the second row of Eqs. 9.13 and 9.16 multiplied by $\bar{m}(a_{\mathrm{BF}})$. Figure 9.10 shows the results for the highest spectral contribution of order U^{BB}. The difference of the bosonic two-body interactions with and without a fermion, $U_{2,1}^{\mathrm{BB}} - U_{2,0}^{\mathrm{BB}}$ (dashed line), decreases monotonically, changing signs at $a_{\mathrm{BF}} = 0$. This behavior is the expected, because the presence of an attractively interacting fermion shrinks the bosonic wavefunction such that $U_{2,1}^{\mathrm{BB}} > U_{2,0}^{\mathrm{BB}}$, while the reverse happens for repulsive Bose-Fermi interactions. The dependence of $2\left(U_2^{\mathrm{BF}} - U_1^{\mathrm{BF}}\right)$ (dotted line) on a_{BF} is plausible as well: For repulsive interspecies interactions the broadening of the wavefunctions is more pronounced for a fermion and two bosons (which also interact repulsively), compared to the case of a fermion and a single boson, such that $U_2^{\mathrm{BF}} < U_1^{\mathrm{BF}}$. In the regime of moderately attractive interspecies interactions, $-|a_{\mathrm{BB}}| \lesssim a_{\mathrm{BF}} < 0$, the addition of a boson leads to a broadening of the bosonic and a shrinking of the fermionic wavefunction. However, the broader σ_{B} overcompensates the shrunk σ_{F}, such that $|U_1^{\text{BF}}| > |U_2^{\text{BF}}|$. On the other hand, for stronger interspecies attraction $a_{\text{BF}} \lesssim -|a_{\text{BB}}|$, the additional narrowing of the fermionic wavefunction upon addition of a boson dominates, and hence $|U_1^{\text{BF}}| < |U_2^{\text{BF}}|$.

The sum of the two contributions (dashed and dotted lines) corresponds to the shift with respect to $U_{2,0}^{\text{BB}}$, shown as a blue solid line in Fig. 9.10a. When additionally the experimentally observed variation of the fermionic filling $\bar{m}(a_{\text{BF}})$ is taken into account, the total shift appears approximately linear for $a_{\text{BF}} \gtrsim -125a_0$ (see Fig. 9.10b, blue solid line).

The results of the variational mean-field model are compared to the experimental data in Fig. 9.9. The vertical extent of the gray shaded areas, that denote the calculated *shifts*, represents the uncertainties in the experimental determination of $\bar{m}(a_{\text{BF}})$. The calculated shifts are added to the values of $U_{2,0}^{\text{BB}}$ (lower panel) and $3U_{3,0}^{\text{BB}} - U_{2,0}^{\text{BB}}$ (upper panel), respectively, which have been measured in a purely bosonic sample. We observe that the matching between experimental and theoretical shifts is remarkably good. However, the agreement must be regarded fortuitous given the simplicity of our theory. These findings quantitatively support the indirect evidence for a renormalization of Hubbard parameters due to interspecies interactions discussed in Chap. 8 [13, 15, 19].

9.3 Excursion: Quantum Revival Spectroscopy with Fermions

It is a natural question, whether the time evolution of the Bose-Fermi few-body system also leaves signatures in the fermionic momentum distribution. Actually, one can expect that the quantum evolution of the spin-polarized Fermi gas is solely determined by the Bose-Fermi interaction energy, because the fermions do not interact directly. The numerous bosonic contributions observed above will not play a role. Therefore, the experimental investigation of the fermionic component has the potential to lead to much cleaner spectra. However, what would be an appropriate observable?

9.3.1 Experimental Results

When the data for this chapter was recorded, there have been no intentions to analyze the fermionic momentum distributions (see Sect. 9.2 for the experimental sequence). Initially, the images of the fermionic cloud have only served to monitor the number of fermions in each experimental run. However, after finding an appropriate observable it has turned out that the recorded images indeed reveal faint signals of fermionic quantum dynamics.

In principle, bosons and fermions can be expected to show similar behavior: At the time of the collapse, bosons are maximally dephased and have a structureless momentum distribution that is determined by the Wannier background. Upon further evolution towards the revival, the characteristic interference pattern appears again as long-range phase coherence is reestablished. In analogy, a metallic state of fermions that corresponds to a partially filled first band (mean local filling $\bar{m} < 1$, see Chap. 6 and Fig. 6.6) shows an interference pattern with broad peaks in the momentum distribution (see Sect. 4.2.2) or a quasi-momentum distribution with a filling in the center of the Brillouin zone (see Sect. 4.2.4). When the quantum evolution leads to dephasing, the interference pattern should wash out corresponding to an homogeneously filled first Brillouin zone.

In analogy to the observables introduced to analyze the bosonic momentum distribution, visibility and the fraction of coherent atoms (see Sect. 4.2.2), we define the fermionic visibility as an observable that measures the homogeneity of the filling in the first Brillouin zone:

$$\tilde{\mathcal{V}} = 1 - \frac{9}{4}\frac{N_{\text{edges}}}{N_{\text{BZ}}}. \tag{9.26}$$

Here, N_{BZ} is the total atom number in the first Brillouin zone and N_{edges} counts the atom number in the four edges as illustrated in Fig. 9.11b. For a homogeneously filled Brillouin zone the ratio $N_{\text{edges}}/N_{\text{BZ}}$ equals $4/9$ and the fermionic visibility vanishes $\tilde{\mathcal{V}} = 0$, while the maximal value $\tilde{\mathcal{V}} = 1$ is reached when the edges are empty, as expected for a metallic state with a very low filling. A related observable is used in the PhD thesis of Tim Rom [27] to measure the ratio of atoms in a band insulating state (compare Sect. 3.3.3). Additionally, theoretical studies on lattice Fermi gases have used the population at the Brillouin zone edge to quantify nonequilibrium dynamics after an interaction quench [28].

All fermionic time-of-flight images that have been recorded alongside with the bosonic data of the previous section are analyzed in terms of $\tilde{\mathcal{V}}$. As shown in Fig. 9.11c an oscillatory behavior is clearly discernible. The amplitude of the oscillations is relatively small, probably due to limited overlap between the bosons and the fermions. Accordingly, the largest amplitudes are observed when strong interspecies attraction improves the overlap between the components. At a hold time of about 2 ms the fermionic visibility shows a pronounced kink that originates from Bloch oscillations that are superimposed on the data due to the effect of gravity. They slowly shift the position of the interference pattern, which is taken into account in the evaluation. At the time when the diffraction pattern fully rolls over a compensation is no longer possible and the kink appears.

The first 1.2 ms of the fermionic quantum evaluation are used for spectral analysis that is carried out in a procedure analogous to Appendix D. The resulting spectra are shown in Fig. 9.12a–e for attractive interspecies interactions, complementing Fig. 9.6 of the bosonic analysis. Due to the shorter length of the traces the spectral resolution is reduced compared to the bosonic data. The strongest peak can be clearly identified to correspond to the direct Bose-Fermi interaction energy $|U^{\text{BF}}|$. Plotting this frequency as a function of interspecies scattering length a_{BF}, we obtain an almost

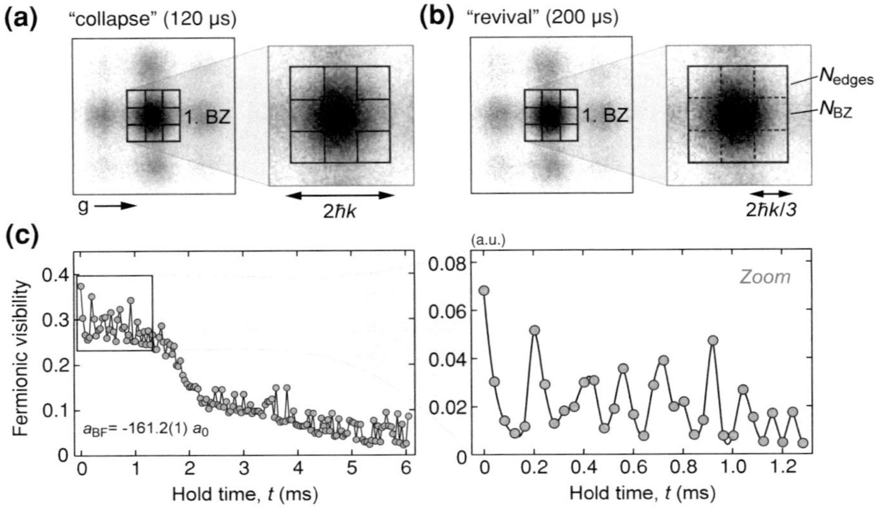

Fig. 9.11 Quantum phase revival dynamics of the fermionic component in a Bose-Fermi mixture. Exemplary momentum distributions recorded after 10 ms time-of-flight expansion show a collapse **a** and a revival **b**, which are hardly discernible with the bare eye. The interspecies attraction is $a_{\mathrm{BF}} = -161.2(1)\ a_0$ and the lattice depth $V_{\mathrm{H}}^{\mathrm{F}} = 18.8(2)\ E_{\mathrm{rec}}^{\mathrm{F}}$. **c** Evaluation of the fermionic visibility $\tilde{\mathcal{V}}$ (see Eq. 9.26) clearly reveals collapse and revival dynamics arising from the interspecies interactions U^{BF}. The kink at about $t = 2$ ms arises due to dephasing induced by superimposed Bloch oscillations [26] along the direction of gravity. Therefore, the analysis is restricted to the first 1.2 ms. The baseline $\tilde{\mathcal{V}} = 0$ corresponds to a homogeneously filled first Brillouin zone

perfectly linear trend (see Fig. 9.12) similar to the results in the previous section (see Figs. 9.6 and 9.7).

A remarkable advantage of evaluating the fermionic momentum distributions is the absence of the numerous bosonic frequencies, which can overlap with the Bose-Fermi peaks. This is strikingly visible when the panels Fig. 9.12d, e are compared to the corresponding spectra in Fig. 9.6. For the $|U^{\mathrm{BF}}|$ peaks, in principle, a comb-like substructure is expected reflecting the dependence of the Bose-Fermi interaction energy on the bosonic occupation (see Sect. 9.2.2). However, the individual comb-teeth cannot be spectrally resolved within the Fourier limit and merge into a broad single peak. In addition to the dominant Bose-Fermi peak, more spectral features are observed (particularly in panels Fig. 9.12b, d). They do not show clear trends as a function of the interspecies scattering length and their physical origin can so far not be identified. Further investigations with higher spectral resolution will be needed for a precise understanding of the spectra.

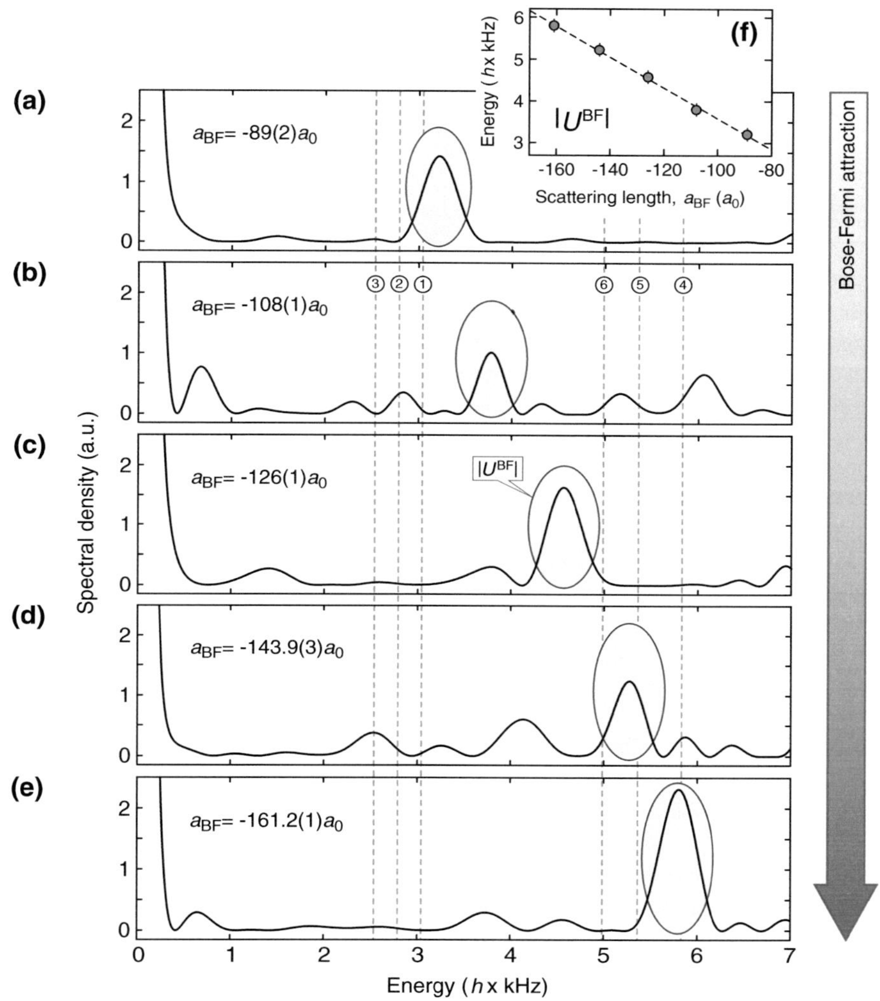

Fig. 9.12 **a**–**e** Fourier spectra of fermionic quantum revival traces for linearly increasing interspecies attraction. The spectral contribution corresponding to the Bose-Fermi interaction energy $|U^{BF}|$ is highlighted by *shaded ovals*. *Dashed lines* (① to ⑥) indicate the locations of the contributions of orders U^{BB} and $2U^{BB}$ that are suppressed in the fermionic dynamics (compare Fig. 9.6). **f** The experimentally measured Bose-Fermi interaction energy $|U^{BF}|$ shows a linear trend as a function of a_{BF} in analogy to our findings in Sect. 9.2

9.3.2 Theoretical Toy Model

A very basic understanding of fermionic quantum phase revivals is gained by a theoretical toy model that is based on our considerations of the previous section.

Similar to the bosons, the time-dependent momentum distribution of the fermions is given by (see Sect. 4.2.2)

$$n_{\mathrm{F}}(\mathbf{k}, t) = |w_{\mathrm{F}}(\mathbf{k})|^2 \sum_{ij} e^{-i\mathbf{k}(\mathbf{r}_i-\mathbf{r}_j)} \langle \hat{c}_i^\dagger \hat{c}_j \rangle(t), \tag{9.27}$$

where $\hat{c}_i$ and $\hat{c}_i^\dagger$ are the fermionic annihilation and creation operators for lattice site i, respectively. As in the bosonic case, $n_{\mathrm{F}}(\mathbf{k}, t)$ is crucially determined by the time-dependent coherences between lattice sites $\langle \hat{c}_i^\dagger \hat{c}_j \rangle(t)$. If all off-diagonal ($i \neq j$) elements of the single-particle density matrix $\langle \hat{c}_i^\dagger \hat{c}_j \rangle$ vanish, the momentum distribution is uniform corresponding to a homogeneously filled first Brillouin zone. The images in Fig. 9.11 clearly reveal that this is not the case in the experiment, implying finite values for off-diagonal elements, at least on a short range. Because such short-range coherences are hard to include in a simple theoretical description, we resort to the Gutzwiller-type on-site wavefunction $|\psi_{\mathrm{BF}}(t)\rangle$ introduced in Eq. 9.1. In this model wavefunction the fermionic coherences are infinitely long-range, which is obviously an exaggeration.

With the on-site wavefunction $|\psi_{\mathrm{BF}}(t)\rangle$ (identical for each lattice site), eigenenergies $E_{n,m}$ as defined in Sect. 9.1 and coherent states for the bosonic component, the dynamical evolution of the fermionic coherences reads

$$|\langle \hat{c} \rangle(t)|^2 = (1 - \bar{m})\bar{m} \cdot e^{2\bar{n}(\cos(U^{\mathrm{BF}}t)-1)}, \tag{9.28}$$

where $\bar{m}$ and $\bar{n}$ denote the mean fermionic and bosonic filling, respectively. Despite the weaknesses of the model this result illustrates two important features: First, the fermionic dynamics are solely determined by the Bose-Fermi interaction energy $|U^{\mathrm{BF}}|$ and its higher harmonics. Second, the strength of the dynamics depends on the mean fermionic filling $\bar{m}$. At vanishing and unity filling the dynamical evolution is fully suppressed, while it is maximal at half-filling. Nevertheless, a more accurate theory of the dynamics beyond the Gutzwiller approximation would be highly desirable.

9.3.3 Summary

In this excursion, we have presented the first observation of collapse and revival dynamics in a fermionic lattice quantum gas. The findings were enabled by the definition of a new experimental observable, the fermionic visibility $\tilde{\mathcal{V}}$, that is sensitive to the evolution of the fermions after an interaction quench. In fact, the findings of this excursion come somewhat unexpected, because the experimental sequence was initially not designed for optimal acquisition of the fermionic data. Therefore, several aspects can be improved in future realizations: First, the quasi-momentum distribution after Brillouin-zone mapping is likely to provide stronger and cleaner signals than the momentum distribution recorded here. Second, the data should be taken in the direction gravity such that the quench dynamics are not obscured by superimposed Bloch oscillations. Third, the dynamics should be recorded in a deeper lattice to suppress relaxation via tunneling during the experimental observation time. These improvements are likely to lead to a much better spectral resolution.

There are numerous applications for the methods presented in this brief excursion. The evolution of fermionic lattice quantum gases in one- and higher dimensions has been dealt with in many theoretical studies. It has been proposed to use nonequilibrium quantum dynamics of fermions for the identification of quantum phases in the Hubbard model [29–31], in chains of spin-polarized fermions with nearest- and next-nearest neighbor interactions [28, 32–34], or in a Heisenberg chain [35, 36]. Furthermore, the quench dynamics studied here are strongly related to questions of thermalization and equilibration in closed quantum systems [30, 33–35, 37, 38], which have so far only been experimentally studied in a bosonic lattice system [39]. Additionally, we envisage the potential to perform interaction spectroscopy in multi-component fermionic mixtures [40], which might enable the detection of direct higher-body interactions.

9.4 Conclusion and Outlook

In this chapter, we have discussed the quantum dynamics in elementary few-body systems formed by a single fermion and a small bosonic field at the sites of an optical lattice. Quantum revival spectroscopy has been demonstrated as a useful technique to measure the interaction energies in this system. While in the case of direct Bose-Fermi interactions an essentially linear dependence on the interspecies scattering length is found, we additionally observe how the interaction among the bosons is modified, mediated by the presence of an interacting fermion. The detection of mediated interactions could further be improved by preparing the bosonic quantum gas on top of a fermionic band insulator, where each lattice site is occupied by exactly one fermion (in contrast to a mean fermionic filling $\bar{m} < 0.5$ in the measurements of this chapter). Under these conditions Bose-Fermi features of order $|U^{\mathrm{BF}}|$, $|U^{\mathrm{BB}} + U^{\mathrm{BF}}|$, etc. are fully suppressed and all Bose-Bose features display the full shift induced by Bose-Fermi interactions.

Because of an interference between Bose-Bose and Bose-Fermi phase dynamics we can infer the mean fermionic occupation of the mixture; intrinsically the fermionic filling is only detected on those lattice sites that are also occupied by bosons. This feature makes quantum revival spectroscopy a useful tool to reveal regimes of phase separation, even if the separation happens on a microscale between neighboring lattice sites, such as in a supersolid [41–43]. Such a scenario could not be conclusively identified via standard in-situ imaging. Generally, a detailed experimental analysis on the redistribution of atoms during lattice loading would be highly desirable, particularly for heteronuclear mixtures (compare Chap. 8). If a shallow lattice V_{L} is loaded at variable interspecies interactions a_{BF}, the subsequent application of quantum revival spectroscopy for fixed values of V_H and a_{BF} may allow to probe the redistribution of atoms. On the technical side, this requires fast switching of the Feshbach field prior to the onset of the quantum evolution. Furthermore, it might be interesting to utilize the fermions as local probes at individual lattice sites: A quantum system of bosons could be prepared in the presence of fermions, which are tuned to transparency by a Feshbach resonance. The fermions would not influence the formation of a bosonic equilibrium quantum phase, but for detection they could be switched to finite interactions.

Miniature impurity systems, as the one presented in this chapter, are suited to study polaron physics in ultracold quantum gases [1–4] and form an ideal test bed for effective field theories [16] that are highly relevant to the description of atomic nuclei [9]. Our measurement technique might further enable thermometry in Bose-Fermi mixtures based on a temperature dependence of the fermionic filling and allow for exact absolute measurements of two- and higher-body interaction energies [15–19] in multi-component quantum systems. Furthermore, the demonstrated control of interatomic collisions, shows that Bose-Fermi systems may qualify for applications in quantum information processing [10].

References

1. A.S. Alexandrov, N.F. Mott, Bipolarons. Rep. Prog. Phys. **57**, 1197 (1994)
2. M. Bruderer, A. Klein, S.R. Clark, D. Jaksch, Polaron physics in optical lattices. Phys. Rev. A **76**, 011605 (2007)
3. J. Tempere, W. Casteels, M.K. Oberthaler, S. Knoop, E. Timmermans, J.T. Devreese, Feynman path-integral treatment of the BEC-impurity polaron. Phys. Rev. B **80**, 184504 (2009)
4. A. Privitera, W. Hofstetter, Polaronic slowing of fermionic impurities in lattice Bose-Fermi mixtures. Phys. Rev. A **82**, 063614 (2010)
5. A. Schirotzek, C.-H. Wu, A. Sommer, M.W. Zwierlein, Observation of fermi polarons in a tunable fermi liquid of ultracold atoms. Phys. Rev. Lett. **102**, 230402 (2009)
6. S. Nascimbène, N. Navon, K.J. Jiang, L. Tarruell, M. Teichmann, J. McKeever, F. Chevy, C. Salomon, Collective oscillations of an imbalanced fermi gas: axial compression modes and polaron effective mass. Phys. Rev. Lett. **103**, 170402 (2009)
7. C. Zipkes, S. Palzer, C. Sias, M. Köhl, A trapped single ion inside a Bose-Einstein condensate. Nature **464**, 388 (2010)
8. S. Schmid, A. Härter, J.H. Denschlag, Dynamics of a cold trapped ion in a Bose-Einstein condensate. Phys. Rev. Lett. **105**, 133202 (2010)
9. L. Platter, Low-energy universality in atomic and nuclear physics. Few-Body Syst. **46**, 139 (2009)
10. A. Albus, F. Illuminati, J. Eisert, Mixtures of bosonic and fermionic atoms in optical lattices. Phys. Rev. A **68**, 023606 (2003)
11. K. Günter, T. Stöferle, H. Moritz, M. Köhl, T. Esslinger, Bose-Fermi mixtures in a three-dimensional optical lattice. Phys. Rev. Lett. **96**, 180402 (2006)
12. S. Ospelkaus, C. Ospelkaus, O. Wille, M. Succo, P. Ernst, K. Sengstock, K. Bongs, Localization of bosonic atoms by fermionic impurities in a three-dimensional optical lattice. Phys. Rev. Lett. **96**, 180403 (2006)
13. T. Best, S. Will, U. Schneider, L. Hackermüller, D. van Oosten, D.-S. Lühmann, I. Bloch, Role of interactions in ^{87}Rb-^{40}K Bose-Fermi mixtures in a 3D optical lattice. Phys. Rev. Lett. **102**, 030408 (2009)

14. O.E. Alon, A.I. Streltsov, L.S. Cederbaum, Zoo of quantum phases and excitations of cold bosonic atoms in optical lattices. Phys. Rev. Lett. **95**, 030405 (2005)
15. D.-S. Lühmann, K. Bongs, K. Sengstock, D. Pfannkuche, Self-trapping of Bosons and Fermions in optical lattices. Phys. Rev. Lett. **101**, 050402 (2008)
16. P.R. Johnson, E. Tiesinga, J.V. Porto, C.J. Williams, Effective three-body interactions of neutral bosons in optical lattices. New J. Phys. **11**, 093022 (2009)
17. H.P. Büchler, Microscopic derivation of Hubbard parameters for cold atomic gases. Phys. Rev. Lett. **104**, 090402 (2010)
18. O. Dutta, A. Eckardt, P. Hauke, B. Malomed, M. Lewenstein, Bose-Hubbard model with occupation-dependent parameters. New J. Phys. **13**, 023019 (2011)
19. A. Mering, M. Fleischhauer, Multiband and nonlinear hopping corrections to the three-dimensional Bose-Fermi-Hubbard model. Phys. Rev. A **83**, 063630 (2011)
20. A. Simoni, M. Zaccanti, C. D'Errico, M. Fattori, G. Roati, M. Inguscio, G. Modugno, Near-threshold model for ultracold KRb dimers from interisotope Feshbach spectroscopy. Phys. Rev. A **77**, 052705 (2008)
21. E.G.M. van Kempen, S.J.J.M.F. Kokkelmans, D.J. Heinzen, B.J. Verhaar, Interisotope determination of ultracold rubidium interactions from three high-precision experiments. Phys. Rev. Lett. **88**, 093201 (2002)
22. N.N. Klausen, J.L. Bohn, C.H. Greene, Nature of spinor Bose-Einstein condensates in rubidium. Phys. Rev. A **64**, 053602 (2001)
23. S. Will, T. Best, U. Schneider, L. Hackermüller, D.-S. Lühmann, I. Bloch, Time-resolved observation of coherent multi-body interactions in quantum phase revivals. Nature **465**, 197 (2010)
24. M. Greiner, O. Mandel, T. Hänsch, I. Bloch, Collapse and revival of the matter wave field of a Bose-Einstein condensate. Nature **419**, 51 (2002)
25. D.-S. Lühmann, *Multiorbital physics in optical lattices*. Ph.D. Thesis, Universität Hamburg, 2009
26. G. Roati, E. de Mirandes, F. Ferlaino, H. Ott, G. Modugno, M. Inguscio, Atom interferometry with trapped fermi gases. Phys. Rev. Lett. **92**, 230402 (2004)
27. T. Rom, *Bosonische und fermionische Quantengase in dreidimensionalen optischen Gittern*. Ph.D. Thesis, Ludwig-Maximilians-Universität München, 2009
28. S.R. Manmana, S. Wessel, R.M. Noack, A. Muramatsu, Strongly correlated fermions after a quantum quench. Phys. Rev. Lett. **98**, 210405 (2007)
29. M. Moeckel, S. Kehrein, Interaction quench in the Hubbard model. Phys. Rev. Lett. **100**, 175702 (2008)
30. M. Eckstein, M. Kollar, P. Werner, Thermalization after an interaction quench in the Hubbard model. Phys. Rev. Lett. **103**, 056403 (2009)
31. M. Schiró, M. Fabrizio, Time-dependent mean field theory for quench dynamics in correlated electron systems. Phys. Rev. Lett. **105**, 076401 (2010)
32. S.R. Manmana, S. Wessel, R.M. Noack, A. Muramatsu, Time evolution of correlations in strongly interacting fermions after a quantum quench. Phys. Rev. B **79**, 155104 (2009)
33. M. Rigol, Quantum quenches and thermalization in one-dimensional fermionic systems. Phys. Rev. A **80**, 053607 (2009)
34. M. Rigol, Breakdown of thermalization in finite one-dimensional systems. Phys. Rev. Lett. **103**, 100403 (2009)
35. P. Barmettler, M. Punk, V. Gritsev, E. Demler, E. Altman, Relaxation of antiferromagnetic order in spin-1/2 chains following a quantum quench. Phys. Rev. Lett. **102**, 130603 (2009)
36. P. Barmettler, M. Punk, V. Gritsev, E. Demler, E. Altman, Quantum quenches in the anisotropic spin-$\frac{1}{2}$ Heisenberg chain: different approaches to many-body dynamics far from equilibrium. New J. Phys. **12**, 055017 (2010)
37. M. Srednicki, Chaos and quantum thermalization. Phys. Rev. E **50**, 888 (1994)
38. M. Rigol, V. Dunjko, M. Olshanii, Thermalization and its mechanism for generic isolated quantum systems. Nature **452**, 854 (2008)

39. S. Trotzky, Y.-A. Chen, A. Flesch, I.P. McCulloch, U. Schollwöck, J. Eisert, I. Bloch, Probing the relaxation towards equilibrium in an isolated strongly correlated one-dimensional Bose gas. Nat. Phys. **8**, 325 (2011)
40. T. Lompe, T.B. Ottenstein, F. Serwane, A.N. Wenz, G. Zürn, S. Jochim, Radio-frequency association of Efimov trimers. Science **330**, 940 (2010)
41. H.P. Büchler, G. Blatter, Supersolid versus phase separation in atomic Bose-Fermi mixtures. Phys. Rev. Lett. **91**, 130404 (2003)
42. E. Kim, M.H.W. Chan, Probable observation of a supersolid helium phase. Nature **427**, 225 (2004)
43. I. Titvinidze, M. Snoek, W. Hofstetter, Supersolid Bose-Fermi mixtures in optical lattices. Phys. Rev. Lett. **100**, 100401 (2008)

Chapter 10
Conclusions and Outlook

The common theme of the experiments described in this thesis is the *realization*, *characterization* and *analysis* of many-body quantum states with bosons, fermions and Bose-Fermi mixtures. In all these aspects novel approaches have been pursued to access new physics:

The aspect of *realization* includes a number of technical innovations. The main accomplishment is the implementation of an optical lattice setup that allows for independent tuning of tunneling, interaction and underlying harmonic confinement. This high degree of flexibility is achieved by the combination of a blue-detuned optical lattice and a red-detuned dipole trap, for which a number of new alignment techniques have been developed. Together with Feshbach resonances great freedom is gained in the realization of Hamiltonians, which is utilized in all equilibrium and nonequilibrium measurements presented in this work.

The aspect of *characterization* comprises two points: First, several new techniques for the characterization of optical potentials are introduced, for example, for the calibration of the lattice depth and the measurement of the anticonfinement in a blue-detuned lattice. The resulting calibration data has been cross-checked with an accurate numerical model of the combined lattice and dipole trap setup. Second, the development of quantum revival spectroscopy has enabled the direct measurement of a central ingredient of many-body Hamiltonians, namely the interaction energy of Fock states at single lattice sites. This novel atom optical technique crucially relies on the capability to instantaneously switch from a confined to a homogeneous lattice potential.

The aspect of *analysis* includes novel methods for the observation of many-body states in the lattice. In the case of bosons, established experimental observables for the coherence of a quantum gas (visibility and fraction of coherent atoms) are applied to detect nonequilibrium dynamics and an analogous observable for fermions is introduced. For the identification of equilibrium phases in fermionic spin mixtures an entirely new approach is developed: High-field phase contrast imaging is used to record the in-situ density distribution as a function of the external confinement. From a sequence of images, the compressibility of the lattice quantum gas can be

S. Will, *From Atom Optics to Quantum Simulation*, Springer Theses,
DOI: 10.1007/978-3-642-33633-1_10, © Springer-Verlag Berlin Heidelberg 2013

extracted, which allows to distinguish compressible and incompressible many-body states.

With the above innovations it was possible to gain new insight into lattice quantum gases on several levels: The Fermi-Hubbard model has been implemented with repulsively interacting spin mixtures of ^{40}K (Chap. 6). By monitoring the in-situ density distribution via phase-contrast imaging, the compressibility of the global many-body state has become an experimental observable and revealed the formation of metallic and insulating phases. Most remarkably, the emergence of a Mott insulating state has been detected, which has also been signaled by a suppression of doubly occupied lattice sites in the relevant regime. In fact, the observation of the Mott insulator has been somewhat fortuitous, because the average entropy per particle has been more than a factor of 2 larger than the maximally allowed value in the Mott insulator phase of a homogeneous system. DMFT calculations have shown that the underlying harmonic confinement induces a highly inhomogeneous entropy distribution that favors the formation of insulating phases, because entropy is largely carried by metallic regions. Similar to previous occasions, nature seems to play into the hands of the ultracold atom physicist. The fundamental impact of entropy in finite-sized strongly interacting systems has also become evident in attractively interacting spin mixtures. Here, the formation of spin up-spin down pairs has been shown to reduce the local entropy capacity so drastically that the in-situ cloud size of the mixture is forced to expand despite of strongly attractive interactions between the atoms.

The development of quantum revival spectroscopy and its application to interacting bosonic ^{87}Rb atoms (Chap. 7) has enabled the most precise measurements of atomic interactions in optical lattices to date. With an accuracy on the per cent level, the energies of atomic Fock states have been measured and the data reveals that Hubbard's interaction parameter U is affected by crucial modifications already at moderate interaction strengths. Interactions induce virtual transitions to higher-lying lattice bands, giving rise to characteristic Fock state energies that can be understood in terms of effective multi-body interactions. This finding shows that multi-orbital effects can have a notable influence on the physics in optical lattices and they provide crucial input for the comparison of optical-lattice quantum simulators with many-body quantum theory.

The measurements on equilibrium Bose-Fermi mixtures (Chap. 8) have systematically analyzed the role of interspecies interactions. It has been found that attractive interactions between bosons and fermions shift the bosonic superfluid to Mott insulator transition towards lower lattice depths, contrary to earlier theoretical predictions. The effective deepening of lattice sites for bosons owing to the presence of attractively interacting fermions has been identified as the dominant reason for this behavior. For repulsive interspecies interactions, the position of the Mott insulator transition remains essentially unchanged hinting at rapid phase separation.

By applying quantum revival spectroscopy to an array of few-body Bose-Fermi mixtures on individual lattice sites (Chap. 9), it has been possible to monitor the absolute value of the Bose-Fermi interaction energy as a function of the interspecies scattering length. Additionally, precise measurements of Bose-Bose interactions have shown that their strength indirectly depends on the Bose-Fermi interaction. The modification of bosonic interactions is mediated by the presence of the fermion, which points to effects beyond the single-band Bose-Fermi Hubbard model and substantiates the indirect evidence for renormalized Hubbard parameters in Chap. 8. In addition, a novel method to measure the fermionic filling has been devised that utilizes an interesting interference effect in quantum revival spectroscopy. The results confirm the hints for phase separation obtained in Chap. 8. Finally, a novel evaluation technique has allowed to transfer the principles of quantum revival spectroscopy to the analysis of fermionic momentum distributions. By this, coherent fermionic quantum dynamics after an interaction quench have been observed for the first time.

Perspectives

Due to the diversity of the experiments in this thesis, specific outlooks are given at the end of each chapter. These make direct connections to the contents and point out specific improvements and extensions. Here, I want to take a somewhat broader perspective on future possibilities that are related to the concepts and findings of this thesis.

The future view on ultracold atoms in optical lattices is likely to be shaped by experiments with single-site and single-atom resolution. So far, this level of control has only been achieved in few setups with bosonic ^{87}Rb [1, 2]. Despite a few extra challenges, the realization of single-site resolution with fermionic species (^{6}Li or ^{40}K) will probably only be a matter of time, given the impetus of the research field and the expected potential gain of insight. Nonetheless, several of the concepts that underlie the experiments of this thesis are likely to remain important:

① Quantum simulation.

② Precision measurements of Hamiltonian parameters.

③ Accurate control of lattice potentials.

④ Nonequilibrium physics.

The concept of quantum simulation ① is currently one of the most important driving forces for the research field. The prospect to solve long-standing open questions of condensed-matter physics with the help of ultracold atoms is highly appealing [3]. For example, the key to a detailed understanding of high-temperature superconductivity might be found in the phase diagram of a doped two-dimensional Fermi-Hubbard model [4]. In the case of success, the impact on research and technology would be immense. Progress towards this goal can probably be made by taking the experimental studies of the Fermi-Hubbard model discussed in Chap. 6 to the next levels. The observation of antiferromagnetic ordering in a repulsively interacting fermionic spin mixture would be the next step and open the door to the exploration of quantum magnetism with ultracold atoms. The equivalent achievement on the side of attractive interactions would be the observation of the BCS-BEC crossover in an optical lattice.

On approach to these goals it will be crucial to reach lower entropies. This is likely to require a whole set of measures: First, the minimization of heating rates during the experimental sequence, for example, by reducing collisions with the background gas of the vacuum chamber and by suppression of three-body losses and inelastic photon scattering. Second, novel cooling schemes to reach lower temperatures inside the lattice, for example, via dimple-type trapping geometries or active entropy removal schemes. Those techniques might profit from the inhomogeneous distribution of entropy revealed in the context of this thesis. Third, the identification of intelligent paths through phase space to minimize nonadiabaticities during thermodynamic changes of state, for example, during lattice loading or adjustment of interactions via Feshbach resonances.

Lower entropies will also be crucial for quantum simulation with Bose-Fermi mixtures. The experiments in this thesis have been among the first to study this intriguing hybrid quantum system. In contrast to the Fermi-Hubbard model, the Bose-Fermi Hubbard model is not directly resembled in real solid state systems. Nevertheless, theoretical studies have predicted interesting zero-temperature phases in one- and two-dimensional systems, such as the formation of charge-density waves and the emergence of a supersolid. Furthermore, Bose-Fermi mixtures may allow to investigate polaron physics [5–8] where the fermions take the role of electrons and bosons resemble the phonons. Along these lines, it may further be possible to observe mediated interactions by realizing an optical lattice potential that acts on the fermions, but is invisible to the bosons. Then the bosons can serve as exchange particles and generate an effective interaction between the fermions. Further possibilities include the simulation of disordered Hubbard models [9] or Kondo lattice models [10].

The precise measurement of Hamiltonian parameters is a prerequisite for accurate quantum simulation ②. When it comes to precision measurements of quantum phase transitions, it will be crucial to quantitatively know the ingredients of the Hamiltonian that is realized in the experimental setup. The development and application of quantum revival spectroscopy has been the first precision study of interatomic interactions in optical lattices revealing the importance of multi-orbital effects. Founded on our results, recent investigations have addressed the shift of the superfluid to Mott insulator transition due to effective multi-body interactions [11] as well as multi-orbital corrections to tunneling [12]. Further experiments will be necessary to understand the impact of higher bands on tunneling [13, 14] and interactions [15], particularly for the case of interacting quantum many-body systems. Multi-orbital physics in optical lattices has already become an active field of theoretical investigations [16–18].

On the frontier of multi-body interactions, it would be exciting to observe not only effective, but also intrinsic three- and higher-body interactions. These might be enhanced close to Feshbach or Efimov resonances. Effective and direct multi-body interactions play an important role in several recent predictions of novel strongly correlated quantum phases [19, 20] with topological order [21] or exotic properties [22, 23].

The tunability of our lattice potential has enabled new classes of experiments ③. Enhanced control of optical lattice potentials is likely to stay on the technological forefront. Being based on traditional experimental techniques, optical superlattices form a meaningful extension to the usual simple cubic lattices. So far, they have only been realized in setups with bosonic ^{87}Rb [24–27], but the prospect to use them in connection with fermions is realistic and thrilling. In addition to studies in arrays of isolated double-well systems or plaquettes, the formation of extended many-body states in this nontrivial lattice structure can be explored. For example, the formation of interaction-induced excited band condensates [28], the artificial creation of antiferromagnetic order [29] or the creation of d-wave resonating valence bond states [30, 31] are some of many interesting possibilities.

Optical lattice setups that rely on the interference of Gaussian beams are typically bound to a harmonic confinement and a fixed lattice constant ($a = \lambda/2$ with λ being the wavelength of the lattice laser). More freedom will be gained by the projection of optical dipole potentials [32], particularly when a high-resolution objective is available [1, 2]. Here the lattice structure is not determined by the laser wavelength, but by the projection pattern. This brings potential landscapes with arbitrary geometry (for example, hexagonal lattices [33]) or flexible underlying confinement (for example, a box potential) into reach and it might even be possible to imitate condensed-matter structures beyond the level of idealized models.

The real-time control over optical potentials allows to expose many-body states nonadiabatically to new Hamiltonians ④. This can give rise to unusual quantum states [34] and allow to investigate transport and relaxation phenomena that are typically inaccessible in real quantum systems [35–37]. Experimental studies on the relaxation of bosonic and fermionic many-body states may allow to address fundamental questions of statistical physics, such as the elusive inner workings of thermalization [38–40], and may develop into a fruitful branch of future research [41–43].

In the past fifteen years the field of ultracold atoms has contributed to physics in several different respects. Many experiments have been wonderfully clean and lucid realizations of phenomena that have previously been theoretical textbook examples. Such experimental results now enter the textbooks themselves and have an enormous educational value. In recent years, the field has set out to experimentally observe complex few- and many-body phenomena that are also at the forefront of theoretical research—the results of this thesis make a humble contribution at this stage of the development. Analytical modeling of current observations is often impossible; the numerical validation is computationally demanding and usually complies with measured data. It is probably not too far-fetched to take this as a sign that synthetic ultracold quantum matter is on the verge to finding answers to unsolved and—even more exciting—to so far unasked questions of many-body quantum physics. The "ultracold" future of quantum magnetism, quantum disorder, nonequilibrium physics and much more is lying ahead of us... and it will be more than thrilling to see where the field has moved in a couple of years from now.

References

1. W. Bakr, J. Gillen, A. Peng, S. Fölling, M. Greiner, A quantum gas microscope for detecting single atoms in a Hubbard-regime optical lattice. Nature **462**, 74 (2009)
2. J. Sherson, C. Weitenberg, M. Endres, M. Cheneau, I. Bloch, S. Kuhr, Single-atom-resolved fluorescence imaging of an atomic Mott insulator. Nature **467**, 68 (2010)
3. A. Cho, The mad dash to make light crystals. Science **320**, 312 (2008)
4. P.A. Lee, N. Nagaosa, X.-G. Wen, Doping a Mott insulator: physics of high-temperature superconductivity. Rev. Mod. Phys. **78**, 17 (2006)
5. M. Bruderer, A. Klein, S.R. Clark, D. Jaksch, Polaron physics in optical lattices. Phys. Rev. A **76**, 011605 (2007)
6. M. Bruderer, A. Klein, S.R. Clark, D. Jaksch, Transport of strong-coupling polarons in optical lattices. New J. Phys. **10**, 033015 (2008)
7. M. Bruderer, W. Bao, D. Jaksch, Self-trapping of impurities in Bose–Einstein condensates: strong attractive and repulsive coupling. Europhys. Lett. **82**, 30004 (2008)
8. A. Privitera, W. Hofstetter, Polaronic slowing of fermionic impurities in lattice Bose–Fermi mixtures. Phys. Rev. A **82**, 063614 (2010)
9. K. Byczuk, W. Hofstetter, D. Vollhardt, Competition between Anderson localization and antiferromagnetism in correlated lattice fermion systems with disorder. Phys. Rev. Lett. **102**, 146403 (2009)
10. M. Foss-Feig, A.M. Rey, Phase diagram of the bosonic Kondo–Hubbard model. Phys. Rev. A **84**, 053619 (2011)
11. M.J. Mark, E. Haller, K. Lauber, J.G. Danzl, A.J. Daley, H.-C. Nägerl, Precision measurements on a tunable Mott insulator of ultracold atoms. Phys. Rev. Lett. **107**, 175301 (2011)
12. R. Ma, M.E. Tai, P.M. Preiss, W.S. Bakr, J. Simon, M. Greiner, Photon-assisted tunneling in a biased strongly correlated bose gas. Phys. Rev. Lett. **107**, 095301 (2011)
13. C. Weitenberg, M. Endres, J. Sherson, M. Cheneau, P. Schauß, T. Fukuhara, I. Bloch, S. Kuhr, Single-spin addressing in an atomic Mott insulator. Nature **471**, 319 (2011)
14. J. Heinze, S. Götze, J.S. Krauser, B. Hundt, N. Fläschner, D.-S. Lühmann, C. Becker, K. Sengstock, Multiband spectroscopy of ultracold fermions: observation of reduced tunneling in attractive Bose–Fermi mixtures. Phys. Rev. Lett. **107**, 135303 (2011)
15. P. Ernst, S. Götze, J. Krauser, K. Pyka, D.-S. Lühmann, D. Pfannkuche, K. Sengstock, Probing superfluids in optical lattices by momentum-resolved Bragg spectroscopy. Nat. Phys. **6**, 56 (2009)
16. D.-S. Lühmann, K. Bongs, D. Pfannkuche, Excitation spectrum of Mott shells in optical lattices. J. Phys. B **42**, 145305 (2009)
17. O. Dutta, A. Eckardt, P. Hauke, B. Malomed, M. Lewenstein, Bose–Hubbard model with occupation-dependent parameters. New J. Phys. **13**, 023019 (2011)
18. D.-S. Lühmann, O. Jürgensen, K. Sengstock, Multi-orbital and density-induced tunneling of bosons in optical lattices. New J. Phys. **14**, 033021 (2012)
19. O.E. Alon, A.I. Streltsov, L.S. Cederbaum, Zoo of quantum phases and excitations of cold bosonic atoms in optical lattices. Phys. Rev. Lett. **95**, 030405 (2005)
20. K.R.A. Hazzard, E.J. Mueller, On-site correlations in optical lattices: band mixing to coupled quantum hall puddles. Phys. Rev. A **81**, 031602 (2010)
21. B. Paredes, T. Keilmann, J.I. Cirac, Pfaffian-like ground state for three-body hard-core bosons in one-dimensional lattices. Phys. Rev. A **75**, 053611 (2007)
22. B. Capogrosso-Sansone, S. Wessel, H.P. Büchler, P. Zoller, G. Pupillo, Phase diagram of one-dimensional hard-core bosons with three-body interactions. Phys. Rev. B **79**, 020503 (2009)
23. S. Pielawa, T. Kitagawa, E. Berg, S. Sachdev, Correlated phases of bosons in tilted frustrated lattices. Phys. Rev. B **83**, 205135 (2011)
24. J. Sebby-Strabley, B.L. Brown, M. Anderlini, P.J. Lee, W.D. Phillips, J.V. Porto, P.R. Johnson, Preparing and probing atomic number states with an atom interferometer. Phys. Rev. Lett. **98**, 200405 (2007)

25. M. Anderlini, P. Lee, B. Brown, J. Sebby-Strabley, W. Phillips, J. Porto, Controlled exchange interaction between pairs of neutral atoms in an optical lattice. Nature **448**, 452 (2007)
26. S. Fölling, S. Trotzky, P. Cheinet, M. Feld, R. Saers, A. Widera, T. Müller, I. Bloch, Direct observation of second-order atom tunnelling. Nature **448**, 1029 (2007)
27. S. Trotzky, P. Cheinet, S. Fölling, M. Feld, U. Schnorrberger, A.M. Rey, A. Polkovnikov, E.A. Demler, M.D. Lukin, I. Bloch, Time-resolved observation and control of superexchange interactions with ultracold atoms in optical lattices. Science **319**, 295 (2008)
28. Q. Zhou, J.V. Porto, S. Das Sarma, Interaction-induced excited-band condensate in a double-well optical lattice. Phys. Rev. A **84**, 031607 (2011)
29. M. Lubasch, V. Murg, U. Schneider, J.I. Cirac, M.-C. Bañuls, Adiabatic preparation of a Heisenberg antiferromagnet using an optical superlattice. Phys. Rev. Lett. **107**, 165301 (2011)
30. S. Trebst, U. Schollwöck, M. Troyer, P. Zoller, d-Wave resonating valence bond states of fermionic atoms in optical lattices. Phys. Rev. Lett. **96**, 250402 (2006)
31. A.M. Rey, R. Sensarma, S. Fölling, M. Greiner, E. Demler, M.D. Lukin, Controlled preparation and detection of d-wave superfluidity in two-dimensional optical superlattices. Europhys. Lett. **87**, 60001 (2009)
32. B. Zimmermann, T. Müller, J. Meineke, T. Esslinger, H. Moritz, High-resolution imaging of ultracold fermions in microscopically tailored optical potentials. New J. Phys. **13**, 043007 (2010)
33. P. Soltan-Panahi, J. Struck, P. Hauke, A. Bick, W. Plenkers, G. Meineke, C. Becker, P. Windpassinger, M. Lewenstein, K. Sengstock, Multi-component quantum gases in spin-dependent hexagonal lattices. Nat. Phys. **7**, 434 (2011)
34. L.E. Sadler, J.M. Higbie, S.R. Leslie, M. Vengalattore, D.M. Stamper-Kurn, Spontaneous symmetry breaking in a quenched ferromagnetic spinor Bose–Einstein condensate. Nature **443**, 312 (2006)
35. U. Schneider, L. Hackermüller, J.P. Ronzheimer, S. Will, S. Braun, T. Best, I. Bloch, E. Demler, S. Mandt, D. Rasch, A. Rosch, Fermionic transport and out-of-equilibrium dynamics in a homogeneous Hubbard model with ultracold atoms. Nat. Phys. **8**, 213 (2012)
36. S. Trotzky, Y.-A. Chen, A. Flesch, I.P. McCulloch, U. Schollwöck, J. Eisert, I. Bloch, Probing the relaxation towards equilibrium in an isolated strongly correlated one-dimensional Bose gas. Nat. Phys. **8**, 325 (2011)
37. D. Chen, M. White, C. Borries, B. DeMarco, Quantum quench of an atomic Mott insulator. Phys. Rev. Lett. **106**, 235304 (2011)
38. M. Srednicki, Chaos and quantum thermalization. Phys. Rev. E **50**, 888 (1994)
39. M. Rigol, V. Dunjko, M. Olshanii, Thermalization and its mechanism for generic isolated quantum systems. Nature **452**, 854 (2008)
40. I.E. Mazets, T. Schumm, J. Schmiedmayer, Breakdown of integrability in a quasi-1D ultracold bosonic gas. Phys. Rev. Lett. **100**, 210403 (2008)
41. C. Kollath, A.M. Läuchli, E. Altman, Quench dynamics and nonequilibrium phase diagram of the Bose–Hubbard model. Phys. Rev. Lett. **98**, 180601 (2007)
42. S.R. Manmana, S. Wessel, R.M. Noack, A. Muramatsu, Strongly correlated fermions after a quantum quench. Phys. Rev. Lett. **98**, 210405 (2007)
43. M. Rigol, Quantum quenches and thermalization in one-dimensional fermionic systems. Phys. Rev. A **80**, 053607 (2009)

Appendix A
Atomic Data

A.1 Potassium Data

^{40}K

Natural abundance		0.0117 %
Mass	m	39.963 999 amu
Nuclear spin	I	4
Electron gyromagnetic ratio ($4\,^2S_{1/2}$)	g_J	2.002 294 21(24) [1]
Electron gyromagnetic ratio ($4\,^2P_{3/2}$) (theoretical)	g_J	4/3
Nuclear gyromagnetic ratio	g_I	+0.000 176 490(34) [1]
Magnetic dipole constant ($4\,^2S_{1/2}$)	A	−285.7308(24) MHz [1]
Magnetic dipole constant ($4\,^2P_{3/2}$)	A	−7.59(6) MHz [1]
Ground state hyperfine splitting	ν_{hfs}	1285.790(7) MHz [1]
Vacuum wavelength D_1-transition	λ_{D1}	770.108 136 5(2) nm [2]
Vacuum wavelength D_2-transition	λ_{D2}	766.700 674 7(3) nm [2]
Line width D_1-transition (^{39}K)	Γ_{D1}	$2\pi \times 5.96(1)$ MHz [3]
Line width D_2-transition (^{39}K)	Γ_{D2}	$2\pi \times 6.04(1)$ MHz [3]
Life time $\lvert 4\,^2P_{1/2}\rangle$ (^{39}K)		26.70 ns
Life time $\lvert 4\,^2P_{3/2}\rangle$ (^{39}K)		26.35 ns
Absorption cross section on D_2 transition ($\sigma^{\pm}$ pol.)	σ_0	$2.8067 \times 10^{-9}\,\mathrm{cm}^2$
Saturation intensity on D_2 transition ($\sigma^{\pm}$ pol.)	I_{sat}	$1.752(3)\,\mathrm{mW/cm}^2$
Triplet scattering length	a_{T}	$+104\,a_0$ [4]
Singlet scattering length	a_{S}	$+174\,a_0$ [4]
Melting point	T_{m}	63 °C
Boiling point	T_{b}	759 °C
Vapor pressure at 25 °C	p_{v}	2.4×10^{-7} mbar

S. Will, *From Atom Optics to Quantum Simulation*, Springer Theses,
DOI: 10.1007/978-3-642-33633-1,

A.2 Rubidium Data

^{87}Rb

Natural abundance		27.835 %
Mass	m	86.909 180 amu
Nuclear spin	I	3/2
Electron gyromagnetic ratio ($5\,^2S_{1/2}$)	g_J	2.002 331 13(20) [1]
Nuclear gyromagnetic ratio	g_I	−0.000 995 141 4(10) [1]
Magnetic dipole constant ($5\,^2S_{1/2}$)	A	+3417.341 306 42(15) MHz [1]
Ground state hyperfine splitting	ν_{hfs}	6834.682 610 904 29(9) Hz [5]
Vacuum wavelength D_1-transition	λ_{D1}	794.978 851 156(23) nm [6]
Vacuum wavelength D_2-transition	λ_{D2}	780.241 209 686(13) nm [6]
Line width D_1-transition	Γ_{D1}	$2\pi \times 5.746(8)$ MHz [6]
Line width D_2-transition	Γ_{D2}	$2\pi \times 6.065(9)$ MHz [6]
Life time ($5\,^2P_{1/2}$)		27.70 ns
Life time ($5\,^2P_{3/2}$)		26.24 ns
Absorption cross section on D_2 transition ($\sigma^\pm$ pol.)	σ_0	$2.906\,693 \times 10^{-9}$ cm^2 [6]
Saturation intensity on D_2 transition ($\sigma^\pm$ pol.)	I_{sat}	1.669 33(35) mW/cm^2 [6]
Triplet scattering length	a_{T}	+98.98(4) a_0 [7]
Singlet scattering length	a_{S}	+90.4(2) a_0 [7]
$\lvert F = 1, m_F = +1\rangle$ scattering length	a_s	+102(2) a_0 [7, 8]
Three-body loss rate	K_3	$5.8(1.9) \times 10^{-29}$ cm^6/s [9]
Melting point	T_{m}	39 °C
Boiling point	T_{b}	688 °C
Vapor pressure at 25 °C	p_{v}	$5.22(27) \times 10^{-7}$ mbar

Appendix B
^{87}Rb and ^{40}K in the Presence of Magnetic Fields

In this Appendix we discuss the Zeeman shifts of ^{87}Rb and ^{40}K in the presence of external magnetic fields. On the one hand, Zeeman shifts are experimentally important, because they are used to accurately calibrate magnetic fields. On the other hand, in this thesis they have particular relevance for high-field phase contrast imaging in the experiments of Chap. 6. Here not only the shifts in the ground state ($4\,^2S_{1/2}$), but also the excited state ($4\,^2P_{3/2}$) of ^{40}K must be considered.

B.3 Formulas

Zeeman shifts can be calculated analytically for the case of electronic angular momentum $J = 1/2$ (implying a total angular momentum $F = I \pm 1/2$) [10]. The analytic solution is given by the Breit-Rabi formula [11]:

$$E_{\mathrm{BR}}(B) = -\frac{\Delta E_{\mathrm{hfs}}}{2(2I+1)} + g_I m_F \mu_B B \pm \frac{\Delta E_{\mathrm{hfs}}}{2} \begin{cases} 1 - x\,\mathrm{sgn}(A) & -m_F\,\mathrm{sgn}(A) = I + 1/2, \\ \sqrt{1 + \frac{4 m_F x}{2I+1} + x^2} & \text{otherwise,} \end{cases} \tag{B.1}$$

where the dimensionless variable x is related to the electronic and the nuclear Landé g-factors, g_J and g_I, according to

$$x = \frac{g_J - g_I}{\Delta E_{\mathrm{hfs}}} \mu_B B. \tag{B.2}$$

The hyperfine splitting $\Delta E_{\mathrm{hfs}} = hA(I + 1/2)$ depends on the magnetic dipole constant A that is typically given in units of a frequency. The atomic data needed to perform the calculations for ^{87}Rb and ^{40}K are provided in Appendix A.

S. Will, *From Atom Optics to Quantum Simulation*, Springer Theses,
DOI: 10.1007/978-3-642-33633-1,

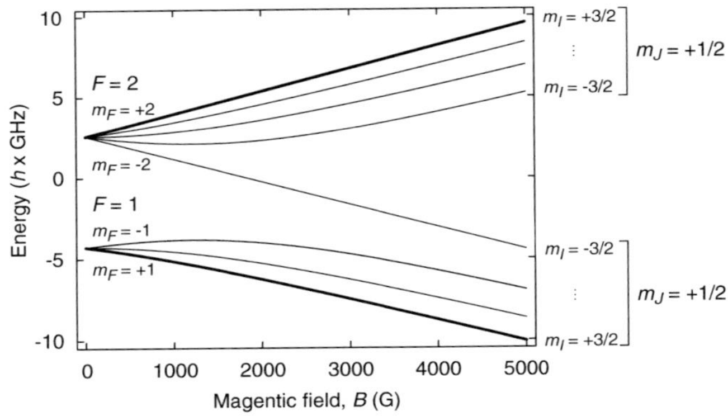

Fig. B.1 Zeeman shifts for ^{87}Rb in the $5^2S_{1/2}$ manifold. In the experiments of this thesis, only the magnetically trappable state $|2, +2\rangle$ and the absolute ground state $|1, +1\rangle$ are used (*bold lines*). The Paschen-Back regime is only reached well above 1000 G, beyond the experimentally relevant range

For high magnetic fields the coupling between the nuclear and the electronic angular momentum becomes negligible. In this case, which is referred to as the Paschen-Back regime, both I and J are quantized along the direction of the magnetic field and m_J and m_I are good quantum numbers. The energies of the Zeeman states are obtained by

$$E_{\mathrm{PB}}(B) = hAm_I m_J + (g_J m_J - g_I m_I)\mu_B B. \tag{B.3}$$

B.4 Rubidium

Figure B.1 shows the Zeeman shifts in the ^{87}Rb ground state manifold $5^2S_{1/2}$ calculated with the Breit-Rabi formula. The Paschen-Back regime is not reached for experimentally relevant magnetic fields below 1000 G. The Breit-Rabi formula offers a convenient way to precisely calibrate magnetic fields: At a fixed current in the coils the transition frequency between Zeeman sub-states in an ultracold sample is measured via radio frequency or microwave spectroscopy. Inversion of Eq. B.1 yields the corresponding magnetic field.

B.5 Potassium

In the experiments of this thesis potassium is used both in fermionic spin mixtures ($|F, m_F\rangle = |9/2, -9/2\rangle$ and $|9/2, -7/2\rangle$) and Bose-Fermi mixtures ($|9/2, -9/2\rangle$ with rubidium in state $|1, +1\rangle$). The interactions are varied using Feshbach resonances

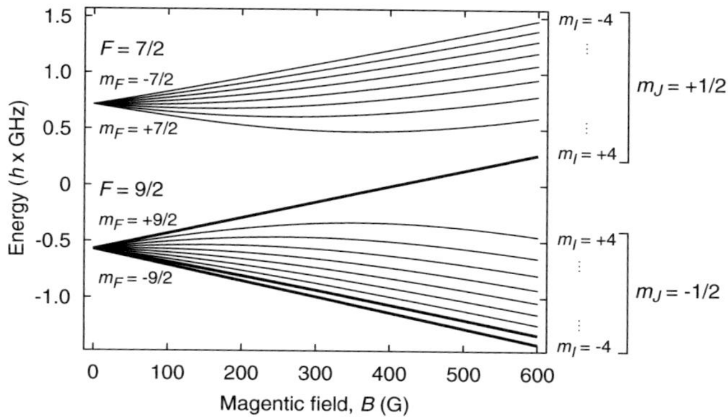

Fig. B.2 Zeeman shifts for ^{40}K in the $4^2S_{1/2}$ manifold. Mainly the magnetically trappable $|9/2, +9/2\rangle$ state and the two lowest lying states $|9/2, -7/2\rangle$ and $|9/2, -9/2\rangle$ are used in this work (*bold lines*). The Paschen-Back approximation is not applicable for the experimentally relevant magnetic fields below 600 G

Table B.1 Shifts of the ^{40}K spin states $|9/2, -9/2\rangle$ (third column) and $|9/2, -7/2\rangle$ (fourth column) at the relevant working points (compare Sect. 2.3.5)

	B (G)	$\Delta_{e1} - \Delta_{g1}$ (MHz)	$\Delta_{e2} - \Delta_{g2}$ (MHz)
①	209.9	−293	−328
②	215.8	−301	−337
③	258.7	−361	−403
④	198.8	−277	−310

All transitions shift towards lower frequencies. Because the imaging laser is tuned to resonance at zero field, it is blue-detuned with respect to the shifted transition

located at around 200 G in the first, and around 550 G in the second case (see Sect. 2.3.5).

In the experiments of Chap. 6 high-field phase contrast imaging is used to image the spin mixture. The laser beam is tuned to resonance at zero field, but the image is taken at a large field in the vicinity of the Feshbach resonance. In order to derive the effective imaging detuning, both the shift in the ground state manifold $4^2S_{1/2}$ and the excited state manifold $4^2P_{3/2}$ are to be taken considered. The ground state manifold can be treated with the Breit-Rabi formula (see Fig. B.2). This is not possible for the excited state because of $J = 3/2$, but numerical diagonalization of the Hamiltonian shows that the Paschen-Back approximation is reasonably accurate already above 100 G [12].

The detunings for high-field phase contrast imaging at the relevant working points are summarized in Table B.1 (see Fig. B.3 for the corresponding level scheme). In the Paschen-Back regime a cycling transition between $|F, m_F\rangle = |9/2, -9/2\rangle$ and $|11/2, -11/2\rangle$ (corresponding to $|m_J, m_I\rangle = |-1/2, -4\rangle$ and $|-3/2, -4\rangle$ in high field) can be driven according to the selection rule $\Delta m_I = 0$, $\Delta m_J = -1$. Accordingly, the $|9/2, -7/2\rangle$ state has a cycling transition with $|11/2, -9/2\rangle$ (corresponding to $|-1/2, -3\rangle$ and $|-3/2, -3\rangle$ in high field).

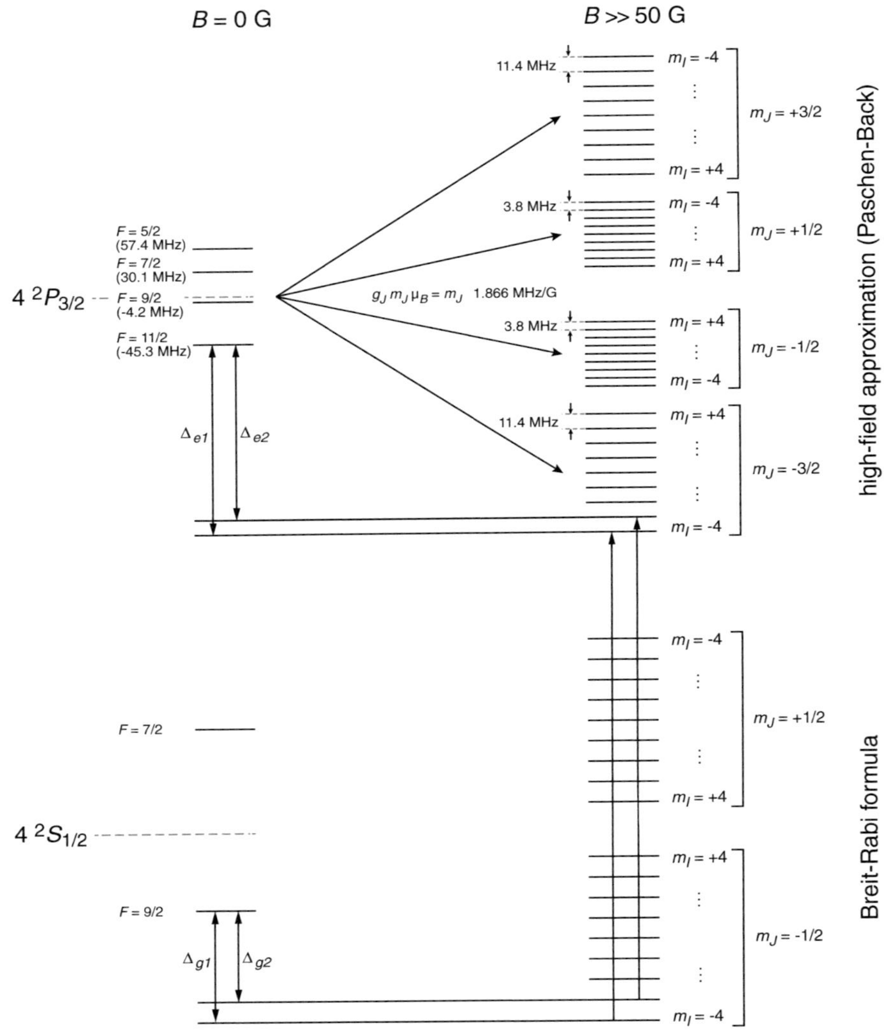

Fig. B.3 Shifts in the potassium $4^2S_{1/2}$ and $4^2P_{3/2}$ manifolds at high magnetic fields that are relevant to high-field phase contrast imaging. The ground state Δ_{g1} and Δ_{g2} are obtained using the Breit-Rabi formula, while the excited state allows to use the Paschen-Back formula (B.3) to calculate Δ_{e1} and Δ_{e2}

Appendix C
Thermometry of Harmonically Trapped Fermi Gases

We determine the temperature of quantum degenerate Fermi gases by fitting a Fermi-Dirac distribution to the recorded atom clouds. The fit model is based on the three-dimensional (3D) Thomas-Fermi distributions that are derived in Sect. 2.1.3. Imaging, however, integrates along the line-of-sight and the image shows a column density. Therefore, the general Thomas-Fermi 3D distribution (Eq. 2.27) must be integrated along one axis, for which the relation [13]

$$\int_{-\infty}^{\infty} dx\, \mathrm{Li}_n(ze^{-x^2}) = \sqrt{\pi}\, \mathrm{Li}_{n+1/2}(z) \tag{C.1}$$

is helpful. Equations 2.28 and 2.29 show the Thomas-Fermi distribution in the two limiting cases of infinite ($T \to \infty$) and zero ($T \to 0$) temperature. Ideally, a fit model should capture both limits equally. We use an interpolation between the cloud sizes $\sigma_\alpha = \sqrt{\frac{2k_B T}{m\omega_\alpha^2}}$ and $R_{\mathrm{F}\alpha} = \sqrt{\frac{2\epsilon_\mathrm{F}}{m\omega_\alpha^2}}$ ($\alpha = x, y, z$) as suggested in Ref. [13]:

$$R_\alpha^2 = \frac{2k_B T}{m\omega_\alpha^2} f(z) \to \begin{cases} \sigma_\alpha, & T \to \infty \\ R_{\mathrm{F}\alpha}, & T \to 0 \end{cases} \quad \text{with} \quad f(z) = \frac{1+z}{z} \ln(1+z)\ , \tag{C.2}$$

where $z = e^{\mu\beta}$ is the fugacity, which determines the shape of the cloud and, consequently, the degree of degeneracy.

The above relations and Eq. 2.27 yield the fit model

$$n_\mathrm{F}^{2\mathrm{D}}(x, y) = n_{\mathrm{F},0}^{2\mathrm{D}} \frac{\mathrm{Li}_2\left(-z \exp\left[-\left(\frac{x^2}{R_x^2} + \frac{y^2}{R_y^2}\right) f(z)\right]\right)}{\mathrm{Li}_2(-z)} \tag{C.3}$$

for the two-dimensional column density (see Fig. 6.4a) and

S. Will, *From Atom Optics to Quantum Simulation*, Springer Theses,
DOI: 10.1007/978-3-642-33633-1, © Springer-Verlag Berlin Heidelberg 2013

$$n_{\mathrm{F}}^{\mathrm{2D,az}}(x) = n_{\mathrm{F},0}^{\mathrm{2D,az}} \frac{\mathrm{Li}_2\left(-z \exp\left[-\frac{x^2}{R_x^2} f(z)\right]\right)}{\mathrm{Li}_2(-z)} \tag{C.4}$$

for azimuthally averaged data (see Fig. 6.4b). These profiles refer to the in-situ density distribution. However, for ballistic expansion of a noninteracting cloud the sizes simply scale as [13, 14]

$$R_\alpha \to R_\alpha b_\alpha(t) = R_\alpha \sqrt{1 + (\omega_\alpha t)^2}, \tag{C.5}$$

where t is the duration of time-of-flight expansion. When the conditions for ballistic expansion are fulfilled, the above models can be applied to time-of-flight images without any changes. The resulting fugacity z is converted to the dimensionless temperature via

$$\frac{T}{T_{\mathrm{F}}} = \left[-6\,\mathrm{Li}_3(-z)\right]^{-1/3}, \tag{C.6}$$

which is readily derived by combining Eqs. 2.23 and 2.24.

To ensure a reliable fitting routine we have done the following: First, before carrying out the Fermi-Dirac fit, the data is fitted by a Gaussian. This helps to find good starting conditions (cloud position, cloud size, peak optical density). Second, when azimuthally averaged data is fitted, the pixels in the cloud center, which are noisy due to poor averaging, are excluded. Third, $\ln(z)$ is used as a fit parameter instead of the fugacity z. This point is essential, particularly, when cold clouds are concerned: The fugacity increases by more than three orders of magnitude, when the temperature decreases from $T/T_{\mathrm{F}} = 0.2$ to 0.1, whereas $\ln(z)$ only rises from 4 to 10. The numerical fitting routine yields more reliable results when the prospective dynamic range of the fit parameters is limited.

Given the typical cloud sizes and the imaging noise in our setup, the reliability limit of thermometry via Fermi-Dirac fits is reached for temperatures around $T/T_{\mathrm{F}} \approx 0.10$. The differences in the wings of the cloud, which form the basis of this thermometry method, become too marginal for lower temperatures. It is interesting to note that typical imperfections of the imaging (out of focus imaging, finite resolution etc.) wash out the characteristics of the Fermi-Dirac distribution and let it appear more Gaussian. Therefore, temperatures T/T_{F} that are measured on the basis of the cloud shape usually constitute a conservative upper bound.

Appendix D
Fourier Analysis of Quantum Phase Revival Traces

The long quantum phase revival traces discussed in Chaps. 7 and 9 are typically recorded in steps of 40 μs with 175 ± 25 points per trace (see Fig. D.1a). The origin of the time axis ($t = 0$) corresponds to the beginning of the lattice ramp from V_L to V_H. Given the total duration of the recorded dynamics of $7\pm$ 1 ms we obtain a Fourier limit on the individual frequency components between 150 ± 25 Hz. In order to increase this resolution by a factor of two and to smoothen the Fourier spectrum, we have devised the following processing scheme:

- The raw data points are interpolated using cubic splines. Typically, the first 30 μs of the interpolated trace are removed to account for the phase evolution that already starts slowly during the 50 μs ramp from V_L to V_H. Otherwise the stretched initial collapse and revival cycle can lead to a distortion of the frequency analysis.
- The damping time scale of the trace is determined by fitting an exponential decay. According to the time scale obtained in the fit, a long exponential decay is smoothly appended to the data for times longer than the original observation time.
- The resulting curve is concatenated to its mirror image, which is obtained upon exchanging times t by $-t$. This smooth trace with doubled length is again sampled in steps of 40 μs (see Fig. D.1b) and a numerical Fourier analysis is performed (see Fig. D.1c).

The advantage of this scheme is two-fold: First, by exploiting the knowledge of the initial phase of the quantum phase revivals [$|\langle\hat{a}\rangle(t)|^2$ is known to start with a maximum at $t = 0$] the size of the data set is legitimately doubled and thereby the Fourier limit is reduced to 75 ± 15 Hz. Second, the Fourier spectrum has a smooth envelope and spectral artifacts due to sharp cut-offs are efficiently avoided.The improvements resulting from the processing scheme can be directly seen by comparing Fig. D.1c to the Fig. D.1d, the later being the direct Fourier transformation of the raw data (Fig. D.1a).

S. Will, *From Atom Optics to Quantum Simulation*, Springer Theses,
DOI: 10.1007/978-3-642-33633-1, © Springer-Verlag Berlin Heidelberg 2013

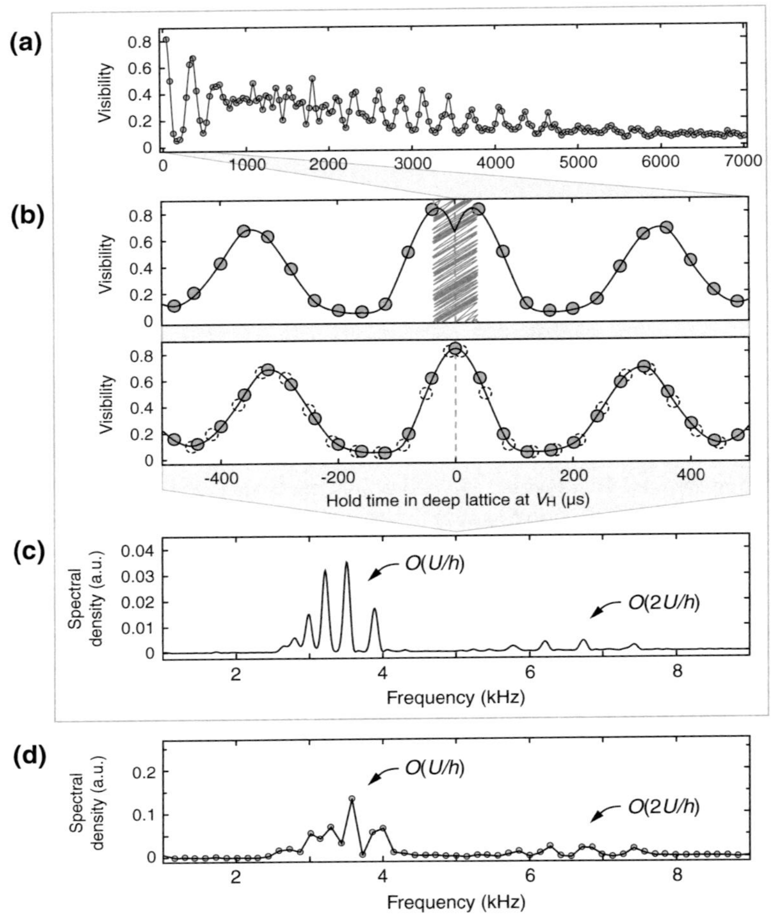

Fig. D.1 Processing of quantum phase revival traces and numerical Fourier analysis. **a** The measured data points, which are typically spaced by 40 μs, are interpolated using cubic splines (*solid line*). **b** Typically, the first 30 μs of the interpolated curve are removed to compensate for the 50 μs ramp from V_L to V_H (*crossed-out area*). Furthermore, a simple exponential decay is appended to the data for times longer than the original observation times (see text). The resulting long interpolation curve is concatenated to its mirror image and the resulting composite curve is again sampled in steps of 40 μs (*solid circles*). The positions of the original data points are shown as *dashed circles*. **c** Numerical Fourier transform is applied to these data, revealing frequencies of orders U/h and $2U/h$. **d** Direct Fourier transform of the data in **a** without further processing

Appendix E
Theory on Bose-Fermi Quantum Phase Revivals

This Appendix provides additional theoretical background on quantum revival spectroscopy in Bose-Fermi mixtures (see Chap. 9). In order to observe quantum phase revivals it is necessary that the bosonic component shows on-site number fluctuations. Therefore, the (nonadiabatic) preparation sequence needs to ensure that the bosons are delocalized in the deep lattice V_{H}. However, for the fermionic component one can think of two limiting scenarios that leave the same signature in the dynamical evolution of the bosons, which we will derive below. In the first scenario (see Fig. E.1a), the fermions are delocalized as well, which is assumed in Chap. 9. In the second scenario (see Fig. E.1b), fermions are localized, but not every lattice site with bosons is occupied by a fermion.

For both scenarios we derive the implications for the bosonic time evolution that is contained in the momentum distribution (see Sect. 7.2.2)

$$n(\mathbf{k}, t) = |w(\mathbf{k})|^2 \sum_{ij} e^{-i\mathbf{k}\cdot(\mathbf{r}_i - \mathbf{r}_j)} \langle \hat{a}_i^\dagger \hat{a}_j \rangle(t), \tag{E.1}$$

assuming a homogeneous lattice with M sites. The coordinate $\mathbf{r}_i = a\mathbf{n}_i$ denotes the location of lattice site i, where $a = \lambda/2$ is the lattice constant and $\mathbf{n}_i \in \mathbb{Z}^3$, and the indices i and j run over all sites. The bosonic momentum distribution $n(\mathbf{k}, t)$ is experimentally observed in time-of-flight imaging after variable hold times t. The dynamics are determined by the expectation value $\langle \hat{a}_i^\dagger \hat{a}_j \rangle(t)$ that is taken with respect to the many-body quantum state $|\Psi(t)\rangle$. Following the arguments of Sect. 7.2, we assume that the global many-body state can be approximated by a Gutzwiller-type product of decoupled on-site states

$$|\Psi(t)\rangle = \prod_{i=1}^{M} |\psi_i(t)\rangle. \tag{E.2}$$

S. Will, *From Atom Optics to Quantum Simulation*, Springer Theses,
DOI: 10.1007/978-3-642-33633-1,

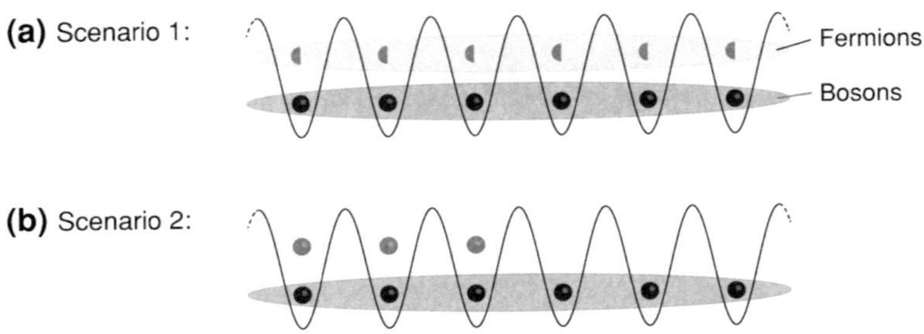

Fig. E.1 Schematic illustrations of the scenarios for the distribution of the fermions discussed in the main text. The *shaded ovals* indicate delocalization

E.1 Scenario 1

In the first scenario, the fermions are delocalized. This means there is a finite probability of finding zero or one fermion at each site. Higher occupations are not possible due to Pauli's exclusion principle. Therefore, we write the on-site state as a superposition of bosonic Fock states without and with a fermion:

$$|\psi_i\rangle = \sum_{n=0}^{\infty} c_n^i e^{-iE_{n,0}t/\hbar}|n, 0\rangle + d_n^i e^{-iE_{n,1}t/\hbar}|n, 1\rangle. \tag{E.3}$$

The eigenenergies $E_{n,m}$ (see Eq. 9.2) are assumed to be identical for each lattice site, because the lattice depth is typically rather uniform across the extent of the atomic clouds. Additionally, we assume identical atom number statistics on each lattice site such that $c_n^i = c_n$ and $d_n^i = d_n$ (compare Sect. 7.2).

We focus on the dynamics of $n(\mathbf{k}, t)$ at the position of constructive interference, where the interference term $e^{-i\mathbf{k}\cdot(\mathbf{r}_i-\mathbf{r}_j)}$ is unity [corresponding to $\mathbf{k} = (2\pi/a)\mathbf{n}$ with $\mathbf{n} \in \mathbb{Z}^3$ in a three-dimensional simple cubic optical lattice]. The dynamics at the interference peak are determined by the time-evolution of the correlations between sites i and j:

$$\begin{aligned}\sum_{ij}^{M}\langle\hat{a}_i^\dagger\hat{a}_j\rangle(t) &= \sum_{i=1}^{M}\langle\psi_i(t)|\hat{a}_i^\dagger\hat{a}_i|\psi_i(t)\rangle + \sum_{i\neq j}^{M}\langle\psi_i(t)|\hat{a}_i^\dagger|\psi_i(t)\rangle\cdot\langle\psi_j(t)|\hat{a}_j|\psi_j(t)\rangle,\\ &= M\langle\hat{a}^\dagger\hat{a}\rangle + (M-1)^2|\langle\hat{a}\rangle(t)|^2,\end{aligned} \tag{E.4}$$

where the sum is split into diagonal and off-diagonal terms. Equating this expression further using Eqs. E.2 and E.3 yields for the first summand

$$M\langle\hat{a}^\dagger\hat{a}\rangle = M\sum_{n=1}^{\infty} n(|c_n|^2 + |d_n|^2) \tag{E.5}$$

and the factors of the second summand read

$$\langle \hat{a} \rangle(t) = \sum_n \sqrt{n+1} \left(c_n^* c_{n+1} \, e^{-i(E_{n+1,0}-E_{n,0})t/\hbar} + d_n^* d_{n+1} \, e^{-i(E_{n+1,1}-E_{n,1})t/\hbar} \right), \tag{E.6}$$

which motivates the abbreviations for the purely bosonic contribution

$$C_n(t) = \sqrt{n+1} \, c_n^* c_{n+1} \, e^{-i(E_{n+1,0}-E_{n,0})t/\hbar} \tag{E.7}$$

and the Bose-Fermi contribution

$$D_n(t) = \sqrt{n+1} \, d_n^* d_{n+1} \, e^{-i(E_{n+1,1}-E_{n,1})t/\hbar} \tag{E.8}$$

already introduced in Chap. 9 (see Eq. 9.3). Hence, Eq. E.4 can be cast in the form

$$\begin{aligned}
\sum_{ij}^{M} \langle \hat{a}_i^\dagger \hat{a}_j \rangle(t) &= M \sum_{n=1}^{\infty} n \left(|c_n|^2 + |d_n|^2 \right) + (M-1)^2 \left| \sum_{n=0}^{\infty} C_n(t) + D_n(t) \right|^2 \\
&\approx (M-1)^2 \left| \sum_{n=0}^{\infty} C_n(t) + D_n(t) \right|^2 \\
&= (M-1)^2 \sum_{n,\tilde{n}}^{\infty} C_{\tilde{n}}^*(t) C_n(t) + D_{\tilde{n}}^*(t) D_n(t) + C_n(t) D_{\tilde{n}}^*(t) \\
&\quad + C_{\tilde{n}}^*(t) D_n(t),
\end{aligned} \tag{E.9}$$

The summand representing the diagonal terms is neglected in the second step. It only scales as $\propto M$ compared to $\propto M^2$ for the off-diagonal terms. It does not play a role for large M.

In the following, we assume that the presence of a fermion does not notably modify the bosonic atom number statistics, which is justified in the limit of small interspecies interactions (see also Sect. 9.1). Accordingly, we introduce the replacement $c_n \rightarrow \sqrt{1-\bar{m}}\, c_n$ and $d_n \rightarrow \sqrt{\bar{m}}\, c_n$. Here $\bar{m} = \langle \hat{m} \rangle$ is the mean fermionic filling that is identical to the probability of finding a fermion at a lattice site (Note that m is either 0 or 1). With the additional assumption $c_n \in \mathbb{R}$, we obtain $C_n \rightarrow (1-\bar{m})\, C_n$ and $D_n \rightarrow \bar{m}\, C_n$, where the abbreviations $C_n \equiv C_n(0)$ and $D_n \equiv D_n(0)$ are used.

Suppression of Revivals

In order to specify the derivation further, we insert the single-orbital eigenenergies following to the Bose-Fermi Hubbard model (see Sect. 9.1)

$$E_{n,m} = \frac{U^{\mathrm{BB}}}{2} n(n-1) + U^{\mathrm{BF}} n\, m. \tag{E.10}$$

This yields for the approximate time-evolution of the correlations

$$\begin{aligned}\sum_{ij}^{M} \langle \hat{a}_i^\dagger \hat{a}_j \rangle(t) \approx (M-1)^2 \sum_{n,\tilde{n}}^{\infty} & C_{\tilde{n}} C_n \left[(1-\bar{m})^2 + \bar{m}^2 \right] e^{-iU^{\mathrm{BB}}(n-\tilde{n})t/\hbar} \\ + \;& C_{\tilde{n}} C_n (1-\bar{m})\, \bar{m}\, e^{-i(U^{\mathrm{BB}}(n-\tilde{n}) - U^{\mathrm{BF}})t/\hbar} \\ + \;& C_{\tilde{n}} C_n (1-\bar{m})\, \bar{m}\, e^{+i(U^{\mathrm{BB}}(\tilde{n}-n) - U^{\mathrm{BF}})t/\hbar}. \end{aligned} \tag{E.11}$$

Without making any further assumptions on the exact atom number statistics of the bosonic component (i.e. the coefficients c_n and C_n) the scaling of the first revival as a function of the fermionic filling $\bar{m}$ can be derived for $U^{\mathrm{BF}}/U^{\mathrm{BB}} = z + 0.5$ ($z \in \mathbb{Z}$). Inserting this condition and the first revival time $t_{\mathrm{rev}} = \hbar/U^{\mathrm{BB}}$ we get

$$\begin{aligned}\sum_{ij}^{M} \langle \hat{a}_i^\dagger \hat{a}_j \rangle(t) &= (M-1)^2 \sum_{n,\tilde{n}}^{\infty} C_{\tilde{n}} C_n \big[(1-\bar{m})^2 + \bar{m}^2 \big] - 2 C_{\tilde{n}} C_n (1-\bar{m})\bar{m} \\ &= (M-1)^2 (1-2\bar{m})^2 \sum_{n,\tilde{n}}^{\infty} C_{\tilde{n}} C_n, \end{aligned} \tag{E.12}$$

showing the $s = (1-2\bar{m})^2$ scaling of the suppression factor discussed in Sect. 9.1.

Coherent States

In the limit of small interactions, the bosonic coefficients can be chosen to correspond to coherent states $c_n = e^{-\bar{n}/2} \alpha^n / \sqrt{n!}$, where $\bar{n}$ is the mean bosonic filling. This gives the dynamical evolution in the single-orbital coherent state model

$$\sum_{ij}^{M} \langle \hat{a}_i^\dagger \hat{a}_j \rangle(t) \approx (N-1)^2 \bar{n}\, e^{2\bar{n}(\cos(U^{\mathrm{BB}} t/\hbar) - 1)} \times \{1 - 2\bar{m}(1-\bar{m})[1 - \cos(U^{\mathrm{BF}} t/\hbar)]\} \tag{E.13}$$

already introduced in Sect. 9.1.

E.2 Scenario 2

In the second scenario, we assume that some of the M lattice sites are occupied by a fermionic atom ($i \in \mathfrak{N}_{\mathrm{BF}}$ where, $\mathfrak{N}_{\mathrm{BF}} = 1, \ldots, N_{\mathrm{BF}}$) and the others are not ($j \in \mathfrak{N}_{\mathrm{BB}}$, where $\mathfrak{N}_{\mathrm{BB}} = N_{\mathrm{BF}} + 1, \ldots, M$). The Gutzwiller many-body quantum state can be ordered:

$$|\Psi(t)\rangle = \prod_{i \in \mathfrak{N}_{\mathrm{BF}}} |\phi_i(t)\rangle \prod_{j \in \mathfrak{N}_{\mathrm{BB}}} |\phi_j(t)\rangle, \tag{E.14}$$

where the time-dependent wavefunctions for sites with a fermion are given by

$$|\phi_i(t)\rangle = \sum_n d_n e^{-iE_{n,1}t/\hbar} |n, 1\rangle \tag{E.15}$$

and those without a fermion read

$$|\phi_j(t)\rangle = \sum_n c_n e^{-iE_{n,0}t/\hbar} |n, 0\rangle. \tag{E.16}$$

As in scenario 1 we make the assumptions $c_n^i = c_n \rightarrow \sqrt{1-\bar{m}}\, c_n$ and $d_n^i = d_n \rightarrow \sqrt{\bar{m}}\, c_n$.

The two kinds of occupation give rise to individual terms in the dynamical evolution. Focussing on the correlations between sites i and j (compare Eq. E.4) the sum can be split into four terms

$$\sum_{ij}^{M} \langle \hat{a}_i^\dagger \hat{a}_j \rangle(t) = \sum_{i=1}^{M} \langle \hat{a}_i^\dagger \hat{a}_i \rangle + \sum_{\substack{i,j \in \mathfrak{N}_{\mathrm{BF}} \\ i \neq j}} \langle \hat{a}_i^\dagger \hat{a}_j \rangle + \sum_{\substack{i,j \in \mathfrak{N}_{\mathrm{BB}} \\ i \neq j}} \langle \hat{a}_i^\dagger \hat{a}_j \rangle + 2\mathfrak{Re}\left[\sum_{\substack{i \in \mathfrak{N}_{\mathrm{BF}} \\ j \in \mathfrak{N}_{\mathrm{BB}}}} \langle \hat{a}_i^\dagger \hat{a}_j \rangle \right]. \tag{E.17}$$

The first summand appears identically in scenario 1 and it scales $\propto M$. The second summand scales as $\propto N_{\mathrm{BF}}^2$, the third as $\propto (M - N_{\mathrm{BF}})^2 \equiv N_{\mathrm{B}}^2$ and the fourth as $\propto 2N_{\mathrm{BF}}N_{\mathrm{B}}$. The time evolution is again determined by the eigenenergies of Eq. E.10 and $\langle \hat{a}_i^\dagger \hat{a}_j \rangle(t) = \langle \phi_i(t)|\hat{a}_i^\dagger|\phi_i(t)\rangle \cdot \langle \phi_j(t)|\hat{a}_j|\phi_j(t)\rangle$ for $i \neq j$.

Further evaluation yields that the first term is constant in time and merely counts the total number of bosons. The evolution of the second and the third term is determined by the energies $U^{\mathrm{BB}}(n - \tilde{n})$. They can interfere with the dynamics of the fourth term, which evolves according to the energies $U^{\mathrm{BB}}(n - \tilde{n}) + U^{\mathrm{BF}}$. At the time of the first quantum phase revival $t = h/U^{\mathrm{BB}}$ the fourth term destructively interferes with the second and the third one when the additional condition $U^{\mathrm{BF}}/U^{\mathrm{BB}} = z + 0.5$ ($z \in \mathbb{Z}$) is met. The modulation depth of the first revival scales as

$$M + \underbrace{\bar{m}^2 M^2}_{\text{2.term}} + \underbrace{M^2 - 2\bar{m}N^2 + \bar{m}^2 M^2}_{\text{3.term}} - \underbrace{(2\bar{m}M^2 - 2\bar{m}^2 M^2)}_{\text{4.term}} \approx M^2(4\bar{m}^2 - 4\bar{m} + 1)$$
$$= M^2(1 - 2\bar{m})^2. \quad \text{(E.18)}$$

where $\bar{m} = N_{\text{BF}}/M$ is the fermionic filling. An overall factor given by the bosonic filling $\bar{n}$ has been omitted for simplicity. The scaling for the suppression of the first revival is identical to the first scenario! Full suppression is achieved when half of the lattice sites are occupied with a fermion, $\bar{m} = 0.5$.

References

1. E. Arimondo, M. Inguscio, P. Violino, Experimental determinations of the hyperfine structure in the alkali atoms. Rev. Mod. Phys. **49**, 31 (1977)
2. S. Falke, E. Tiemann, C. Lisdat, H. Schnatz, G. Grosche, Transition frequencies of the *D* lines of ^{39}K, ^{40}K, and ^{41}K measured with a femtosecond laser frequency comb. Phys. Rev. A **74**, 032503 (2006)
3. H. Wang, J. Li, X.T. Wang, C.J. Williams, P.L. Gould, W.C. Stwalley, Precise determination of the dipole matrix element and radiative lifetime of the ^{39}K 4p state by photoassociative spectroscopy. Phys. Rev. A **55**, R1569 (1997)
4. C. Ticknor, C.A. Regal, D.S. Jin, J.L. Bohn, Multiplet structure of Feshbach resonances in nonzero partial waves. Phys. Rev. A **69**, 042712 (2004)
5. S. Bize, Y. Sortais, M.S. Santos, C. Mandache, A. Clairon, C. Salomon, High-accuracy measurement of the 87 Rb ground-state hyperfine splitting in an atomic fountain. Europhys. Lett. **45**, 558 (1999)
6. D.A. Steck, Rubidium 87 D Line Data. http://steck.us/alkalidata (2009)
7. E.G.M. van Kempen, S.J.J.M.F. Kokkelmans, D.J. Heinzen, B.J. Verhaar, Interisotope determination of ultracold rubidium interactions from three high-precision experiments. Phys. Rev. Lett. **88**, 093201 (2002)
8. N.N. Klausen, J.L. Bohn, C.H. Greene, Nature of spinor Bose-Einstein condensates in rubidium. Phys. Rev. A **64**, 053602 (2001)
9. E.A. Burt, R.W. Ghrist, C.J. Myatt, M.J. Holland, E.A. Cornell, C.E. Wieman, Coherence, correlations, and collisions: what one learns about Bose-Einstein condensates from their decay. Phys. Rev. Lett. **79**, 337 (1997)
10. W. Ketterle. Physics 8.421 (Atomic Physics I). Lecture Notes, Massachusetts Institute of Technology (2006)
11. J.M. Goldwin, *Quantum Degeneracy and Interactions in the* ^{87}Rb-^{40}*K Bose-Fermi-Mixture*. Ph.D. Thesis, University of Colorado, 2005
12. T. Tiecke. Properties of Potassium. http://staff.science.uva.nl/tgtiecke/PotassiumProperties.pdf (2010)
13. W. Ketterle, M.W. Zwierlein, in *Making, Probing and Understanding Ultracold Fermi Gases*, ed. by M. Inguscio, W. Ketterle, C. Salomonin. Ultracold Fermi Gases, Proceedings of the International School of Physics "Enrico Fermi", Course CLXIV, Varenna, 20–30 June 2006, (IOS Press, Amsterdam, 2008)
14. C. Regal, *Experimental Realization of BCS-BEC Crossover Physics with a Fermi Gas of Atoms*. Ph.D. Thesis, University of Colorado, 2005

Curriculum Vitae: Dr. Sebastian Will

Personal

Birthday:	September 5, 1980 in Limburg an der Lahn
Citizenship:	Germany

Employment and Education

since 11/2011	Postdoc: Massachusetts Institute of Technology, Cambridge, MA, Group of Prof. Dr. Martin Zwierlein
11/2011	Graduation (Dr. rer. nat.): Johannes Gutenberg University Mainz, Thesis: "Interacting bosons and fermions in 3D optical lattice potentials" Grade: summa cum laude
2006–2011	Graduate Research Assistant: Johannes Gutenberg University Mainz and (from 2009) Ludwig-Maximilians-University Munich, Group of Prof. Dr. Immanuel Bloch
11/2006	Diploma (Dipl.-Phys.): Johannes Gutenberg University Mainz, Thesis: "Atom optical experiments with ultracold sodium atoms" Grade: with distinction
2005–2006	Research visit: Massachusetts Institute of Technology, Cambridge, MA, Group of Prof. Dr. Wolfgang Ketterle
2001–2006	Undergraduate studies in physics: Johannes Gutenberg University Mainz
1991–2000	Secondary education: Tilemannschule, Limburg an der Lahn, Grade: 1.0

Honors and Awards

since 2007	Junior fellow of the Gutenberg-Academy Mainz, Academic mentor: Prof. Dr. Jürgen Gauss (University of Mainz), External mentor: Prof. Dr. Jürgen Kluge (Fritz Haniel & Cie.)
2006–2009	Fellowship of the Materials Science Graduate School MAINZ
2002–2006	Fellowship of the Studienstiftung des deutschen Volkes

S. Will, *From Atom Optics to Quantum Simulation*, Springer Theses,
DOI: 10.1007/978-3-642-33633-1, © Springer-Verlag Berlin Heidelberg 2013

MIX
Papier aus verantwortungsvollen Quellen
Paper from responsible sources
FSC® C105338

FSC
www.fsc.org

If you have any concerns about our products,
you can contact us on
ProductSafety@springernature.com

In case Publisher is established outside the EU,
the EU authorized representative is:
Springer Nature Customer Service Center GmbH
Europaplatz 3, 69115 Heidelberg, Germany

Printed by Libri Plureos GmbH
in Hamburg, Germany